Pre-industrial Cities & Technology

The Cities and Technology series

Pre-industrial Cities and Technology
(C. Chant and D. Goodman, eds)

The Pre-industrial Cities and Technology Reader
(C. Chant, ed.)

European Cities and Technology: industrial to post-industrial city
(D. Goodman and C. Chant, eds)

The European Cities and Technology Reader: industrial to post-industrial city
(D. Goodman, ed.)

American Cities and Technology: wilderness to wired city
(G.K. Roberts and J.P. Steadman)

The American Cities and Technology Reader: wilderness to wired city
(G.K. Roberts, ed.)

Pre-industrial Cities & Technology

Edited by
Colin Chant and David Goodman

in association with

Published by Routledge   Written and produced by The Open University
11 New Fetter Lane   Walton Hall
London EC4P 4EE   Milton Keynes MK7 6AA

Simultaneously published in the USA and Canada by Routledge
29 West 35th Street
New York, NY 10001

Edited, designed and typeset by The Open University
Printed in the United Kingdom by Scotprint Ltd, Musselburgh, Scotland

A catalogue record for this book is available from the British Library
A catalog record for this book is available from the Library of Congress

ISBN 0 415 20075 X hardback
ISBN 0 415 20076 8 paperback

at308bk1i1.1

This book is part of the Cities and Technology series listed on the back of the first page. The series has been prepared for the Open University course AT308 *Cities and Technology: from Babylon to Singapore*. Details of this and other Open University courses can be obtained from the Course Reservations and Sales Office, PO Box 724, The Open University, Milton Keynes MK7 6ZS, United Kingdom; tel: + 44 (0)1908 653231.

Much useful course information can also be obtained from The Open University's website: http://www.open.ac.uk

Contents

Part 3   Pre-industrial Cities in China and Africa

Introduction

This textbook is part of a series about the technological dimension of one of the most fundamental changes in the history of human society – the transition from rural to urban ways of living. The series[1] is intended first and foremost as a contribution to the social history of technology; the urban setting serves above all as a repository of historical evidence with which to interpret the historical relations of technology and society. The main focus, though not an exclusive one, is on the social relations of technology as exhibited in the physical form and fabric of towns and cities.

The main aims of the series are twofold. The first is to investigate the extent to which major changes in the physical form and fabric of towns and cities have been stimulated by technological developments (and conversely how far urban development has been constrained by the existing state of technology). The second aim is to explore within the urban setting the social origins and contexts of technology. To this end the series draws upon a number of disciplines involved in urban historical studies – urban archaeology, urban history, urban historical geography and architectural history. In so doing, it seeks to correct an illusion created by some past historical writing – the illusion that all major changes in urban form and fabric might be sufficiently explained by technological innovations. In brief, the series shows not only how towns and cities have been shaped by applications of technology, but also how such applications have been influenced by, for example, politics, economics, culture and the natural environment.

The wide chronological and geographical compass of the series serves to bring out the general features of urban form which differentiate particular civilizations and economic orders. Attention to these differences shows how civilizations and societies are characterized both by their use of certain complexes of technologies, and also by the peculiar political, social and economic pathways through which the potentials of these technologies are channelled and shaped. Despite its wide sweep, the series does not sacrifice depth for breadth: case-studies of technologies in particular urban settings form the bulk of the material in the Readers associated with each textbook.

Definitions

The series calls upon a diverse set of interpretations, models and approaches from the social history of technology, from urban historical and geographical studies, and from archaeology and architectural history. In a wide-ranging series such as this, it seems appropriate to introduce theoretical issues when called for by a particular topic. There is, however, a place in this introduction for discussion of the series' two main variables.

In this series, 'technology' is interpreted broadly to cover all methods and means devised by humans in pursuit of their practical ends, thereby including relevant developments in science, mathematics, public health and medicine. But however broad this interpretation, there are still distinctions to be drawn between, say, technology (the means of building) and the built environment (the product of building technologies); or between technology (the means of

---

1   The three textbooks and Readers in this Cities and Technology series – all published by Routledge, in association with The Open University – are listed on the back of the first page of this volume.

achieving human ends) and society (perhaps the most ambiguous term of all, but in one sense the summation of human ends in the form of a set of religious, moral and political values). It may turn out to be difficult to draw a sharp line between these concepts; but the same can be said of night and day, or indeed, the modern city and the countryside, and we cannot do without such concepts in our dealings with the world. Similarly, without the distinctions proposed between technology, society and the built environment, we cannot hope to think clearly about the questions this series raises. In practice, the series focuses on the implications for urban form and fabric of a well-defined range of innovations, above all those in: agriculture and food-processing; military technology; energy; materials; transport; communications; water supply, sanitation and other developments in public health and social medicine; production processes; and building construction, including representational and measurement techniques and engineering science.

The contribution of these technologies is analysed at differing angles of relevance to the history of urban form and fabric. Innovations in agricultural, military, industrial and transport technology are linked with broad developments in the history of urbanization, including the origins of urban settlements, the changing relationship between town and country, and the increasingly specialized nature of cities within systems of cities. At the core of the series are the technologies most intimately involved in the processes of city-building – construction techniques, intra-urban transport, energy systems, water supply and sanitation, and communications networks. The series also attends to developments in science, technology and public health stimulated by a variety of urban crises.

If 'technology' is a slippery term, so too is 'city', the other great historical variable of this series. It should be emphasized at the outset that no particular store is set by the distinction between cities and towns, a distinction often made in a culturally specific way – for example, the peculiarly British criterion that cities have cathedrals. In this series, cities and towns are seen as part of an urban continuum, and references to cities should usually be read as covering all settlements of an urban degree of complexity and specialization. This needs to be stressed, because in a series as wide-ranging as this, the largest cities of a given period and region tend to claim the most detailed attention. But this is already to presuppose a distinction between cities and villages, the type of permanent human settlement which developed alongside agriculture, beginning some 11,000 years ago. What differentiates towns or cities from villages, apart from the brute facts that the former are physically bigger and contain more people?

The very fact that 'cities' have been transformed throughout history is what makes them compelling objects of historical investigation; but it also makes any abstract definition elusive. In many earlier periods and places, the city, densely populated and built up within its defining defensive walls, was sharply demarcated from the surrounding countryside and its small agricultural settlements. In the modern era, the old fortification rings which defined the city, especially in Europe, became redundant, partly through developments in artillery. They sometimes metamorphosed directly into circular roads, becoming part of the transport system which facilitated the outward diffusion of the city into the countryside, to the point where some recent commentators have begun to question whether the term 'city' is becoming outmoded. Cities, clearly, are dynamic entities, subject to great changes, not only expanding but declining and contracting, and even at times being reduced to rubble or ashes. For all we know, they may be a phenomenon of certain successive configurations of human society only, and be destined to disappear, just as there were none in human history, it seems, until little more than 5,000 years ago. There is no neat definition of a city, as the following quotation indicates:

> the city may be defined initially as a community whose members live in close proximity under a single government and in a unified complex of buildings, often surrounded by a wall. Since, however, this definition would also cover many villages, military camps, religious communities and the like, the city may further be described as a community in which a considerable number of the population pursue their main activities within the city, in non-rural occupations. But other communities, such as a monastery or small factory surrounded by the dwellings of its workmen, might be similarly characterized. A third characterization may therefore be that the city is a community which extends at least its influence and preferably its control over an area wider than that simply necessary to maintain its self-sufficiency.
>
> (Hammond, 1972, p.8)

The problem with this process of definition and redefinition is that in order to cover marginal exceptions, the definition gets loaded with various historical attributes of certain cities that are not part of the *meaning* of words such as 'city' or 'urban'. The search for a perfect definition, intended not only to cover all types of city in history but also to expose those settlements with bogus urban credentials, is surely a futile one, and this series will eschew it. The main point that will be emphasized is that a significant or dominant proportion of the population is engaged in activities other than agriculture: government, religion, administration, law, education, finance, manufacture, commerce, entertainment and so on. As it turned out, those involved in the non-agricultural occupations made possible by improved agricultural productivity generally clustered in relatively densely populated and built-up settlements; the reasons for this are considered at appropriate points in the series.

But a criterion based on economic specialization seems too disembodied, too dissociated from the physical urban reality, to capture the full meaning of 'city'. It is important therefore to add that the activities which distinguish towns and cities from villages become associated with particular spaces and structures: forums, squares and parks, streets, bridges and railways, markets, factories and shopping malls, temples, town halls and theatres; and that these spaces and structures often become emblematic of a given town or city. Always remembering that the urban built environment is designed by human beings for human purposes, we suggest that it is, apart from geographical location itself, the most definitive aspect of a city's identity, however dramatically it may be transformed by war or disaster. Emphasis on the changing spatial form and physical fabric of cities, and the ways in which they emerge from human activities, is therefore no arbitrary selection from the multiple historical phenomena of urbanization.

Conventions and acknowledgements

Short numbered passages – readily identifiable by the different typeface used and the grey rectangle beside the page number – have been printed at the end of some chapters of this book. These are extracts from published works, and in some cases bibliographical references have been edited out of them without ellipses in the text; ellipses do appear where other portions of text have been omitted. Where references have been retained in such extracts, the full source is included in the bibliographic list at the end of the chapter.

Chapter authors have sought to avoid what are now widely accepted as sexist expressions in their own writing, but these remain without editorial comment where they occur in quotations from existing sources. As a general rule, measures have been given in metric rather than imperial units; where non-metric measures appear in existing sources, metric equivalents have been added. It may be that many readers have until now negotiated life's vagaries without having at the forefront of their minds the difference between the metric 'hectare' and the imperial 'acre'. It might therefore be helpful to point out that a hectare is equivalent to a square with sides of 100 metres. A hectare is nearly two and a half times an acre, which is equivalent to a rectangle with sides of the old furlong (220 yards – about 200 metres) and the old chain (22 yards – about 20 metres).

The editors and authors of a book such as this, which provides part of the print backbone of a mixed-media Open University course, are indebted to an unquantifiable degree to innumerable colleagues in all areas of the institution. Special thanks are due to Denise Hall, the course manager, without whose unsurpassed ability in her role, and total commitment to the course, there would be no series of books; to Linda Camborne-Paynter and her colleagues in Course Management, without whose expertise in the conversion of authors' drafts to first-rate electronic text, the textbook deadlines could never have been met; to Jonathan Davies and Sarah Hofton of The Open University's Design Studio for their skill and imagination in the exterior and interior graphic design, and to their colleagues Ray Munns and Michael Andrew Whitehead for their painstaking and creative work on numerous maps and illustrations; to picture researchers Celia Hart, Anne Howard and Paul Smith, for their unstinting dealings with the appropriately variegated cultures and technologies of illustration archives; and last but by no means least, to the dedicated team of Open University editors, Hazel Coleman, John Pettit and Jane Wood, whose close and rigorous reading of the textbook chapters has saved readers from numerous opacities, solecisms and awkward expressions. The rigorous and constructive comments of colleagues on a variety of university and Arts Faculty committees have informed the thinking of the Course Team during the course's gestation. Thanks are due in particular to Tim Benton, Tony Coulson, Colin Cunningham, Janet Huskinson, Anne Laurence and Bernard Waites for comments and advice at this stage. Help given on individual chapters has been acknowledged *in situ*. The Course Team has been fortunate indeed to have the benefit of the appropriately broad knowledge and experience of its external assessor, Anthony Sutcliffe, who has been a most constructive critic, always tolerant of academically respectable interpretations that diverged from his own.

Colin Chant
The Open University

Reference

HAMMOND, M. (1972) *The City in the Ancient World*, Cambridge, Mass., Harvard University Press.

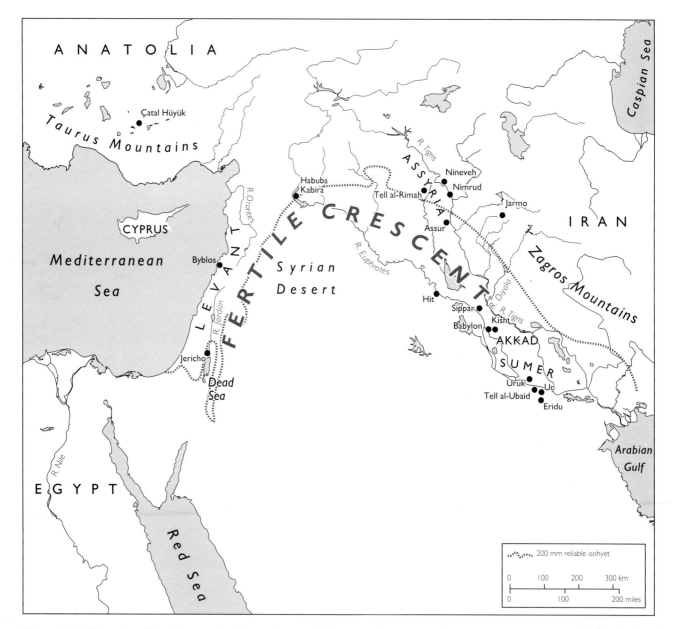

Figure 1.1 The ancient Near East; north of the isohyet, rainfall was usually at least 200 mm a year – enough to support agriculture without recourse to irrigation

Part 1
ANCIENT CITIES

Chapter 1: THE NEAR EAST

by Colin Chant

1.1 Introduction

The first settlements of an urban degree of complexity emerged, probably during the late fourth millennium BCE,[1] in the Near Eastern region often referred to as the 'Fertile Crescent' (see Figure 1.1). An ambiguous geographical term, the Fertile Crescent in its fullest sense denotes the sequence of river valleys and alluvial plains stretching from the Tigris and Euphrates in Mesopotamia (modern Iraq and eastern Syria), along the Orontes and Jordan in the Levant, to the Nile in Egypt. The trading activity of the first cities stimulated urbanization elsewhere during the later third millennium, notably in the Indus Valley to the east, and in the Mediterranean region to the west – where the first European cities, those of the Minoans and Mycenaeans, developed. This chapter on the Near East, therefore, logically introduces Chapters 2 and 3 on early European urbanization. As one authority on the region has put it, 'the history of the Western world begins in the Near East, in the Nile Valley and in Mesopotamia, the basin of the Tigris and Euphrates' (Postgate, 1992, p.xxi).

In a chapter dealing with the vicissitudes of urban civilization over more than two millennia, comprehensive coverage is out of the question. Instead, some of the themes of this series of three books are examined through the lens of a period that raises problems of historical interpretation in an acute form. The main difficulty is the fragmentary and uneven nature of the data. Written evidence is available from very soon after the emergence of the first cities, but its survival depends on the materials used and the conditions under which they happen to have been kept. The preservation of artefacts also depends on materials and conditions: biodegradable materials such as wood, textiles and leather survive only in special environments such as waterlogged peatbogs, arid Egyptian tombs and the permafrost of the Altai Mountains; whereas building stones, pottery and glass are eminently durable. But many durable materials, especially precious stones and metals, have been plundered by history's asset-strippers, from ancient tomb-robbers to modern colonialists, and partly for these reasons the recovery of entire settlements is rare. Another reason is that many important sites have been continuously occupied and developed, and as a result the early history of many of the locations evidently most conducive to urbanization is irretrievably buried.

The author of this chapter is grateful for detailed advice and corrections from Roger Moorey, and for the comments of colleagues at The Open University, including Olwen Williams-Thorpe. The overall approach, however, is the author's own.

[1] In this series of books, BCE (Before Common Era) and CE (Common Era) are used instead of BC and AD. All of the historical periods covered in Chapters 1 and 2 are BCE.

1.2   The emergence of cities: a technological revolution?

What part has technological innovation played in the development of cities? The simplicity of the question is deceptive: any serious attempt to answer it will require the unravelling of complex historical relationships. An appropriate way to start is to consider the relevance of technological change to the *origins* of urban settlements. In Section 1.2, the focus is on technological developments arguably necessary for the *emergence* of urban civilization, rather than those deployed in the process of city-building.

Any introduction to this issue falls under the long shadow cast by the archaeologist V. Gordon Childe (1892–1957), and his concept of a technologically driven 'Urban Revolution' in Mesopotamia in the fourth millennium, persuasively presented in *Man Makes Himself*, first published in 1936. Both his use of the term 'revolution', and his privileging of technology in the transition from village to city life, have been challenged ever more vigorously in recent years. However, Childe's work has been so influential that it must be addressed in any contemporary account of the relations between technology and the emergence of urban society.

The beginnings of technology

Technology is a likely focus for any analysis of changes in human behaviour leading to urbanization, because so much of the prehistoric evidence consists of artefacts. At least two and a half million years before the emergence of cities, early African hominids began with their chipped pebble tools to demonstrate the technological aptitude of the human family – the physical and intellectual capacity to adapt to a given environment by making use of it. This aptitude enabled early humans to colonize diverse environments throughout the Old World (Africa, Europe and Asia). A series of prolonged glacial periods during the later Palaeolithic (Old Stone Age) in Europe and Asia encouraged developments in the control of fire, and the use of permanent shelters. These early humans had their domestic technologies – hearths for cooking and heating, and lamps in the form of small stone bowls (presumably fuelled with animal fat). They shaped a variety of tools and weapons, probably from wood but also – as we know from surviving artefacts – from stone and bone; they also developed communication, probably by gestures as well as spoken language. These attributes helped them to supplement their diet by trapping large animals, and also to protect themselves from harsh climates with animal skins. A notable technological extension of human capacities was the augmenting of human muscle power by hunting-weapons such as the spear-thrower (a long rod with a hook at one end) and the bow, described as 'perhaps the first engine man devised' (Childe, 1966, p.59).

It is technology that traditionally defines this mind-numbingly protracted 'Old Stone Age', but there is also evidence of a burgeoning culture expressed in the representation of daily life (as in the cave paintings at Lascaux in southern France), and the ritual burial of the dead. It seems, then, that it was a blend of technological and social characteristics that enabled small groups, bound by ties of kinship, to survive thousands of millennia by hunting and gathering. Indeed, the successive human species not only survived, but prospered: by the end of the Palaeolithic, *Homo sapiens* was the 'only animal with a near-global distribution', having moved out of the African hominid heartland as far afield as Australia and the Americas (Gamble, 1995, p.1).

In some Palaeolithic habitats unusually well stocked with game and fish (as in central France), there were caves that were settled permanently. On the plains of Russia and central Europe, large camp sites – with some substantial, often semi-subterranean dwellings (pit houses) – were associated with the hunting of herds of mammoth, bison and reindeer. But any sizeable, permanent human settlement was out of the question because the population had to be mobile and, generally, was thinly spread: it has been estimated that in hunter-gatherer societies, population density rarely exceeded one person per ten square kilometres (Morris, 1979, p.3).

Childe's Neolithic Revolution

For cities to be sustainable, a dramatic change in this ratio had to become possible. The key to this change was agricultural technology. For prehistorians of the Old World, the division between the Palaeolithic (Old Stone Age) and the Neolithic (New Stone Age) is marked by the transition from the nomadic business of hunting animals and gathering plants in the wild, to the settled occupation of agricultural food-production. The Neolithic began around 9000, as the ice sheets of the most recent Ice Age receded – though, as with all of these technologically defined periods, the timing for a given region varies with the rate of diffusion of the innovations. Moreover, in any such region, the adoption of agriculture was a gradual process, taking place during a period lasting two or three millennia. This transitional period is sometimes identified as the Mesolithic (Middle Stone Age), and is characterized by innovations in hunting (the domestication of the dog) and fishing (harpoons, nets and paddles). Even where agriculture was fully adopted, not all settlements were permanent (nomadic agriculture is still practised, as in the soil-exhausting slash-and-burn methods of nomadic farmers in the Amazon basin). And even when, after the introduction of irrigation and manuring, permanent plantations became possible, cultivators of plants co-existed with more mobile pastoralists or herdspeople, and nomadic hunters. Periodic tensions and competition for resources between these groups, though often exaggerated by historians, go some way to explain the ebbing and flowing of ancient empires and civilizations.

[handwritten margin note: Hardly a 'revolution' if it took so long!]

Despite these qualifications, the invention of agriculture, when viewed against the long Palaeolithic struggle for survival, appears as a radical change in the technological relation between humans and the natural environment. For this reason, as early as the 1920s Childe identified the transition to agriculture, and the relative flood of innovations it seemed to trigger, as a Neolithic Revolution. It was the first of three sets of fundamental change in human history, each consisting of revolutionary innovations in the economic sphere leading to population growth; to follow were the Urban Revolution and the Industrial Revolution. He saw the establishment of cities during the Urban Revolution as driven by the cumulative technological impetus of the Neolithic Revolution, which led to a surplus of food, occupational specialization and the development of trade.

[handwritten margin note: – TMA 01]

One defining characteristic of agriculture, the basis of Childe's Neolithic Revolution, was that plants were cultivated on sites of human choosing, and then of human artifice – rather than gathered where they naturally occurred. The creation of plantations was accompanied by the development of implements, such as hoes, sickles and sandstone grain mills. By the fourth millennium, the animal-drawn plough was beginning to replace the hoe as the main tool for cultivation:

> Of all the devices created by mankind up to the end of the fourth millennium [BCE] there can be little doubt that the plough had overall the greatest effect. More probably than not it was mainly responsible for the rise in population in the small Mesopotamian and Egyptian cities.
>
> (Hodges, 1971, p.77)

The other main agricultural innovation was the domestication of animals. Goats were tamed in the Near East around 8500, followed by sheep (*c.*8000), pigs (*c*.7000) and cattle (*c*.7000) (Roaf, 1990, p.36). To begin with, such animals were a controllable source of meat, skins and bones, but their captors came to appreciate the value of live animals, first as sources of milk, manure and woollen cloth, and later of motive power replacing humans as drawers of ploughs and carts. These subsequent uses have been called a 'secondary products revolution'. Placed among the technological developments precipitating Childe's Urban Revolution, this particular complex of innovations is seen as 'a major biotechnological shift', leading to the denser occupation of deserts, steppes and mountains, and in the longer term to urban systems, animal-powered machinery and, eventually, to industrialization (Sherratt, 1981, 1996).

Associated with the establishment of agriculture was the invention of household crafts – notably pottery, carpentry and textile-manufacture. Like agriculture, these crafts involved the manipulation of easily accessible earth, plants and animals to meet human needs. The need both to prepare and to store cereal foods was met by pottery, which replaced containers made of skins, gourds and wood. Making pots was, after cooking, the first human project in practical chemistry: it required knowledge of appropriate mixes and colours of clay, the control of the firing conditions, and skill in applying pigments and varnishes. The seminal principle of the pottery kiln, which may have been established as early as the sixth millennium, was the separation of the vessels from the fire by a perforated clay floor (see Figure 1.2). In the often thickly wooded environments of prehistoric agriculture, timber was a readily available and versatile material, suitable for making tool hafts, ploughs, wheels and houses. Woodworking skills were stimulated by the Neolithic innovation of stone axes (polished with hard, quartz-bearing sandstone), along with adzes, chisels and bow-drills. Another craft associated with mixed farming was the manufacture of woven fabrics of linen or wool for clothing, replacing the hunters' dressed skins. Textile-manufacture required the invention of spindles and of the loom, 'one of the great triumphs of human ingenuity' (Childe, 1966, p.95). Pottery, carpentry and textiles were all crafts which the inhabitants of a self-sufficient agricultural village would practise alongside their often seasonal agricultural tasks. It was not until the rise in productivity associated with the potter's wheel in the fourth millennium that pottery became the kind of specialized occupation that would characterize urban settlements. The distinction between non-specialized villages and specialized towns and cities, however, can be exaggerated: there is some evidence from northern Mesopotamia of interdependence and specialization among villages, at least one of which may have been a community of hunters and tanners processing onagers, or wild asses (Moorey, 1994, p.2).

Where did the transition to agriculture take place? It is important to recognize that agriculture and its related crafts, and the urban civilizations which followed, must have developed independently in more than one region of the world – for example, in the New World in Peru and Mesoamerica, and probably also in China. The agricultural roots of the first Near Eastern cities trace back to a set of climatological, geological and biological features specific to the fertile land stretching from the Levant to the foothills of the Zagros Mountains in Iran (see Figure 1.1). The ending of the most recent of the Ice Ages was crucial: once the glaciers had receded, this region offered rainfall adequate for farming, or else river valleys and oases with naturally moist soils. Equally important, it became a habitat for varieties of plant suitable for cultivation, such as the wild grasses that were bred into wheat and barley. There were also species of animal that lent themselves to domestication (domesticates), such as cattle, goats, sheep and pigs. Indeed, the human pioneers of agriculture would have found the cattle and sheep feeding off the grasses.

Figure 1.2 Restored pottery kiln, Neo-Assyrian Period (Moorey, 1994, p.145; by permission of Oxford University Press)

first?

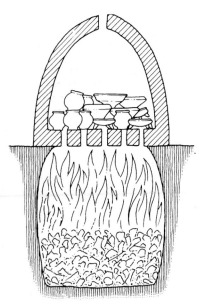

Çatal Hüyük and Jericho

Under these favourable conditions, agriculture enabled a surplus of food to be produced, allowing the human population to expand. It also stimulated the building of permanent storage facilities for the surplus, and these were often more complex than the dwellings. Some quite large agricultural villages have been excavated in this region. Jarmo (in modern Iraq) in the foothills of the Zagros Mountains, and Çatal Hüyük in Anatolia (modern Turkey), are examples with a long history of settlement, but which never became major centres of non-agricultural activities (they are given in Figure 1.1). Çatal Hüyük, first excavated in the early 1960s by James Mellaart, flourished from about 6300 to 5500. Covering more than twelve hectares, its mound is the largest Neolithic site yet identified (*hüyük* means 'mound' in Turkish). The excavated area reveals a mixture of shrines and distinctive mud-brick and timber dwellings, all of them entered by ladder from the roof (see Figure 1.3). More than a thousand such houses were packed together, implying a population of 5,000–6,000. Luxury goods were traded for obsidian, a local volcanic glass that could be chipped to make tools. Pottery and weaving were practised, and copper is found in the later levels of settlement, but no workshops have been unearthed, nor roads nor market-places. There were no separate boundary walls – merely the blank walls of the outermost houses and storehouses (Mellaart, 1967). Despite Mellaart's vigorous advocacy of its urban character, the consensus among prehistorians is that Çatal Hüyük was no more than a large, culturally developed village. Its position on a plain to the north of the Taurus Mountains isolated it from the technological, economic and political flow that reinforced urbanization elsewhere (Redman, 1978, pp.236–7).

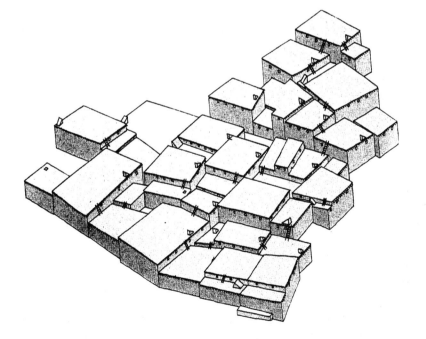

Figure 1.3 Çatal Hüyük: a reconstruction of part of one of the settlement's excavated building levels, dated early sixth millennium. The single-storey buildings were entered by the roof, and the roofs rose up in terraces toward the centre of the settlement (Roaf, 1990, p.44; by permission of Andromeda Oxford)

The assertion that the first cities emerged in Mesopotamia has also been challenged by excavators of Jericho, another settlement in the original zone of agriculture (the Fertile Crescent). Situated north of the Dead Sea in the Jordan Valley, Jericho dates from around 8000 as an agricultural community, with a population of about 2,000. Even before the onset of agriculture, the copious perennial spring at the site, together with the post-glacial profusion of wild cereals and nuts, led hunter-gatherers to establish a permanent settlement.

During the eighth millennium, stone houses were built over the remains of dwellings of sun-dried bricks. There are traces of the domestication of animals, and of an irrigation system – an indication of the progressive drying of the Levantine climate. Dame Kathleen Kenyon (who excavated Jericho's earliest layers) unearthed a stone wall, some two metres wide, that adjoined a substantial circular tower with an internal staircase (see Figure 1.4). The wall established for her the urban character of the settlement (Kenyon, 1979, p.26); but her inference that it was defensive has been challenged. A more recent interpretation is that the wall protected the settlement against flooding (Roaf, 1990, p.31). There is, in any case, no evidence that a significant proportion of Jericho's Neolithic population specialized in occupations other than agriculture, and so it is not now generally considered to be the first city, or one of the first cities. At this very early point in its development it was essentially a large agricultural village. Its expansion during its Bronze Age, and the construction of extensive fortifications, took place after the emergence of cities elsewhere in the Near East.

Figure 1.4 Jericho. *Left* a stone tower, *c.*10,000 BCE, preserved to a height of more than eight metres. The square opening halfway between the two standing figures leads to a stairway of twenty-two stone steps in the centre of the tower (Kenyon, 1981, plate 5; by permission of The British School of Archaeology in Jerusalem). *Right* a cross-section of the same tower, showing also the wall and ditch (Redman, 1978, p.80)

Villages such as Jarmo, Jericho and Çatal Hüyük were situated in the region most immediately favourable to rain-fed mixed farming after the end of the most recent Ice Age. But no system of cities emerged; one of the reasons is that the light soils of this region were easily exhausted, and agricultural settlements were often relatively transient, with farmers becoming increasingly remote from the forest resources on which they also depended. The first settlements of an urban degree of size and complexity were actually established further south, on the fertile alluvial plain of the rivers Tigris and Euphrates.

Childe's Urban Revolution

In this region, designated here by the historically somewhat ambiguous term 'Mesopotamia', an arid climate prevailed after the most recent Ice Age. But because the soil was constantly replenished with river-borne, mineral-rich silts, the plain had great potential to provide generous food surpluses. Before the fertility of the alluvium could be fully exploited, a technological effort was required to clear a great tract of swamps between the rivers Euphrates and Tigris, which was covered with reeds and date-palm groves. Channels were dug to drain the swamps, and ground was raised above flood levels: for this purpose, a platform of reeds underlay the agricultural settlement which developed into the city of Uruk. Thereafter, systematic control of the natural order was needed, in the form of irrigation channels and flood-control works. Although the pattern of inundations of both rivers was unpredictable, the main threat of flooding was posed after the melting of mountain snows in the spring, near or during the time of harvest. Consequently, dikes and basins had to be constructed to protect the crops from flood-water, as well as canals to irrigate them. The Euphrates was more important than the Tigris as a focus of settlement in antiquity (see Figure 1.1), being slower-moving and therefore easier to control.

The direct evidence for ancient irrigation works is scarce, but it seems that the first were local and village-based, and located on the margins of the flood-plain. More extensive cultivation of the plain required a much more systematic and large-scale approach to the challenge posed by great volumes of water. This presupposed the deployment of large teams of workers, and an organized division of labour, and was therefore arguably both a condition and a stimulus for the emergence of more and bigger settlements. According to a controversial thesis put forward by Karl Wittfogel, these great hydraulic projects also served to concentrate power in the social class that administered them, and thereafter the ruling elites were able to mobilize their teams of conscripted labour to construct the great defensive walls, temples, palaces, government buildings and granaries that dominated their cities (Wittfogel, 1957).

A particular technological innovation probably stimulated by irrigation works was the *shaduf* (see Figure 1.5). This was a simple device, based on the lever, one of the principal ancient mechanical devices augmenting muscle power. It raised water from an irrigation channel, or from a river to an

Figure 1.5 Raising water from a river by a pair of *shadufs*. The *shadufs* were poised on mud uprights, and were operated by men standing on brick platforms. As the counter-weight pulled the bucket up to each man's waist, he emptied it into an irrigation trough behind him. From the palace of the Assyrian ruler Sennacherib at Nineveh, seventh century (Singer *et al.*, 1954, p.524; by permission of Oxford University Press)

irrigation channel, by means of a vessel attached to a wooden beam with a clay counter-weight at the other end; each person operating a *shaduf* could raise over 2,500 litres per day (Singer *et al.*, 1954, p.525). The main agricultural crops fed by such irrigation techniques were barley and wheat. Other cultivated plants which afforded storable food were date-palms, fig trees, olive trees and vines: even more than irrigated land for grain cultivation, such plantations implied long commitment to place. Animal products and animal traction were supplied from the pastoralism practised between the cultivated, irrigated areas.

The first urban settlements based on irrigated agriculture were the independent city states of Sumer, the region of southern Mesopotamia just above the marshy area of the delta (see Figure 1.1). Many of these cities, such as Ur and Eridu, grew out of agricultural settlements established at least as early as the sixth millennium. A necessarily arbitrary date of 3200 has been suggested for the start of Sumerian urbanization (Hammond, 1972, p.35), though some recent estimates push the emergence of urban characteristics further back in the fourth millennium. A case in point is Habuba Kabira, a fully excavated mid-fourth millennium site in northern Syria; though only eighteen hectares in area, it had monumental architecture and a defensive wall (van de Mieroop, 1997, p.39). Even though urbanization implies a greater commitment to particular locations, the fortunes of individual cities over the long span of ancient Mesopotamian history were notoriously variable. One of the reasons for this volatility was political and military rivalry between the independent city states. Another was geographical: the Tigris and Euphrates frequently changed course, and as a result locations that once had plentiful water were left literally high and dry. The famous ancient sites of Ur and Uruk, for example, are several kilometres (west and east respectively) from the present course of the Euphrates.

Two features of the region were crucial to its priority in the history of cities: first, its great potential to produce an agricultural surplus, provided that immense organized effort was invested in hydraulic engineering; and second, the paucity of many of the other resources likely to be demanded by any society committed to this organized effort, such as timber and stone suitable for building. Also lacking were materials other than flints for tools and weapons, such as metals and obsidian. To make good these deficiencies, the settlers turned to trade. Relatively early in the growth of these Sumerian settlements, specialization had developed within agriculture as more and more of the alluvium was turned over to cultivation. The cultivators probably found it easier to barter their surplus for the products of neighbouring hunters, fishers and pastoralists than to practise mixed farming themselves. Obtaining industrial raw materials required a commitment to trading over longer distances. From Neolithic villages in the region, there is evidence of trade in special materials such as shells and obsidian, and coloured minerals such as green malachite. Some of these came from as far afield as Egypt and the eastern Mediterranean, and were probably collected by nomadic hunters, whose contribution to many ancient trading systems should not be overlooked. Some of these exotic materials were used for decoration: malachite, for example, gave a green pigment used for cosmetics. Their availability initially stimulated crafts such as gem-cutting, and in the longer term opened up the can of metallurgy; malachite, in particular, was discovered to be a rich copper ore. Some materials, when made into beads and amulets, were probably invested with magic properties. Technology and magic, it should be stressed, were alternative – and indeed overlapping – ways of seeking to turn the natural environment to human advantage; it would be anachronistic to read our modern contempt for magic back into the ancient period.

The technologies of the Urban Revolution

Examples such as Jarmo show that it took more than ecology conducive to agriculture for cities to develop. Childe painted a vivid picture of the subsequent innovations deemed necessary for the Urban Revolution:

> The scene of the drama lies in the belt of semi-arid countries between the Nile and the Ganges. Here epoch-making inventions seem to have followed one another with breathless speed, when we recall the slow pace of progress in the millennia before the first revolution or even in the four millennia between the second and the Industrial Revolution of modern times.
>
> Between 6000 and 3000 [BCE] man has learnt to harness the force of oxen and of winds, he invents the plough, the wheeled cart, and the sailing boat, he discovers the chemical processes involved in smelting copper ores and the physical properties of metals, and he begins to work out an accurate solar calendar. He has thereby equipped himself for urban life, and prepares the way for a civilization which shall require writing, processes of reckoning, and standards of measurement – instruments of a new way of transmitting knowledge and of exact sciences. In no period of history till the days of Galileo was progress so rapid or far-reaching discoveries so frequent.
>
> (Childe, 1966, p.105)

The rise of metallurgy was described by Childe as 'a dominant factor in the second revolution' (Childe, 1966, pp.115–16), though he later recognized that the Mayan civilization of Mesoamerica developed without metal tools. Extracting metals from their ores (smelting), and then either beating them into shape (forging) or pouring them into moulds (casting), necessitated developments in heating equipment (pyrotechnology); metallurgical furnaces may well have been a progression from the pottery kiln. An intermediate stage may have been the manufacture of a substitute for the mineral turquoise, known as blue faience – a composite material consisting of sintered quartz (powdered quartz fused by heating) and a glaze, obtained initially by heating copper ores such as malachite and blue azurite with powdered talc-stone. Invented in Mesopotamia before 4000, faience-manufacture spread to Egypt, and was widely practised in the Near East and Europe over the next two millennia.

The first copper artefacts were made from limited amounts of naturally occurring or 'native' deposits; but when it was discovered that certain less uncommon minerals heated with charcoal yielded the metal, its advantages over stone were pursued with vigour. As well as being forged (a process analogous to the working of implements in stone or bone), copper could also be melted and cast using clay moulds. It could take an edge or point as good as any in stone, and was more durable; unlike stone or bone, it could be endlessly recycled. But copper metallurgy was a process involving many stages: a repertoire of technical knowledge and skill had to be built up for mining and smelting the ore, and casting or forging the extracted metal. Heat (and therefore timber fuel) was needed at almost every stage, and much more than that generated by a domestic fire. Casting, for example, required a furnace with a forced draught in order to attain the 1,083°C temperature necessary to melt copper, let alone the craft skill and knowledge necessary for successful castings. The main development in metallurgy in the urban era was the replacement of the furnace blowpipe by animal-skin bellows, operated by hand or foot (see Figure 1.6 overleaf).

The occupations of the miner, smelter and smith were clearly specialized, presupposing a surplus of food. This growing specialization of occupation, and the increase in trade necessary for the development of metals, were also preconditions for the emergence of cities. Although it was not until after the Industrial Revolution that metals became a significant part of the physical structure of cities, it can still be argued that the development of the craft of

Figure 1.6 Copper metallurgy: foundrymen cast bronze doors, depicted above the figures to the right. In the centre, molten metal is poured into cups leading into the clay mould. At the top-left, the skin-covered drums of the bellows are alternately depressed by foot and raised by string; from an Egyptian tomb-painting (Hodges, 1971, p.121)

metalworking and of a variety of metal tools was a further stimulus to urbanization, even though most of the primary developments in metallurgy took place in the ore-bearing highlands of the Near East, well away from cities. Once cities were established as centres of exchange and consumption, metalworking could flourish as an urban occupation, mostly for the production of luxury goods and weapons.

In Mesopotamia, ore deposits of copper and other metals such as lead, tin, gold and silver were rarely found on the alluvial plains so suited to Neolithic farming. But the region was surrounded by mountains and highland plateaux richly endowed with mineral resources; tapping them required only an organized commitment commensurate with that needed to clear the swampy river valley. The mountains adjoining Mesopotamia have, with good reason, been called a Mineral Crescent, complementing the Fertile Crescent in which the transition to agriculture took place (Hodges, 1971, p.11).

The working of copper, the most widely used metal, appeared in Mesopotamia probably in the first half of the fourth millennium. The commitment of the first cities to long-distance trade is nowhere more evident than in their pursuit of the benefits of metals. Copper ores were obtained from Oman in the Arabian Gulf region and from the Zagros Mountains of Iran, and later from Cyprus and Anatolia. The most common form of copper in the Near East from the fourth to the second millennia was an apparently natural alloy of copper and arsenic. Bronze, an artificial alloy of copper and tin, which is harder than arsenical copper and has a lower melting temperature, was in use from early in the third millennium. This new material has been immortalized in the term 'Bronze Age', which in the Near East traditionally covers the period c.3400–1400; but the term is a misnomer because the alloy only became prevalent in the second half of the second millennium – at the very end of this period. 'Bronze Age' is also a simplification, even where materials are concerned, because a wide range of other metals was in use. The Taurus Mountains had rich deposits of silver-bearing lead ores, and silver started to be recovered from them during the fourth millennium – before, indeed, the lead itself was put to use. From the third millennium, the main silver-trading route to Mesopotamia was down the Euphrates from Anatolia. Gold was brought from Egypt and Nubia, as well as from Iran and Anatolia. Iron artefacts are found throughout the Bronze Age, though the higher melting temperature of iron meant that casting was out of the question with ancient equipment, and it remained a difficult and expensive metal to smelt and work (Moorey, 1994).

Exploiting remote mineral resources required not only the occupational specialization and food surplus that have been identified as necessary for urbanization, but also systems of trade, and therefore of transport, so that the often geographically separated operations of metallurgy – mining, smelting, forging and casting – could be linked. Another advantage of the Near East in this respect was relatively favourable natural communications: the upper valley of the Euphrates, in particular, linked distinct areas of economic activity in and around the Black Sea and the eastern Mediterranean. The valley acted as a natural land highway, with drinking-water at hand, even before rivercraft came to be used for the lower, navigable waters.

Developments in transport involved the harnessing of animals on land, and of human muscle and wind power on water. For developments in land transport, the availability of suitable domestic animals, as well as the nature of the terrain, was crucial. In central and south Asia and Europe, oxen had been taught in Neolithic times to draw ploughs, and probably provided an alternative to dogs for drawing sledges. It was in this region that wheeled transport diffused, an innovation celebrated by Childe in a paroxysm of hindsight:

> The wheel was the crowning achievement of prehistoric carpentry; it is the precondition of modern machinery, and, applied to transport, it converted the sledge into a cart or wagon – the direct ancestor of the locomotive and the automobile.
>
> (Childe, 1966, p.124)

The first wheels were solid discs, most commonly made of three pieces cut from a single plank (see Figure 1.7). Most scholars agree that the first two-wheeled carts and four-wheeled wagons appeared in Mesopotamia during the crucial fourth millennium. Wheeled transport spread throughout those parts of the world where there were suitable domesticates, but was unknown in south-east Asia, sub-Saharan Africa, Australasia and the Americas, where the llama was an effective pack-animal but had little tractive power. But the economic significance of draught-animals and wheeled carts in the Near East was limited. Because of the terrain, the sparsity of good roads and near absence of bridges, long-distance overland trading systems continued to rely on human porters, and on pack-animals such as donkeys (domesticated c.4000) and mules (third millennium onwards). The mule is a horse–donkey hybrid, which is about fifty per cent more effective than a donkey in speed and carrying capacity (Moorey, 1994, p.12).

What of the horse itself, so well known in the modern era as a draught-animal? A simple yoke and pole were sufficient to attach the squat frames of oxen to wagons, but were ill-suited to the equine physique. Special collars and shafts would be needed to exploit its tractive capacity, and horses were

Figure 1.7 Cart with solid wheels; limestone relief from Mesopotamia, about 3000 (Hodges, 1971, p.74; photograph: University of Pennsylvania Museum, neg.S4–134588)

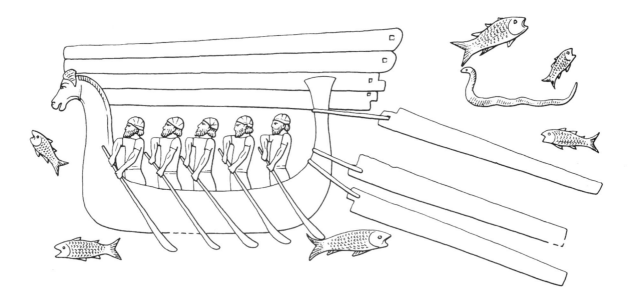

Figure 1.8 The transport of timber by water: the shaped timbers are stacked in the boat and drawn behind it. The freshwater fauna, and the fact that the rowers are standing upright, indicate river rather than sea transport. From an eighth-century BCE Neo-Assyrian relief (Moorey, 1994, p.353; by permission of Oxford University Press)

consequently restricted in antiquity to military and ceremonial uses, starting from the earlier second millennium. An important economic consideration favouring oxen and donkeys was that the rich pasture needed for horses was scarce in the ecologies of the first urban civilizations. One final point to note is that pack-camels – far superior to donkeys, mules and ox-drawn carts for desert carriage – were rarely used in the Near East before the first millennium (Piggott, 1983).

Wheeled transport required level, durable roads: in the semi-arid ancient Near East, packed dirt usually sufficed. There are occasional examples of thoroughfares paved with stone; these usually carried processions from a city to a major religious site. A key transport consideration throughout antiquity (and thereafter until the advent of railways) was the greater efficiency of water transport for carrying bulk goods. The clustering of Near Eastern cities by the main rivers testifies not only to their dependence on irrigated agriculture, but also to the primacy of water transport in supplying urban wants. The use of skin-boats and dug-out canoes for fishing must go back well before the oldest-known north European fragments of a skin-boat, dating from the ninth millennium. The first depicted Mesopotamian vessels were canoes and rafts made of reeds lashed together, though there were wooden boats from the third millennium (see Figure 1.8 for a later example). The greater efficiency of river transport needs to be qualified: boats and rafts in Mesopotamia had to be rowed upriver, or towed by a file of hauliers. By the third millennium, there were Mesopotamian ships trading in the Arabian Gulf, and in due course they would cover the eastern Mediterranean, the Red Sea and the Indian Ocean as far as the Indus Valley (Casson, 1974). The commitment to trade was a great stimulus to urbanization:

> The fertility of lands gave their inhabitants the means for satisfying their need of imports. But economic self-sufficiency had to be sacrificed and a completely new economic structure created. The surplus of home-grown products must not only suffice to exchange for exotic materials; it must also support a body of merchants and transport workers engaged in obtaining these and a body of specialized craftsmen to work the precious imports to the best advantage. And soon soldiers would be needed to protect the convoys and back up the merchants by force, scribes to keep records of transactions growing ever more complex, and State officials to reconcile conflicting interests.

(Childe, 1966, pp.141–2)

Historians have continued to argue that the driving force of Sumerian expansion was the difference in the resource endowments of the lowlands of Mesopotamia, and the surrounding highlands (Algaze, 1993). Trade was also a vital conduit of technical skills and knowledge. There is good evidence that early in the history of Sumerian urban civilization, there were trading contacts with Egypt, and later with the Indus Valley, and it is very likely that craft specialists moved around the whole area. Again, technological and urban developments are seen as mutually reinforcing.

Emphasis on trade and the concomitant voluntary transfer of technologies could result in a somewhat roseate view of the early history of cities. Goods, techniques and the ubiquitous ancient technological resource of slave-labour were often obtained through conquest; this raises the question of the relation between military technology and urbanization. In addition to the direct bearing of fortifications on the form and fabric of many ancient cities (see below), military technology may also have had a more indirect influence on urbanization by facilitating greater social stratification, which many historians have seen as one of the defining characteristics of ancient cities. There has been much speculation about the relationship between military technology, the production of an agricultural surplus and the emergence of a stratified society. One controversial view of the origins of kingship and aristocracies put forward by Lewis Mumford – one of the pioneers of twentieth-century urban history – is that peasants had to be forced to produce an agricultural surplus, in return for the protection of a tribe of pastoralists or nomads (Mumford, 1966, pp.32–5).

Mumford's view probably exaggerates the differences between hunter-gatherers, pastoralists and farmers on the eve of the Urban Revolution. On balance, it seems that cities stimulated developments in military technology, rather than the other way round. Cities as stores of agricultural surplus and traded goods were obvious prizes, and needed to be defended; their rulers increasingly took the same view about sources of raw materials and trading routes. As centres of aristocratic or monarchic power over the producers of food, cities were arsenals and sites of weapons-manufacture. Here is one of the strongest bonds between metallurgy and urbanization, as it was in weaponry that the superior durability of copper and bronze, compared with that of stone and bone, conferred a decisive advantage in combat.

A more difficult, though intriguing, set of relations is that between urbanization and methods and forms of communication – what is now called information technology. Even the compact and centralized physical form and structure of ancient cities may reflect limitations on the flow of detailed information; such information was laborious to copy and could be transmitted only as fast as the fleetest horse-rider or foot-runner. Childe, Gideon Sjoberg and others have argued that a literate elite's system of writing was the main criterion distinguishing the city from early rural settlements (Sjoberg, 1960, pp.32–3). Number signs, writing, and standardized weights and measures are another Mesopotamian cluster of innovations dating from the period immediately preceding the emergence of cities. In Mesopotamia, local technologies furnished the physical medium of this information; in the mature form of their writing system, practised from about 3000, scribes pressed reed styluses into clay tablets to make the wedge shapes of their cuneiform script, and the tablets were sun-dried or baked (see Figure 1.9 overleaf). The earliest clay tablets from Uruk have an already well developed system of ideographic signs (those representing abstract concepts and relations), suggesting that simple pictographic symbols, or picture signs, had been invented in the pre-urban period. If the view is taken that numbers and writing, and related technologies, originated because some permanent record of the agricultural surplus was required, then these intellectual inventions take their place alongside irrigation and other agricultural innovations among the necessary technological preconditions of urbanization.

Figure 1.9 Clay tablet: this administrative text from Uruk, dating from *c*.3000, contains a transaction concerning metal objects. The deep impressions are numerical signs made with a round reed stylus held either vertically (giving a circular impression) or obliquely (giving a crescent shape). Such signs developed into the distinctive wedge-shaped impressions of cuneiform script (Nissen *et al.*, 1993, p.22; by permission of Bildarchiv Preussischer Kulturbesitz)

1.3   The emergence of cities: a social revolution?

In the spirit of Childe, the foregoing account of developments culminating in the emergence of cities in the Near East has highlighted technological changes. Childe, an Australian whose main academic posts were in Edinburgh and London, was a Marxist, and reputedly delighted in reading the *Daily Worker* in London's upper-class Athenaeum club (Harris, 1994, p.121). But emphasis on technology was by no means confined to Marxists. A post-war generation of urban sociologists and historians in the USA saw technological innovations as at least one set of the main determinants of change in urban form and structure. With regard to the period of this book, Sjoberg regarded technology as the 'key independent variable' differentiating pre-industrial and industrial cities (1960, p.7). But as Sjoberg recognized, the priority given to technology in Childe's account of the Urban Revolution was soon challenged by other historians, anthropologists and sociologists. They insisted, often from the perspective of the sociologist Max Weber, that the precondition of the technological and cultural innovations involved in urbanization was an increasing scale and complexity of social organization, which led first to new political and religious institutions (Weber, 1958).

Childe's concept of a technologically based Urban Revolution continues to inspire debate (Harris, 1994). Much of the criticism has focused on points of accuracy in the light of subsequent archaeological work, and on Childe's emphasis on the diffusion of technologies from the Near East; contemporary archaeologists are more inclined to believe that technologies have more than one place of origin. Another scholastic concern is the word 'revolution'. Many contemporary historians share the reservations others have about the Scientific and Industrial Revolutions, stressing the long duration of the transition in question, and the continuities between the periods that such terminology sharply demarcates. This kind of objection was foreshadowed by Mumford, who took issue with Childe on the grounds that all the elements of the city pre-existed in rural societies, albeit in scattered form (Mumford, 1966, p.42). But the issue of prime concern in this section is the priority Childe gave to technology in his explanation of urbanization. This is the issue addressed by the US archaeologist Robert McCormick Adams, who was not hostile to the term 'Urban Revolution' as such:

> In balance, the insights engendered by the term seem to outweigh its drawbacks …
>
> Usefully to speak of an Urban Revolution, we must describe a functionally related core of institutions as they interacted and evolved through time. From this viewpoint, the characteristics Childe adduces can be divided into a group of primary variables, on the one hand, and a larger group of secondary, dependent variables, on the other. And it clearly was Childe's view that the primary motivating forces for the transformation lay in the rise of new technological and subsistence patterns. The accumulative growth of technology and the increasing availability of food surpluses as deployable capital, he argued, were the central causative agencies underlying the Urban Revolution.
>
> This study is somewhat differently oriented; it tends to stress 'societal' variables … Perhaps in part, such an approach is merely an outgrowth of limitations of space; social institutions lend themselves more easily to the construction of a brief paradigm than do the tool types or pottery styles with which the archaeologist traditionally works. But I also believe that the available evidence supports the conclusion that the transformation at the core of the Urban Revolution lay in the realm of social organization. And, while the onset of the transformation obviously cannot be understood apart from its cultural and ecological context, it seems to have been primarily changes in social institutions that precipitated changes in technology, subsistence, and other aspects of the wider cultural realm, such as religion, rather than vice versa.
>
> (Adams, 1966, pp.10–12)

It should be stressed that for Adams and for Childe, both social and technological changes had to be included in any explanation of the origins of cities. The issue, as posed by Adams, was whether the one is more fundamental than the other. Approaches to this complex issue cannot be conveniently grouped into two opposing camps; two extracts from later archaeological writing underline the point. The first, in which Joan Oates deals directly with ancient Mesopotamia, leans towards Childe's emphasis on technology; it can be found as Extract 1.1 on p.43. In the second extract (Extract 1.2 on p.44), Kent Flannery considers whether Childe's Urban Revolution can be applied to the New World (discussed by David Goodman in Chapter 7). In this case, Flannery's reservations echo those of Adams. The author uses the expression 'Nuclear America' for the period in which the first substantial, 'nucleated' settlements emerged.

Technological determinism

The point of looking at the scholarly debate over the Urban Revolution, a debate that is far too wide-ranging to bring to a conclusion here, is that distinctive views on the historical significance of technological changes have been at stake. Some, like Childe, have seen technology as the engine of historical change. This ascription to technology of a fundamental role in the dynamics of historical change is commonly labelled 'technological determinism'. In its strongest form, this is the view that all historical change is ultimately dependent upon developments in the autonomous realms of science and/or technology. The point about their autonomy is crucial. Technological determinism presupposes, whether implicitly or explicitly, that scientific discoveries and technological inventions prevail, not because they serve or embody some dominant social interest but because they are shown to be objectively better – either at explaining the natural world or at turning it to human use.

Given Childe's political affiliation, it is worth noting the continuing debate about whether Karl Marx was a technological determinist. Following the lead of Marx's principal collaborator, Friedrich Engels, many Marxists have distinguished between an economic base (including technology), and a social, cultural and intellectual superstructure determined by that economic base. This model of society has undoubtedly been influential in the history of technological determinism, even if it is debatable whether Marx's own writings warrant it (MacKenzie, 1984). But there should not be too much emphasis on Marx; technological determinism is at least implicit in a great deal of non-Marxist, especially US, scholarship. One of the few technological determinists actually to identify himself as such was Leslie A. White, who was influenced by Childe's writings (White, 1959). On a more popular level, implicit technological determinism has underpinned much analysis of the 'impact' of technology on twentieth-century society – for example, Alvin Toffler's *Future Shock* and *The Third Wave*; and the writings of Marshall McLuhan on the communications revolution. These are all heady scenarios of startling social changes made 'inevitable' by technological innovations.

Technological determinism has also featured in more sober historical writing. For example, Lynn White, Jr saw the introduction of the heavy plough as the ultimate cause of industrialization and urbanization in northern Europe (White, 1962; discussed in Chapter 4). Well before White, Lewis H. Morgan in *Ancient Society* (1877) had devised a typology of societies, progressing from savagery, through barbarism, to civilization; although Morgan set great store by the development of human institutions in this process, he at least implicitly presented inventions and discoveries as fundamental. Childe was quite explicit about the relation between the technological and the social:

> The archaeologist's divisions of the prehistoric period into Stone, Bronze, and Iron Ages are not altogether arbitrary. They are based upon the materials used for cutting implements, especially axes, and such implements are among the most important tools of production. Realistic history insists upon their significance in moulding and determining social systems and economic organization.
>
> (Childe, 1966, p.8)

There are a number of objections to this way of looking at things. First, it can be argued that the technologies upon which so many consequences are being loaded are not autonomous, but are themselves shaped by social, economic, political, cultural and other conditions. As we shall see in this series, some historians have seen the social mix of cities themselves as part of the set of conditions that stimulate technological innovations. So even if it could be shown that a given technology, once adopted, inevitably brought about a certain set of urban changes, this would not amount to technological determinism in the strictest sense, which requires that technology be autonomous. This kind of objection is exemplified in the quotation from Adams on p.15, and in Extract 1.2 by Flannery on p.44. Contemporary social constructionists, who will shortly be discussed, also insist that technological innovations have social origins.

Second, it can be objected that technologies are not the only cause of historical change: they are *necessary*, but not *sufficient*. The second objection, like the first, invokes the wider context, but it is in principle a different *kind* of objection in that it is compatible with the autonomy of technology. It is arguing that technologies, even if autonomous, are only one part of the explanation of historical change. Belief in the autonomy of technology, it should therefore be stressed, is not the same as belief in technological determinism. Henry Hodges, in his popular work on ancient technology, implicitly distinguished between necessary and sufficient conditions: having identified abundant raw materials and adequate communications as necessary for the development of 'centres of technological advance', he insisted that they are not sufficient:

> In effect, one of the limitations to technological evolution in antiquity seems to have been imposed by man himself, for certain social conditions were apparently inimical to further technological innovation. Indeed authoritarian governments aiming at stable social conditions appear to have been those under which there was least technological advance.
>
> (Hodges, 1971, p.241)

Many authors have been branded as technological determinists when they are only arguing (implicitly or explicitly) that certain technologies are a necessary condition of certain historical changes.

Third, there are objections which question the strength of the causal links posited between technology and historical change. For example, it might be argued that technologies, however *influential*, cannot *determine* social change, because human beings have free will. Technological determinism, as some historians would have it, overlooks *human agency*. Again, this kind of objection is quite compatible with belief in the autonomy of technology. It may even allow that technologies are among the most potent forces for historical change, but reserve for human beings and societies some capacity to resist or channel them in distinctive ways; Oates makes this point at the end of Extract 1.1. The point was also made by Wittfogel. Although he ascribed a fundamental role to irrigation works in shaping ancient oriental societies, a view characterized as 'hydraulic determinism', he nevertheless insisted that there was a 'genuine choice' for peoples about whether to practise large-scale irrigation agriculture (Wittfogel, 1957, p.18).

Technology and society: some alternative approaches

Technological determinism was the main quarry of an important collection of articles, *The Social Construction of Technological Systems*. In place of technological determinism, the editors promoted a view of technology and society as a 'seamless web' (Bijker, Hughes and Pinch, 1987). One of the approaches falling under this general rubric was social constructionism, exemplified in Trevor Pinch and Wiebe Bijker's interpretation of the history of bicycle design. Briefly, they argued that the emergence of the modern safety bicycle from earlier designs was not a linear process of the discovery of the most efficient form of human-powered wheeled transport. What is seen as technically the best solution was actually the result of negotiation between all social groups interested in the bicycle, and the design stabilized only when all social interests were reconciled. There is nothing inevitable about any design: it is always socially constructed. How could this approach be applied to the debate over the Urban Revolution? It might, for example, be argued that some of the agricultural innovations associated with the emergence of cities – such as the ox-drawn plough, or large-scale irrigation networks – were those that reconciled the interests of the social groups concerned, rather than those that met some objective criterion of greater agricultural productivity. It is even clearer, however, that in the ancient context some social groups are much more powerful than others – a political fact that Pinch and Bijker have been criticized for overlooking in their analysis.

In fact, more than one viewpoint was represented in the collection. Another centred on the notion of 'technological systems', a concept developed by Thomas P. Hughes in his study of electricity supply. Hughes insisted that no technology should be treated in isolation, since it operates at the core of a large social system: technology is therefore socially shaped, as well as shaping society. He believed that such systems acquire 'technological momentum', but this is not the same as the autonomy of technology, precisely because the momentum is due to the increasing number of public and private organizations that have a stake in these systems, rather than the unfolding power of their component technologies. Again, this concept has been developed through the analysis of an industrial society, and it is not self-evident that it can be applied to ancient technologies. Nevertheless, at first glance it seems appropriate to the labour-intensive and highly organized ancient projects of irrigation, and of the construction of monumental public buildings – and perhaps even to that great technological system, the ancient city itself.

In a more recent collection, some leading US historians of technology have attempted to reconcile the opposite poles of technological determinism and social constructionism (Smith and Marx, 1994). Hughes, who was one of the editors of the 1987 collection, now argues that the relations of technology and society change over time: social constructionism is more applicable to young technological systems (which are open to sociocultural influences), and technological determinism is more applicable to mature technological systems (which can be seen as powerful enough to shape society in their own interests). Again, he insists that his own concept of technological momentum is a better alternative to technological determinism, giving equal weight to the social and technological; but he now makes it clear that he rejects social constructionism as tending to replace technological determinism with social determinism (Hughes, 1994, p.104). According to Thomas Misa (another of the contributors), **technological determinism** characterizes macro-level vistas of technology and social change (a grand synthesis like Childe's *Man Makes Himself* would fit the bill), whereas social constructionism is a feature of micro-level studies (a study of the water supply of an ancient city might fall into this category). He argues for a 'meso-level' approach synthesizing the two, and attending to 'the institutions intermediate between the firm and the market or between the individual and the state' (Misa, 1994, p.139). However fruitful this approach proves to be, it is clear that many social historians of technology recognize the need to transcend the crudities both of technological determinism, and of accusations of technological determinism. They are ready to move towards a synthesis of technological determinism and social constructionism, whereby socially shaped technologies are accorded their due weight as instruments of social change. To achieve this synthesis, social historians of technology will need to go beyond a preoccupation with the social *roots* of technology, and consider anew, and in addition, its social *effects*.

The Wittfogel thesis

The debate over the Urban Revolution introduces in a broad way the opposing poles of technological determinism and the social construction/shaping of technology. It might help to focus the issues if the relevance of a particular technology to urbanization is considered. Irrigation and flood-control systems, and other water-management techniques, have already been mentioned among the technologies that have a direct bearing on the origins of cities. According to Wittfogel's thesis, the need in Mesopotamia to drain the swamps and control the rivers Tigris and Euphrates was the prime cause of the autocratic or 'agrobureaucratic' nature of its government. While not a thesis specifically about urbanization, it supports by implication the view that hydraulic engineering and the need for centralized water management prompted the social organization necessary for the emergence of cities, rather than the other way about (Wittfogel, 1957).

Ever since *Oriental Despotism* appeared in 1957, Wittfogel's concept of a hydraulic society has been roundly attacked. But such is the enduring appeal of grand monocausal theories – theories that seek to explain historical change by attention to one key variable – that it continues to be discussed and repudiated (Smith and Marx, 1994, pp.172–3). Much of the debate concerns evidence that cities emerged without the prior existence of elaborate irrigation works. In the light of such evidence, it may be that the crucial consideration in Sumer was not irrigation as such, but an ecological context notable for its unpredictable floods, changing watercourses, dust storms and diseases – a context that made institutionalized social cohesion essential for self-preservation. This difficult ecology, moreover, resulted in great inequalities in the value of landholdings, depending in part upon their strategic position in the irrigation network; this was a circumstance which may have encouraged the formation of a class-based society (Redman, 1978, pp.226–7, 232). Mason Hammond attempted to resolve deep differences of interpretation over the relationship between irrigation technology and the origins of cities:

> The planning, construction, and control of even such small-scale irrigation would obviously have entailed group effort and some form of government. Extensive swamps, such as early man found in southern Mesopotamia and in the lower Nile Valley … obviously presented a much greater challenge and would have necessitated much more social and political control to construct the channels and dikes necessary for irrigation and to organize the fair distribution of water. Some scholars, including Robert M. Adams, think that although early farming villages were capable of establishing irrigation on a small scale, major projects would have required the previous existence of a fully developed social and political structure and that, therefore, the emergence of the city in Sumer, and of a centralized monarchy in Egypt, preceded the achievement of irrigation on a large scale. However, it appears more likely that the gradual enlargement of irrigation projects required increasingly elaborate co-operation and management, and that, therefore, response to this challenge was at least one of the causes for the emergence of incipient cities in Sumer, as it seems to have been of a unified monarchy in Egypt.
>
> (Hammond, 1972, p.27)

Karl Butzer, author of a standard work on the 'hydraulic civilization' of ancient Egypt, has assessed the stimulus to archaeological research given by the grand theorizing of Wittfogel, and of Childe. (See Extract 1.3 at the end of this chapter.)

The function of cities

The debate about the causes of the emergence of cities overlaps with another about the primary function of the first cities: did they originate for environmental, technological, economic, military or religious reasons (Carter, 1983, pp.3–9)?[2] The view that ancient cities were religious in origin and primary function has a distinguished pedigree (Mumford, 1966, pp.44); apart from the physical dominance of the sacred precinct, the fact that so many were aligned with the cardinal points of the compass is evidence for some that the city was intended as an earthly representation of heaven. One historian of ancient technology has rejected such interpretations:

> By many writers [the development of the city in the Near East] has been described as essentially the growth of a complex of dwellings around a central temple, shrine or place of worship; and this indeed is often the superficial appearance given by such cities upon excavation. But in fact most cities were primarily centres of particular technologies. Throughout the whole period covered by this book [*Technology in the Ancient World*] most cities were renowned simply for the presence within their walls of one or two main technologies such as the making of pottery or glassware, or the manufacture of jewellery, and there is very good reason to believe that more often than not it was these industries, and not the presence of the temple, that brought the city into being. Of course, where a technology developed there had to be a means for the import of its raw materials and the export of its finished products so that

---

[2] An edited version of Carter can be found as the first item in the Reader associated with this volume (Chant, 1999).

the city became equally a centre of trade and a market-place for the buying and selling of agricultural products and livestock to provide food for its citizens.
(Hodges, 1971, p.212)

And yet Hodges did not go so far as to embrace technological determinism, as is evident in his attempt to explain why cities failed to develop in parts of Europe, despite their adoption of, for example, copper and iron metallurgy:

> Even in those areas from which raw materials were acquired in bulk, as for example around copper- and iron-workings, salt mines or tin mines, no cities seem to have developed, and it would appear that in Barbarian Europe raw materials were acquired and distributed in the same *ad hoc* way as that in which the craftsmen moved from one community to another. Yet the volume of trade was often sufficiently large to warrant urban development and at times it would seem almost as though there were amongst the prehistoric people of Europe a deliberate attitude of thought that ran counter to building cities.
> (Ibid., p.214)

Another approach to the relations of cities and technology, one that is more consonant with social constructionism than with technological determinism, is the view that cities served to stimulate innovations. According to Hodges, the absence of cities may explain the relative lack of technological development of northern Europe, compared with the urban civilizations of the Near East and the Mediterranean. Nevertheless, the example of the Indus Valley civilization suggests that cities were not a *sufficient* condition for technological innovations:

> the great cities of Mohenjo-daro and Harappa were as large and as industrious as any known from the ancient world at this date; and while we may blame the way in which their industries were organized, in all probability the root of the trouble lay in the difficult communications between the Indus Valley and the rest of the civilized world. As it is, it would seem that these two great cities and their dependent towns and villages became technologically introspective. They had achieved a level of technological development that was satisfactory from their point of view, and there appears to have been inadequate contact with the rest of the civilized world to act as a stimulus to further development.
> (Ibid., p.217)

The debate over the relation of technological innovations to the origins of cities will remain hard to resolve, not least because of the nature and limitations of evidence for early urbanization. The bias of the evidence towards material culture, such as masonry structures, pottery and the contents of graves, has already been noted. This bias may well have fostered technology-based interpretations and periodizations; but it also favours religious or military interpretations of the functions of ancient cities, since religious or military structures were most likely to be built of durable stone.

The point of this discussion of the emergence of cities is not to adjudicate between the opposing views presented, but to set up two apparently antithetical notions – the power of technological innovations to change human institutions, and the power of human institutions to influence the implementation of those innovations. The main theme of this series is the consideration of this antithesis in a variety of urban contexts. One conclusion, however, which may be drawn from this section, and which seems to weigh against the technological-determinist interpretation of the Urban Revolution, is that there was no preordained pattern of agricultural innovation leading inexorably to urbanization. The agricultural regime of Sumer was distinctive to that region, and a distinctive pattern of urbanization emerged. Agriculture could become established in another environment, with no immediate prospect of any kind of urban development – as in northern and central Europe, or in the areas where rain-fed agriculture first developed. There seems to have been some kind of technological, social and ecological equilibrium in these areas, of a kind which did not exist in Mesopotamia, or indeed among the nomadic hunters. External forces, whether military or commercial, had to disturb this equilibrium before the move to an urban society could take place. It can never be too strongly emphasized that for most of human history and for most humans, respect for the traditional ways was the norm, and the embracing of innovation the exception.

1.4   *Technology and city-building in Mesopotamia*

The focus now changes from the origins of Mesopotamian cities to their form and fabric, and to the technologies of city-building. For this section, a broad sense is needed of the main periods of ancient Mesopotamian history (see Table 1.1). Also relevant are certain analytical concepts used by urban geographers. One such is *urban morphology*, which is the study of the layout and physical form of a city. Another fundamental concept is *land-use* – the question of whether the function of a building or district of a city is, for example, residential, commercial, industrial or public.

What was the layout of the first Mesopotamian urban settlements – the city states of Sumer? It should be stressed that even though the temple was intended to dominate the Sumerian city, secular functions were always reflected in the built environment:

> According to Sumerian texts, cities usually had three main parts. An inner city housed within its walls the temples of the city's gods, the palace of the ruler, and private houses, distributed in quarters which often had each their own city gate. The suburbs contained houses and gardens and cattle pens, for the immediate support of the population. Finally, there was a commercial section which, though called a harbor, also handled overland commerce and might house foreign as well as native merchants. The harbor quarters seem to have enjoyed a certain administrative self-government and a special legal status.
>
> In short, the cities provided their inhabitants with material security, prosperity, and efficient government and with an assurance of divine protection and favor. They grew to a considerable size; the built-up area of Ur, for instance, occupied one hundred acres [about forty hectares] during [the Early Dynastic] period and might have accommodated twenty-four thousand inhabitants. Its territory, some four square miles [about 1,000 hectares], may have been occupied by half a million people … The walls of Uruk (Erech) surrounded two square miles. Other cities approached the same size. The surrounding fields were intersected by dikes and by canals which not only served for irrigation but gave access to shipping from the rivers and probably from the Persian [Arabian] Gulf.
>
> (Hammond, 1972, pp.37–8)

There is, however, nothing in these early cities identifiable as an open forum or market-place, in which merchants bargained over prices, and it may be inferred that the exchange of goods was controlled by the city's ruler. There is little evidence either of specialized manufacturing districts: workshops seem to have been disseminated throughout the urban fabric. In any case, by no means

Table 1.1   Selected periods and (in italics) reigns of ancient Mesopotamian history (adapted from Postgate, 1992, p.22 and Oates, 1986, pp.199–202)

Dates (BCE)	Period	Dates (BCE)	Period
5000–4000	Ubaid	1950–1600	Old Babylonian
4000–3200	Uruk	*1792–1750*	*Hammurabi*
3200–2350	Early Dynastic	900–650	Neo-Assyrian
2350–2150	Dynasty of Akkad	*704–681*	*Sennacherib*
2334–2279	*Sargon*	650–539	Neo-Babylonian
2150–1950	Third Dynasty of Ur	*604–562*	*Nebuchadnezzar II*

all crafts were located within a city. Some metalworking, especially smelting, took place where the ore was mined; the resulting ingots were then sent to be worked at the point of distribution of the finished goods. Itinerant craft specialists were common, especially potters and smiths.

The built environment of ancient Mesopotamian cities might be expected in some way to reflect their non-agricultural occupations, including crafts such as textiles, carpentry, pottery, stonemasonry, metalworking and jewellery – even if not so dramatically as in the modern industrial city. But excavations have provided a scant basis on which to reconstruct the layout and equipment of craft workshops. Many would have been quite distinctive, as is shown by the large clay beehive ovens in the temple bakeries of Ur; in other urban manufacturing premises, there were pottery kilns, and vats for dyeing cloth or brewing beer. The nature of the craft processes, however, was such that 'it was the workers and their tools which mattered more than any location or fittings' (Moorey, 1994, p.16).

Which technologies did the Mesopotamians apply to their cities? Although this series will become increasingly concerned with global technologies and their urban contexts, it needs to be emphasized that in antiquity, technologies were frequently of a small scale and local character. At the simplest level, the inhabitants of the marshlands of the Tigris–Euphrates delta fashioned their houses from reeds. The process of city-building in antiquity was also heavily dependent on human muscle power, extended in some instances by mechanical devices such as levers and pulleys, or supplemented by animal muscle power.

Building technology and building types

In Sumer, where timber and building stone were scarce, the physical structure of cities testifies to its main natural building resource – earth, an inexhaustible resource on the Mesopotamian flood-plain. The mud-brick was invented in Syria or Mesopotamia well before the urban era, and would allow builders vast scope for architectural experiment, including the construction of monumental buildings of unprecedented size. The basic Sumerian brick was made from clay mixed with water, with chopped straw or dung trodden in. A more dilute form of the mixture was used for mortar and plaster. At first, mud-bricks were hand-shaped and oval in section, perhaps in imitation of rounded stones, but bricks formed in wooden moulds soon appeared. Bricks evidently made in moulds have been found in Anatolia dating from as early as the seventh millennium; the process of moulding is depicted in an Egyptian tomb-painting of about 1500 (Hodges, 1971, p.32). Most Sumerian bricks were dried in the sun, because of the scarcity of timber fuel for baking them. Sun-dried bricks were prone to erosion, especially from moisture rising from the ground, and mud-brick structures had continually to be repaired and rebuilt. The difference this makes to the survival of ancient city sites is apparent if Sumerian city remains are compared with the well preserved baked-brick citadel, buildings, granary, walls and embankments of the ancient Indus Valley city of Mohenjo-daro, dating from about 2000. There was evidently plentiful timber fuel in this region to fire bricks in a kiln.

It may have been experimentation with combinations of bricks that led to the development of the arch, though Childe speculates that the tunnel-like roofs of reed huts inspired its invention (Childe, 1966, p.111). The true arch is held together by the compressive forces exerted by its brick or stone components. Together with its spatially extended forms, the vault and the dome, it was one of the key developments in ancient building technology. For the Mesopotamians, it meant that spaces could be spanned without the expense of importing stone lintels. By the end of the third millennium, the Mesopotamian builders had learned to make 'pitched-brick' vaults, by a construction method that obviated the expense of the temporary wooden support known as 'centring' (Oates, 1986, p.48; see Figure 1.10).

Sir Leonard Woolley excavated Ur between 1922 and 1934. Talking of the Royal Cemetery of Ur in the mid-third millennium, and specifically of the tombs of the *lugal*, or king, Meskalamdug and the *nin*, or queen, Puabi, he had this to say about Sumerian building:

> Not the least surprising aspect of the civilization which the tombs illustrate is the advance it had made in architecture. The doorway of RT789 was capped with a properly constructed brick arch, and its roof was formed by a brick barrel vault [see Figure 1.11] with apsidal ends; Puabi's (RT800) tomb was similarly vaulted, others had vaults of limestone rubble masonry; we find a complete rubble dome built over a timber centring … In these underground buildings no columns were required, but since the column was … freely used in the immediately succeeding period it must have been known in the cemetery age also. Summing this up, we can say that all the basic forms of architecture used today were familiar to the people of Ur in the early part of the third millennium before Christ.
>
> (Moorey, 1982, p.102)

Figure 1.10 Pitched-brick vault, viewed from below; Tell al-Rimah, Assyria, *c.*2100. Successive rings of bricks were laid, starting from each end until they met in the middle (Oates, 1986, p.48; photograph: D. Oates)

Figure 1.11 The south-east wall of the tomb chamber of RT789 at Ur, seen from the outside, showing the brick arch of the door, the door-blocking of brick and stone, the wall of limestone rubble plastered with clay and, behind, the remains of the baked-brick vault (Moorey, 1982, p.70; photograph: University of Pennsylvania Museum, neg.S4–142567)

Figure 1.12 Excavation of the heavily eroded ziggurat at Eridu, dating from the Third Dynasty of Ur, revealed – under one corner – these walls of earlier temples of the Ubaid Period (Oates, 1986, p.20)

Because sun-dried bricks were so widely used and not very durable, it is problematic for archaeologists to reconstruct Sumerian cities. Dotted around the alluvial plain of the Euphrates are a number of mounds ('tells'), which are the accumulated detritus of centuries – and in some cases millennia – of city-building with sun-dried brick. Excavation of one such mound revealed the city of Eridu, the dominant settlement of the Ubaid Period, and by Sumerian literary tradition the first agricultural settlement to become a city. It is typical in that each layer of settlement is centred on a religious building; the earliest mud-brick shrine dates from the sixth millennium. The growth of a settlement into a city was marked by increasingly large and elevated temples, as each predecessor was razed to provide the platform of the next (see Figure 1.12). The impermanence of these individual temple complexes contrasts with their enduring centrality within the physical and social structure of cities throughout the vicissitudes of ancient Mesopotamian history. From about 2000, these temple platforms were raised still higher, with steps or terracing, and became known as ziggurats. The original meaning of the word – a mountain summit – hints at the stunning visual impact of these artificial eminences on the flat Mesopotamian flood-plain (see Figure 1.13).

The sequence of building innovations characterizing the transition to an urban settlement is best seen at Uruk, the dominant city of the Uruk Period. The oldest level of the tell at Uruk is composed of the remains of the reed huts and mud-brick houses of an agricultural village. The transition to a city is dramatically revealed, as there next appear the foundations of a substantial temple complex (see Figure 1.14), and a terrace nearly twelve metres high built of lumps of mud with layers of bitumen in between. Bitumen, though relatively expensive, was readily available in Mesopotamia from the middle Euphrates, and was widely used for mortar and damp-proofing in building

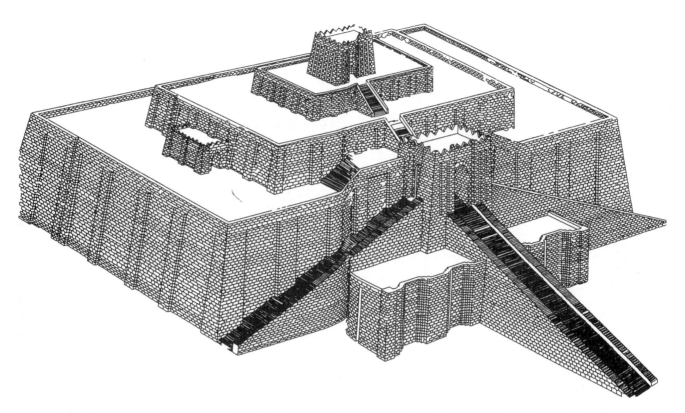

Figure 1.13 Reconstruction by Sir Leonard Woolley of the ziggurat at Ur (dating from the Third Dynasty of Ur), part of a building programme initiated by King Ur-Nammu (see Figure 1.17). Woolley's visualization of the upper stages, including the vaulted gateways, is speculative but technologically feasible (Kuhrt, 1995, p.65; by permission of Routledge)

(Moorey, 1994, pp.332–5). Layers of reed matting also bound the mud in such platforms. Other methods of consolidating the mud are evident at Uruk: the buttresses and recesses of one platform, which was nearly 1,000 square metres in area at the top, were consolidated with countless pots forming decorative patterns. On top of the platform was a staircase and whitewashed shrine (the 'White Temple'), dedicated to the god of the city. In the shrine, excavators found a clay tablet inscribed with numerals, indicating the role of the priestly servants of the gods in administering the food surplus. The careful planning of this monument is attested by the orientation of its corners to the cardinal points.

The numerous temples at Uruk were rebuilt and enlarged several times during the city's 5,000-year history, each new complex reflecting changes in building materials and techniques. Successive platforms were now built of limestone, and of kiln-fired brick, at least where erosion was most likely. The expanding technical capacity of the city's craft specialists is revealed in the

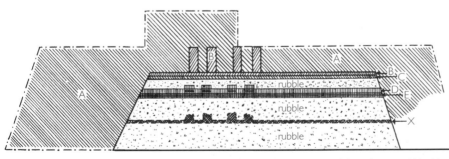

Figure 1.14 Section through the tell at Uruk. X marks the supposed foundation of Uruk's first temple; later foundations are E, D and C. B marks the foundation and walls of the 'White Temple'. All were later encased in brickwork: A (Singer *et al.*, 1954, p.48; by permission of Oxford University Press)

decorative reinforcement of the platform walls. Pots give way to mosaics of baked clay or stone cones, painted red, white and black (see Figure 1.15); next, mother-of-pearl and the decorative stone cornelian are inlaid on bitumen. Inside the sanctuary, figures of animals are found first in clay, then in stone or shell friezes, and then in copper over a bitumen core.

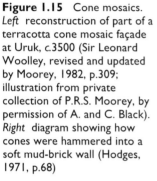

Figure 1.15 Cone mosaics. *Left* reconstruction of part of a terracotta cone mosaic façade at Uruk, *c.*3500 (Sir Leonard Woolley, revised and updated by Moorey, 1982, p.309; illustration from private collection of P.R.S. Moorey, by permission of A. and C. Black). *Right* diagram showing how cones were hammered into a soft mud-brick wall (Hodges, 1971, p.68)

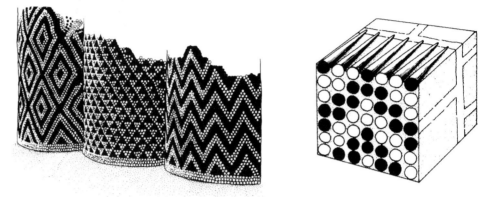

Like religion, warfare and military technology had a direct bearing on the form and fabric of early settlements. For example, Mellaart suggests that the compact layout and roof-entry of the houses at Çatal Hüyük were for defensive purposes (Mellaart, 1967, p.69). But this is nothing compared with the Sumerian city's massive mud-brick fortifications. Such defences reflected not only the threat of nomadic incursions, but the frequent dynastic wars between independent city states locked in commercial rivalry, and in disputes over land and water rights. They also testify to the importance in warfare of the siege, which led to the development of such devices as the wheeled scaling-ladder and the battering-ram (see Figure 1.16). In response, ever higher walls and stronger arched gateways were built, initiating a 'continuous dialogue' between walls and weaponry which would continue for most of the history of the Eurasian urban built environment (Ashworth, 1991, p.17).

These Sumerian urban structures are undoubtedly eloquent about the properties of sun-dried brick in particular; they also bear witness to the capacities and constraints of Sumerian technology in general:

> There is little need to connect these structures [ziggurats] with the Egyptian pyramids ... the similarities can be adequately accounted for by the exigencies of primitive architectural technology.

(Postgate, 1992, p.110)

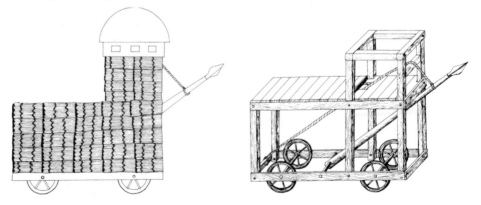

Figure 1.16 *Left* Siege-engine from a Neo-Assyrian relief of *c.*800 BCE. *Right* diagram showing how it may have operated. In its most developed form, the siege-engine combined a battering-ram and siege-tower mounted on wheels. The battering-ram appears to have been a pointed arm that prised the brick from the city wall; archers on the platform above gave protection to those attacking the wall (Hodges, 1971, p.150)

Figure 1.17 Detail from the 'Ur-Nammu Stela', found at Ur, showing King Ur-Nammu (c.2112–2095) officiating at a temple-building ceremony. He carries the tools of a builder – axe, basket, builder's dividers, ladle for bitumen mortar, and a flat wooden trowel (Moorey, 1994, p.303; by permission of Oxford University Press)

But they also raise questions about the relationship of the social structure of the Sumerian city to the built environment. There is some dispute about the nature of the original Sumerian city government, but it is clear that there soon developed in its various city states and empires a very strong form of monarchy, in which the king (*ensi* or *lugal*) occasionally assumed divine status, though more typically acted as the non-divine deputy or vicegerent of a given city's patron deity, on whose behalf the ruler built the city's temple. Building programmes clearly carried high prestige: there are monumental depictions of kings dressed as bricklayers, masons or architects, receiving the divine plan for a temple (see Figure 1.17). Unfortunately, little is known about the social and political groups responsible for building these cities. There are references in texts to gangs of labourers, but it is not known whether these were specialists, or temporary conscripts for particular projects. Slaves existed, but probably in insufficient numbers to have played a substantial role in city-building (Kuhrt, 1995, pp.28–9).

The centralization of power is a recurrent feature of the earliest urban civilizations, and surely a very powerful shaper of the urban built environment. It explains the dominance of temple complexes as locations of religious, political and economic power, though it should always be remembered that domination of what remains can say more about the permanence of the materials used, and indeed the preferences of excavators, than the actual built environment in its time. We also see the beginnings of a great application of building effort and resources into providing the ruler with a suitably lavish and well appointed gateway into the next world. An early example of this, around 2500, is the royal cemetery at Ur, in which the ruler is buried in an underground chamber. Close by are found the bodies of his now redundant attendants and courtiers, along with the animals and equipment needed for a royal after-life.

Transport

One of the themes of this series will be the ways in which developments in transport influenced the physical form and fabric of cities. The example of Çatal Hüyük, with its lack of streets and its closely packed mass of dwellings entered by the roof, is perhaps the point of departure. Here, transport technology seems to have had no direct effect on the morphology of the settlement, unless ladders are defined as a kind of vertical transport! A little more may be ventured about Sumerian cities. A characteristic morphological feature of the Sumerian city is its network of winding unpaved streets (see Figure 1.18). In addition to ordinary pedestrian movement, they accommodated human porters, and pack-animals such as donkeys. Woolley,

Figure 1.18 Plan of a
residential area, early second
millennium Ur. For Woolley,
the site showed the absence of
town-planning, with winding
lanes, blind alleys, and irregular
plots determined by accidents
of land ownership (Sir Leonard
Woolley, revised and updated
by Moorey, 1982, p.194; by
permission of A. and C. Black)

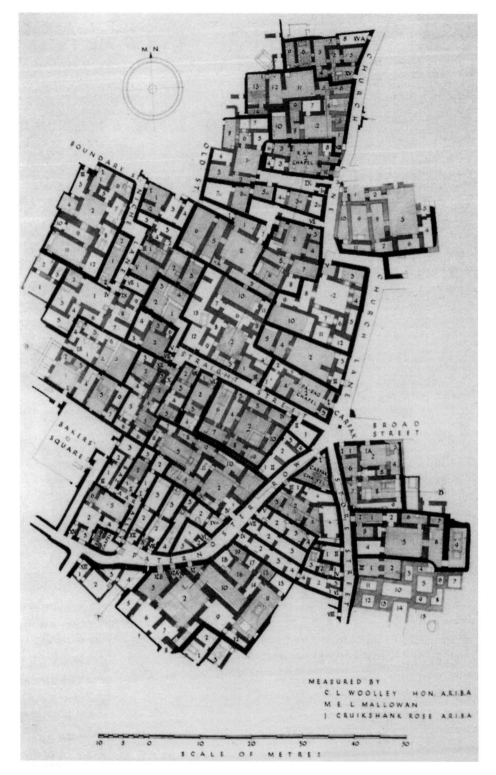

one of the main excavators of Ur, identified a low flight of brick steps against a
house wall as 'clearly a mounting-place for the convenience of riders'. He
assumed, though, that wheeled traffic was debarred from the city:

> Ur was, in fact, a typical Middle East town; its narrow winding lanes are the
> prototypes of those of old Baghdad, and in Aleppo no more than a century ago
> the sight of a wheeled cart or carriage in the streets was so rare as to draw a crowd.
>
> (Moorey, 1982, p.193)

Figure 1.19 Neo-Assyrian aqueduct, carried on arches visible at the right. It supplied water for the park, which was divided by a path leading up a hill to an altar and a royal pavilion. From a relief in the palace at Nineveh, seventh century BCE (Singer *et al.*, 1954, p.553; by permission of Oxford University Press)

Water supply

The need for a natural perennial supply of drinking-water placed a fundamental limitation on the size and spread of ancient urban settlements. The arrangements for the first Sumerian cities were much simpler than the hydraulic engineering required for cultivation. Essentially each city mound was located by a freshwater spring; by various simple means (including pulleys after *c.*1500 BCE), the water was raised from wells, which were usually lined with baked bricks. There was also storage capacity in the form of cisterns. Some developments in water supply and sewerage in the later history of the ancient Near East (earthenware pipes, masonry sewers, water-closets and drains) anticipate the more systematic investment in such facilities made later by the Greeks and, above all, by the Romans. Sennacherib (*c.*704–681), ruler of Assyria in the north of Mesopotamia, built an aqueduct over a distance of eighty kilometres to increase the supply of water to his capital, Nineveh (see Figure 1.19). The main construction materials were blocks of limestone, waterproofed with bitumen, and at one point its 1 in 80 gradient was maintained by an arched structure across a river valley.

The rise of Babylon

In the Mesopotamian region, the political predominance of Sumerian city states such as Ur and Uruk lasted through the Early Dynastic Period, until about 2350. Power then began to shift to the north, to the Semitic peoples who lived in the region known as Akkad (subsequently, as Babylonia). This process began with Sargon of Agade, who founded the first Mesopotamian territorial empire. Agade, Sargon's lost capital, was evidently an administrative centre created for an empire, rather than a self-sufficient entity such as Ur and Uruk – which numbered among Sargon's Sumerian conquests. This move from trading activity to imperial dominion marks a more aggressive stance by the Mesopotamian elite towards meeting their needs for remote resources. Sargon's economic motivation is attested in inscriptions, which list among his conquests the Silver Mountains, the Cedar Forest and the Tin Country (Oates, 1986, p.32).

Sargon's dynasty lasted until around 2150, after which there was a revival of the cities of Sumer during the Third Dynasty of Ur. A more permanent ascendancy of the north, known as the Old Babylonian Period (see Table 1.1), was established in the first half of the second millennium. The best-known ruler of this period was Hammurabi (c.1792–1750), originator of the most famous law code of early Mesopotamia. The code revealed a more complex urban society: private ownership of property and private enterprise were now regularized. This diversification of land ownership and enterprise was momentous in the early history of cities. It reflected the development of the city beyond its original functions as an independent cult centre, and as controller of the irrigation works and food surplus of its agricultural hinterland.

Hammurabi established Babylon as the administrative and military centre of an empire. Though not the earliest, Babylon was probably the biggest ancient Mesopotamian city, its ruins extending over some 850 hectares (Oates, 1986, p.144). It was well placed with regard to river transport, situated as it was on the Euphrates, and in the region where the Tigris and Euphrates ran closest. It was also well placed with regard to overland transport, including the route from Iran to Turkey.

The ascendancy of Babylon was undermined by invasions from the north, and eventually hegemony in Mesopotamia in the first millennium moved to the northernmost region, Assyria, with its capitals of Assur, Nimrud and Nineveh. The Assyrians were a militaristic and agricultural people with a highly autocratic system of government, whose cities were resplendent places of royal residence and centres of imperial control. In this respect they continued the tradition of the Babylonians, diminishing the economic and political independence of the former city states. In order to subdue the city states, the Assyrians developed the technology of siege warfare; they sometimes combined their battering-rams and wheeled siege-towers (see Figure 1.16).

The Babylonians, aided by expert Mede cavalry, managed to reassert their supremacy in the region following the fall of the Assyrian capital, Nineveh, in 612. The outstanding ruler of the New Babylonian or Chaldaean dynasty was Nebuchadnezzar II (604–562), who was responsible for the great rebuilding of

Figure 1.20 Plan of Babylon, incorporating revisions made in the light of textual evidence and excavations in the 1970s and 1980s; some new locations have been suggested for quarters of the city (such as 'Newtown' and 'Kullab'), and for gates in the wall other than those which have been located by excavation. The expression 'TE.E ki' is made up of equivalents of cuneiform signs; the full name of the quarter is unknown (George, 1993, p.739; by permission of Andrew R. George)

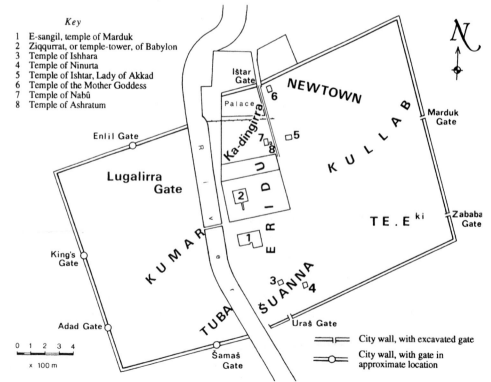

Key

1 E-sangil, temple of Marduk
2 Ziqqurrat, or temple-tower, of Babylon
3 Temple of Ishhara
4 Temple of Ninurta
5 Temple of Ishtar, Lady of Akkad
6 Temple of the Mother Goddess
7 Temple of Nabû
8 Temple of Ashratum

Babylon, to a gridiron plan. What has been excavated of the ancient city was mostly built in this period, which marked the height of the city's power and prestige.

Nebuchadnezzar's city was divided by the Euphrates into two sectors, linked by a bridge, and its inner rectangular area (nearly 400 hectares) was enclosed by a double wall of sun-dried mud-brick (see Figure 1.20). A military road ran between the double walls, and there was a moat on the outside (not visible in the figure), connecting with the Euphrates. The banks of the river were strengthened with masonry bulwarks, and steps led down to landing-stages. In the older eastern sector were the main palaces and temples, and a nearby district of large houses. As was typical in such a hot, arid climate, each house was built around a courtyard, providing cool and shade. The size and relative complexity of the city is shown by another great triple wall beyond the inner city; some eight kilometres long, this afforded protection for the city's suburban districts.

The main temple, dedicated to the city's god Marduk, was called Esagila (E-sangil, in Figure 1.20, location 1). From Esagila, a Processional Way, paved with limestone flags and slabs of the composite stone breccia, ran through the Ishtar Gate to another temple outside the city. The Ishtar Gate, one of the great bronze gates in the city wall, was one of the outstanding building features of the city; it was finished with moulded glazed brickwork decorated with dragons and bulls in relief. Across the Processional Way from Esagila was the city's ziggurat, Etemenanki (location 2 in Figure 1.20), known in the Bible as the Tower of Babel. This is now only a mass of eroded mud-brick, the baked-brick facing having long since been robbed and reused. There was a palace complex near the Ishtar Gate, thought to house the fabled Hanging Gardens (Oates, 1986, pp.144–59). A part of the complex known as the Vaulted Building, supposedly the undercroft of the Hanging Gardens, used hewn stone in substantial quantities. Such usage remains atypical: 'it is not possible to speak of stone architecture in Mesopotamia'; nevertheless, its appearance in Babylon demonstrates that 'the centralized Neo-Babylonian state was geared to providing a much wider range of stone for temple-building through use of river and canal transport' (Moorey, 1994, pp.335, 340).

1.5 Egypt: a civilization without cities?

This Near Eastern introduction to the study of technology and urbanization continues with a brief look at Egypt, partly because at a critical point in the development of a complex society in the Nile Valley, contact was established – initially for commercial reasons – with Sumerian outstations ('colonies') in western Syria. Egypt is also relevant to Chapters 2 and 3 on Athens and Rome because it went on to play a substantial part in a Mediterranean economy successively dominated by the Hellenistic and Roman Empires.

There are striking geographical, political and cultural contrasts between Mesopotamia and Egypt, contrasts which invite consideration of the influence of local conditions on the technologies involved in urbanization. As with the Tigris and Euphrates, the sources of the Nile lay in mineral-rich mountains, but the torrid equatorial climate of this region was a deterrent to exploitation. For much of the period, the Egyptians relied upon the mineral resources of Sinai (copper, turquoise) and the eastern desert (building stone, gold); see Figure 1.21 overleaf. The Nile Valley cut through limestone and sandstone, and so building stone was abundant and conveniently situated for river transport – quite different from the situation in Mesopotamia.

A critical difference between Egypt and Mesopotamia was the flood/harvest cycle. Unlike the broad Mesopotamian alluvial plain on which the Euphrates and Tigris were prone to split and change course, the Nile Valley was a long

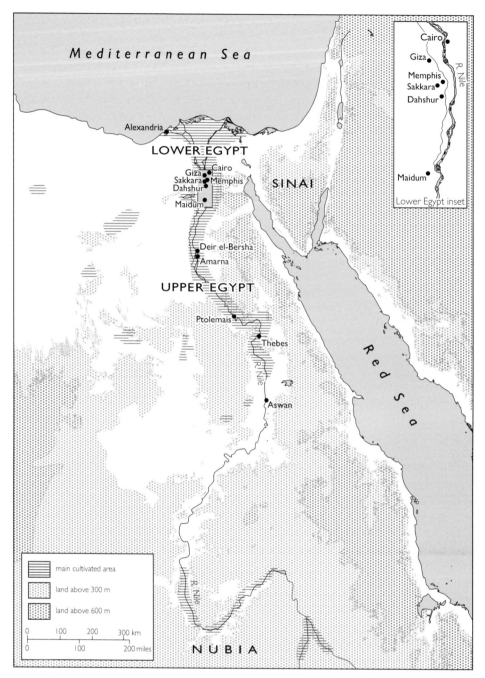

Figure 1.21 Ancient Egypt

strip of fertile land set in the desert, no more than fifty kilometres wide at any point. The flow of the Nile was relatively gentle, and even though it burst its banks every year, the inundation was relatively predictable: indeed it has been seen as the inspiration for the first calendar based on the solar year. The flood was the basis of a sustainable agricultural cycle: the valley floor was under water from mid-July to September; grain was sown after the flood in the autumn, with sufficient moisture in the soil to support most of the growth until the spring harvest; and a period of drought ensued until the next flood. By contrast, in Mesopotamia flooding was a threat to the harvest, rather than a welcome part of a benign cycle. In Egypt, the leaching of the soil by the flood, and its aeration as it cracked during the drought, diminished the build-up of salts through evaporation (salinization); such a build-up was a much bigger problem in Mesopotamia, where artificial irrigation was more prevalent. In Egypt the main hydraulic concern was constructing irrigation canals to supplement the natural supply of water in the flood area, and to extend cultivation beyond it. As in

Mesopotamia, however, a great deal of organized effort was required to bring the swampy Nile Valley under cultivation.

Egypt had more natural barriers to foreign aggression than Mesopotamia: deserts to the east and west of the river; to the north, the marshy Nile delta, which in antiquity had seven outlets and was scarcely navigable; and to the south, the gorges and cataracts of Nubia. By comparison, the urban settlers of Mesopotamia were far more exposed to the raids of neighbouring tribes and of nomads from the north. Some historians have appealed to these considerations to explain a contrast between the 'open' (unwalled) cities of Egypt and the fortified cities of Mesopotamia (Mumford, 1966, p.79). The contrast, however, should not be pressed too far, as there is archaeological and pictorial evidence of fortified settlements at various stages in Egypt's ancient history (Kemp, 1989).

Egypt and Mesopotamia make an intriguing comparison when the theme is the relationship between technological and urban change, because many scholars have denied that there were any true cities in the Nile Valley, especially those who, like Hammond, include economic self-sufficiency in their definition (Hammond, 1972, p.76). Whether or not Egypt's more substantial settlements count as cities, it is clear that the stable, Nile-dominated Egyptian ecology, and the ready availability of many mineral resources, were less of a stimulus to urbanization than in Mesopotamia – where the problem of flood control was more challenging, and long-distance trade in minerals essential. For much of the Pharaonic era, imports were restricted to luxuries such as malachite, gold, gems and spices. Copper and turquoise were brought across the desert by armed convoys from mines in Sinai, and expeditions went up the Nile in search of gold and ivory. However, increasing scarcity of timber eventually brought the Egyptians into the eastern Mediterranean in wooden ships, powered by sails and oarsmen, in order to trade with North Syria through the port of Byblos (see Figure 1.1).

The path of Egyptian civilization was quite different from that of Mesopotamia, in that a unified nation emerged before any substantial built-up settlements. This may be attributed in part to the uniformity of the Nile ecology and inundation pattern, which arguably encouraged the co-ordination of water control over the entire valley, rather than the more unpredictable and variable environment in which the system of independent Sumerian city states developed. Childe argued that this uniformity and predictability also made more likely a monarchical regime that claimed magical powers to influence the natural order (Childe, 1966, pp.157ff.).

The geographical uniformity of the Nile Valley can be overstated. There is a natural break between the delta region of Lower Egypt and the long valley of Upper Egypt; and, initially, two separate kingdoms of Upper and Lower Egypt were formed from the original Neolithic tribes. These kingdoms were unified around 3200, under a ruler who subsequently took the name of Pharaoh. The framework of Egyptian history under the Pharaohs is still essentially that proposed by an Egyptian priest in the third century BCE (Kemp, 1989, p.14; see Table 1.2).

According to tradition, the first Pharaoh organized a great system of canals, and the first damming of the Nile. For well over two millennia, Egypt was governed in a much more uniform way than Mesopotamia. Although there were intermediate periods of relative weakness at the centre, and of foreign rule, the main eras of Pharaonic rule (the Old, Middle and New Kingdoms) had the same fundamental character – the absolute rule of a monarch regarded as an incarnate deity, who owned all land and agricultural surplus, directly controlled all trade, and oversaw the activity necessary to control the river. This personal rule operated through a territorial organization and not through cities, the existence of which

Table 1.2 Egypt under the Pharaohs (based on Kemp, 1989, p.14)

Dates (BCE)	Period*
3050–2695	Early Dynastic (1–2)
2695–2160	Old Kingdom (3–8)
2160–1991	First Intermediate (9–11)
1991–1785	Middle Kingdom (12)
1785–1540	Second Intermediate (13–17)
1540–1070	New Kingdom (18–20)
1070–712	Third Intermediate (21–24)
712–656†	Kushite/Assyrian rule (25)
†664–525	Saite (26)
525–332	Late (27–31)

* The numerals in brackets indicate the dynasty number(s)

† These overlapping dates reflect the complexity of late Pharaonic Egyptian history, when there were sometimes rival dynasties based in different parts of the country

depended entirely on the Pharaoh's will. For Hammond, therefore, there were
no true cities, as they were not economically self-sufficient. But according to
Childe, political unification led to a state based on secondary industry and
commerce, as well as on food production; the conditions were therefore in
place for his Urban Revolution. It is surely no accident that Childe, a Marxist
writing originally in the inter-war period, was much more persuaded than
Hammond of the benefits of a powerful state. He celebrated the internal peace
it provided, and the protection it afforded from foreign aggression and the
incursions of desert tribes (Childe, 1966, pp.157, 161).

Childe did distinguish between cities and the kind of estate (depicted in
paintings in Old Kingdom tombs) that produced a permanent supply of
offerings for the dead ruler's after-life. Although potters, smiths, jewellers and
carpenters worked on them, these estates were more like medieval manors
than cities in size and function. Like Hammond, Childe also distinguished the
economically independent Sumerian city state from the Egyptian city, which
was entirely dependent on the function assigned to it within the monarchical
state; without this function, it would quickly relapse into a self-sufficient
agricultural settlement (Childe, 1966, pp.166–7). He simply drew a different
conclusion about the urban status of the Egyptian type of settlement.

The relationship between technology and urbanization needs to be
considered more closely. How did Egyptian technology compare with that of
Mesopotamia, and do such comparisons throw any light on the nature of
Egyptian urbanization?

> The idea that early Egypt achieved a higher state of technological skill than other
> countries of the Near East is purely illusory and stems from the better
> preservation of many of the materials in that country than elsewhere. More
> commonly than not, we shall find that Egypt received her technologies from her
> neighbour countries at a rather later date than elsewhere.
>
> (Hodges, 1971, p.45)

Indeed, the transition to agriculture in Egypt probably took place some four
millennia later than in the original rain-fed areas of the Fertile Crescent, but not
much later than on the comparable Sumerian flood-plain. It is now thought
that the Nile only began to deposit the silts of the delta region, through which
agricultural practice would have diffused, from the middle of the seventh
millennium. The Egyptian Neolithic people then had to set about clearing the
marshy areas of the lower reaches; this would have removed some natural
barriers to the annual flood, and therefore intensified it. This was a further
stimulus to the control of the flood by dikes and channels, though such works
were never on the same scale as many in Mesopotamia.

Stone implements continued to be used in Egypt long after metal tools had
become common in Mesopotamia. Copper appeared during the fourth
millennium, but because no tin was discovered in ancient Egypt, copper was
probably replaced by bronze later than in Mesopotamia. However, all bronze
objects were luxury items in the Near East until the second millennium.

If transport innovations are deemed necessary for urbanization, then Egypt
makes an intriguing case-study: depictions of Egyptian papyrus rivercraft
constitute the very first evidence of boats with sails. The oldest known plank-
boat, dated about 2650, was uncovered beside the Pyramid of Khufu (Cheops).
But the introduction of wheeled vehicles may have happened later than in
Mesopotamia: there is no record in Egyptian art of carts with solid wheels.
Military chariots with spoked wheels were introduced from the Levant in the
second millennium. This notable difference underlines the importance of the
local natural environment in the technological development of ancient
civilizations. River transport was vastly easier on the north-flowing Nile than
on the south-flowing Tigris and Euphrates, because the prevailing wind blew

from the north. This meant that, on the Nile, boats could travel north with the current, and use the wind to travel south – against the current. On the Tigris and Euphrates, on the other hand, both the current and the wind pushed boats in a southerly direction, making it difficult to travel north (see p.12). This benign circumstance of the Nile Valley surely favoured the invention in Egypt of sails for travelling upstream; for land transport, the Egyptians could make do with pack-donkeys, rather than invest effort and resources into the technological system of wheeled transport, including suitable draught-animals and roads. The dominance of water transport strongly influenced the system of cities in Egypt. It would, however, be hard to show that it affected the internal layout of cities; as has already been seen in the case of Ur (see p.28), even where wheeled transport existed in the Near East, it seems scarcely to have influenced urban morphology.

Building technology

Early settlement in the Nile Valley was confined to the cliffs above the high level reached by the annual flood. Thereafter, settlement on the valley floor was effected by raising villages on mounds. This much is comparable with the situation in Mesopotamia, but there were notable divergences in the building practices of the two regions. As in Mesopotamia, sun-dried brick was the main building material for residential and utilitarian structures in Egypt, and its use from around 3600 seems to have been a direct Sumerian influence: walls appeared with details similar to Sumerian structures. Since stone was near at hand for buildings that needed to be durable and weatherproof, there was no corresponding transition to kiln-fired bricks; instead, the Egyptians developed the arts of quarrying and stonemasonry. The state had sole power over the use of stone, thus ensuring an even more conspicuous social and political shaping of the built environment in Egypt than in Mesopotamia. Despite the availability of stone, arches were used sparingly, and often in the form of a false or corbelled arch, rather than the so-called true arch in which stone wedges, or voussoirs, are held together in compression (Clarke and Engelbach, 1990, pp.181ff.; see Figure 1.22).

The most striking Egyptian building types are funerary structures, beginning with mastabas (from the Arabic for 'bench'), which typically consisted of an underground burial pit surmounted by a rectangular, bench-like, mud-brick structure containing provisions for the after-life. The burial pits were dug ever

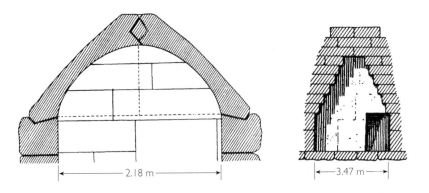

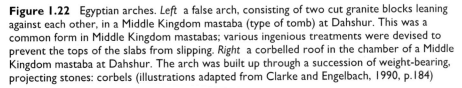

Figure 1.22 Egyptian arches. *Left* a false arch, consisting of two cut granite blocks leaning against each other, in a Middle Kingdom mastaba (type of tomb) at Dahshur. This was a common form in Middle Kingdom mastabas; various ingenious treatments were devised to prevent the tops of the slabs from slipping. *Right* a corbelled roof in the chamber of a Middle Kingdom mastaba at Dahshur. The arch was built up through a succession of weight-bearing, projecting stones: corbels (illustrations adapted from Clarke and Engelbach, 1990, p.184)

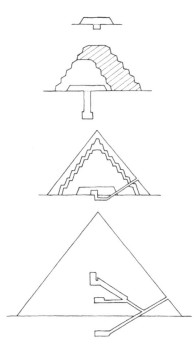

deeper to guard against robbery. Mastabas were built of mud-brick and timber, until about 2600 when stone began to be used. The arrangement of underground tomb and house of the dead developed into the pyramid, which at first had a stepped profile, until the familiar smooth limestone face was introduced (see Figures 1.23a and b). The first stone pyramid, the Step Pyramid of Djoser at Sakkara overlooking the city of Memphis, was built about 2700. The three great pyramids at Giza date from the Fourth Dynasty (2600–2500). At this time, mummification techniques were also developed, giving rise to the specialized occupation of embalming. After the pyramids, tombs – with complex networks of galleries and shafts – were cut out of rock, as the Pharaohs sought even greater security for their posthumous possessions.

How were the pyramids built? They were prodigies of administration, engineering and labour: 'It was the scribe's pen, as much as the overseer's lash or the engineer's ingenuity that built the pyramids' (Kemp, 1989, p.129).

Figure 1.23a The development of the pyramid, from the mastaba (top) to the Pyramid of Khufu (bottom). The second diagram from the top is of the Step Pyramid of Djoser: the original mastaba was built over in two stages. Third from the top is the Pyramid of Seneferu at Maidum (2700 BCE), which retained the mastaba and step structure but was given a smooth facing of limestone. Finally, in the Pyramid of Khufu, the mastaba and steps were eliminated (Hodges, 1971, p.104)

Figure 1.23b The Giza group of pyramids, with the Great Pyramid, or Pyramid of Khufu, in the foreground. This building rose to 147 metres, a building height unsurpassed until the skyscrapers of the twentieth century (Hodges, 1971, p.104; by courtesy of The Museum of Fine Arts, Boston)

They were meticulously planned, on the basis of practical geometry: the base of the Pyramid of Khufu at Giza, for example, is impressively near to a perfect square, with a linear discrepancy of less than twenty centimetres and a maximum angular error of three and a half minutes of arc – an overall level of accuracy better than 1 in 1,000 (Edwards, 1985, p.99). Such accuracy was achieved with little theoretical basis and the simplest of instruments, such as knotted measuring-cords, wood or stone cubit rods, and set squares for the sighting of right angles (Clarke and Engelbach, 1990, pp.65ff.; see Figure 1.24). The achievement attests to the accumulated experience of Egyptian land-surveyors in practical geometry. According to the Greek historian Herodotus, this was acquired through the need, under the prevailing system of land registration and taxation, to remeasure field boundaries after each annual flood.

To assemble the Pyramid of Khufu, huge blocks of limestone and granite were cut from quarries near the Nile, using no more than copper chisels or diorite stone picks, the expansionary power of wetted wooden wedges inserted into cuts in the rock face, and an untold input of human labour. By means of levers, sledges, rollers and ropes, great teams of labourers shifted the blocks to the river, where they were put on a barge and transported downstream (see Figure 1.25). At the site of the pyramid, another team of men using massive papyrus ropes dragged the blocks by sledge up a ramp to a level some thirty metres above the river. An estimated two and a half million

Figure 1.24 Surveyors, equipped with measuring-cords and cubit rods, measuring a field. From a New Kingdom tomb in Thebes, c.1400 (Singer et al., 1954, p.541; by permission of Oxford University Press)

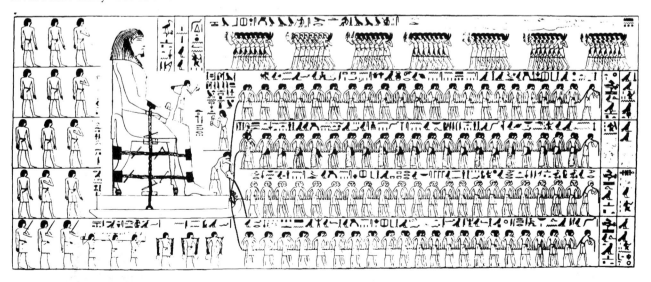

Figure 1.25 Transporting a statue of the Pharaoh Djehutihotep. The men below the statue are carrying wood for the track on which the statue was dragged. A man is pouring water in front of the sledge; this has been seen as a method of lubrication, but is more likely to have been a purification ritual. From a wall-painting in the tomb of Djehutihotep at Deir el-Bersha, dated about 1880 BCE (Cotterell and Kamminga, 1990, p.22, after a lithograph in Layard, 1853; by permission of Cambridge University Press)

blocks were used, averaging about two and a half tonnes each. According to Herodotus,[3] 100,000 men worked on the Pyramid of Khufu for three months a year for twenty years, presumably the three months when the annual Nile flood prevented agricultural work. Although modern scholars reckon that the number of workmen is greatly exaggerated, the fact that so many men dedicated so much of their lives to housing a dead body speaks volumes about the social context of Egyptian building. The Egyptians have been portrayed as being especially preoccupied with the after-life, though this perception is partly shaped by the fact that tombs and their contents are virtually all that survives of ancient Egyptian civilization down to about 2000. The main reason for this is that for residential buildings, even including royal palaces, the Egyptians used perishable sun-dried brick.

The occupational specialization and technical development emphasized by Childe in the process of urbanization are readily inferred from the contents of tombs. Everything thought necessary for a comfortable after-life was included: excavation has revealed furniture, pots containing food and beer, and weapons and ornaments (many made with imported materials such as cedarwood, copper, obsidian and lapis lazuli, and embodying the craft skills of carpenters, smiths, masons and jewellers). More is known about craft technologies in Egypt than in Mesopotamia, because of the depiction – in the wall-paintings of tombs – of Egyptian methods of carpentry, shipbuilding, pottery, weaving, brewing and winemaking, and manufacture of stone vessels; this suggests that crafts carried more social prestige in Egypt. Stone-working was especially well developed. In the manufacture of stone vases, a wooden-shafted, flint-edged drill with heavy weights attached was used to hollow out the block of stone. Multiple drills enabled craft specialists to work on more than one stone bead at a time; this practice has been described rather fancifully as an ancient example of factory mass-production (Stocks, 1989; see Figure 1.26). The abundant papyrus reeds of the Nile delta provided material not only for boats and ropes but, when beaten out, for paper – another example of a local technological solution to the urban imperative to record agricultural production and trade.

In what ways were the pattern of urbanization (if it warrants the name) and the urban built environment shaped by the social and political structure of ancient Egypt? This was a uniform and hierarchical society, with the bulk of the population living in agricultural villages, organized into administrative districts called 'nomes', with governors appointed by the Pharaoh. Each nome had its capital. Memphis emerged as an overall capital for Lower Egypt, but several sites served as capitals in Upper Egypt. Were these capitals true cities? The problem of the perishability of mud-brick structures, including city walls, obtrudes here. Since the Pharaohs tended to build their capitals on new sites, there are fewer Egyptian locations like the many-layered Mesopotamian tells of accumulated building relics and rubbish. Moreover, many ancient settlements are buried beneath modern cities, or too deep under the Nile alluvium to be brought to light. But excavations revealing more planned settlements, often with enclosure walls, have led archaeologists to place Egypt more in the mainstream of classical urban civilization (Kemp, 1977, 1989). This is a corrective to the conclusion most readily drawn from the durable remains of Egyptian urban civilization – that the Pharaohs' absolute power channelled a disproportionate amount of building effort into providing for the royal after-life, instead of creating a living city.

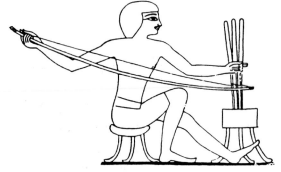

Figure 1.26 A craftsman spinning three drill-rods simultaneously. In this case, the cutting power of the drill-rod (probably bronze, with a wooden handle) is augmented, not by the attachment of heavy weights but by the extra speed of rotation imparted by the bow-string. From a wall-painting in the New Kingdom tomb of Rekhmire at Thebes (Stocks, 1989, p.527; by permission of Denys A. Stocks)

---

[3]   The relevant extract from Herodotus' *History* is given in the Reader associated with this volume (Chant, 1999).

Thebes

An Egyptian settlement arguably worthy of the label 'city' is Thebes. This is the Greek name for a city the Egyptians called Waset or No-Amun, not to be confused with the city of Thebes on the Greek mainland. Its remains fuel the debate above about whether Egyptian Pharaonic administrative centres were 'true cities'. Was the essence of Thebes its necropolis (city of the dead) on the west bank of the Nile, a community of officials and workmen connected with the palace, temples and tombs dedicated to the after-life of the Pharaoh and his courtiers?

Thebes is located well up the Nile, some 500 kilometres south of Memphis (see Figure 1.21); it only came to prominence in Egyptian history at the end of the Old Kingdom, after a period of weakened central authority, perhaps associated with low water levels and famine. It was the princes of Thebes who reunited the country and inaugurated the Middle Kingdom. The ensuing period of renewed prosperity and building activity resulted in 'the first known major city to be laid out on axial lines in ancient times' (Uphill, 1988, p.54). The remains of this city on the east bank of the Nile constitute the mound on which the New Kingdom Karnak temple-complex stands; excavations have revealed part of the Middle Kingdom city wall, some six metres thick (Kemp, 1989, p.160).

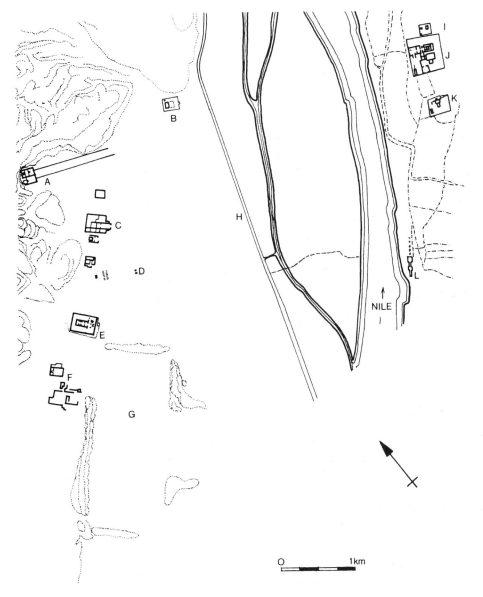

Figure 1.27 Plan of Thebes, showing the principal urban areas on both sides of the river

Key A: Deir el-Bahri; B: Qurna temple; C: Ramesseum; D: colossi of Memnon; E: Medinet Habu; F: Tehen Aten complex; G: harbour; H: canal; I: Temple of Monthu; J: Karnak, Temple of Amun; K: Temple of Mut; L: Luxor temple (Uphill, 1988, p.53; drawing: Helena Jaeschke)

It was also a noble Theban family that expelled the Hyksos from Egypt. The Hyksos were the rulers of a Palestinian desert people who periodically infiltrated the delta for grazing and, from about 1700 BCE, established political supremacy in Lower Egypt – partly through the use of their horse-drawn military chariots. In about 1540 the victorious Theban family founded the New Kingdom, in the course of which the Egyptian Empire expanded into Palestine and Syria. Hammond was prepared to concede that there may have been two true cities during this period – Thebes and Amarna. He still, however, inclined to the view that Thebes was a town for bureaucrats and officials rather than a city in his preferred sense (Hammond, 1972, p.73). Amarna (now known as Tell el-Amarna) was a new capital, built further down the Nile during the first half of the fourteenth century by the Pharaoh Akhenaten. In addition to a stone temple and two palaces, the site reveals a mud-brick workmen's quarter – an example of Egyptian town-planning, with its parallel streets enclosing regular blocks within a rectangular wall.

Apart from the rock tombs in the nearby Valley of the Kings, the remains of New Kingdom Thebes show multi-storey buildings, and a broad street connecting the temples at Karnak and Luxor on the east bank of the Nile (see Figure 1.27). Typical of Egyptian monumental building was the colossal Hypostyle Hall, with 134 columns, and pylon gate, built in front of the temple at Karnak (see Figure 1.28). These temples demonstrate both the power of Egyptian building technology and the constraints under which it operated. Because there was no true arch, columns could not be far apart: when the lintels were of limestone, the space between columns could be no more than three metres, though the widespread use of sandstone during the New Kingdom enabled spans to be tripled. The colossal dimensions of the columns at Karnak were achieved by building up a succession of drums.

Even though the remains of the city are dominated by the cult of the royal dead, there are clues to the size and activity of the ancient living city. Attached

Figure 1.28 Part of the 19th Dynasty Hypostyle Hall in the Temple of Amun at Karnak. 'Hypostyle' means that the roof is supported on columns. The way in which the columns were constructed with a succession of drums is clearly shown (photograph: Heritage and Natural History Photography, Dr John B. Free)

to the Ramesseum, the stone mortuary temple of Rameses II on the west bank, is a series of large, vaulted mud-brick granaries, with the capacity to support up to 20,000 people (Kemp, 1989, p.195). There is also in the immediate vicinity of the western necropolis another surviving example of a workmen's town, containing tools and other everyday objects, as well as a rich store of written records yielding evidence of private production and enterprise conducted alongside state work for the necropolis. This is Deir el-Medina (see Figure 1.29). It dates from the New Kingdom, and was originally surrounded by a mud-brick wall over six metres high. Despite its modesty in the face of the awesome religious complexes, the urban character of Thebes depends to some extent on the significance attached to such remains (David, 1986, pp.56–7).[4]

Several points of similarity and of difference have emerged from this brief comparison of urbanization in Mesopotamia and Egypt. In Egypt, as in Mesopotamia, there were substantial settlements with monumental buildings and occupational specialization, and sometimes with walls and orthogonal layouts. Only historians attached to a clearly culturally-bound definition, emphasizing economic function, would deny that any were cities.

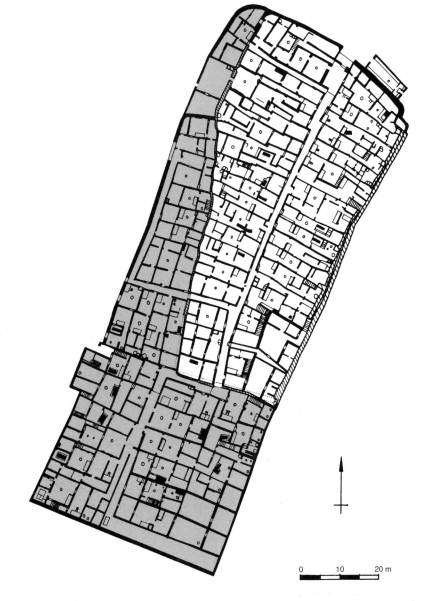

Figure 1.29 Plan of Deir el-Medina. The original mud-walled settlement, shown here as the unshaded area, had one main street – with side alleys. The settlement expanded to the south and west, as indicated by the shaded area, and was enclosed by a stone wall. There were no wells; instead, a reservoir outside the north gate was filled by water carriers from the Nile Valley (adapted from Uphill, 1988, p.22; drawing: Helena Jaeschke)

0   10   20 m

---

[4] There is an extract from David (1986) in Chant (1999) – the Reader associated with this volume.

1.6 Conclusion

The emergence of cities in the Near East depended, at least in part, on the application of humanity's technological aptitude within a region of critical resource imbalance. With great organized effort, sufficient agricultural surplus could be extracted to support other members of that society in the similarly prodigious effort required to obtain raw materials lacking in that area. It seems that the discipline necessary for these efforts engendered a hierarchical and authoritarian political system, characterized by rulers who insisted that the surplus, in the form of the raw materials for which it was exchanged, should be invested in structures built to reinforce their authority. The question as to which came first – the technological effort, or the social structures necessary to see it through – may never be satisfactorily answered, because of the problematic nature of the evidence for ancient settlements. The relative paucity of evidence does not, however, deprive the debate of its value. Apart from the intrinsic importance in world history of its substance, the debate poses in polarized form the issues which have to be addressed in any discussion of the relationship between the urban and technological variables. It also supports a general proposition which is not always self-evident – that however objective and discriminating historians seek to be, their predispositions can have a decisive influence on the conclusions they construct from their evidence.

   This chapter has raised another issue of relevance to the series as a whole. It might be accepted from the discussion of Mesopotamian and Egyptian cities that certain technologies were necessary for the emergence of cities. But it might also be inferred that the geographical setting, rather than technology, is the variable that most influences the pattern of urbanization in the two regions. It is doubtful, though, whether the contributions of geography and technology can be uncoupled: technology, after all, is about the relationship between human aspiration and the natural environment. The reason why the geography of an urbanizing region is seen by many to be decisive in antiquity is surely because technologies were more localized during that period. According to this argument, any polarity between technological and geographical determinism is illusory; the main issue, as far as the physical form and fabric of cities is concerned, continues to be the social relations of technology, as expressed in the urban built environment.

Extracts

1.1   Oates, J. (1986, rev. edn) *Babylon*, London, Thames and Hudson, pp.14–15

With this short summary we may perhaps attempt to assess the possible formative role played by these various environmental factors in the development of Mesopotamian society, although to do so briefly must lead to gross over-simplification. Indeed any such comments must be highly speculative since the diversity of processes that can be observed begin well before the advent of writing. Nonetheless it seems incontrovertible that the peculiar physical environment of southern Mesopotamia, with its exceptional agricultural potential but its lack of natural resources, was a positive stimulus, although certainly not the only one, towards the growth of the world's first cities. It is well known that some of the world's oldest farming communities lie to the north of Babylonia, within the mountainous regions of Iran, Iraq, the Levant and Turkey and on the adjacent rain-fed grasslands, where those plants and animals that were to lend themselves to early and successful domestication – sheep, goat, cattle, pig, barley, emmer and einkorn wheat – were readily available in the wild state. Although it is axiomatic that the development of farming and herding preceded the growth of cities, it is significant that the first cities not only in the Tigris and Euphrates Valleys but also along the Nile and Indus arose outside those 'nuclear' areas where agriculture was first practised. Clearly, more intensive agricultural techniques were required in areas marginal to those endowed with adequate wild resources, and most of all in those regions, such as Sumer, that were climatically unsuited to simple rain-fed farming. Not only did technical innovations such as irrigation and the plough, both attested before the dawn of history, make possible the intensive occupation of these otherwise unproductive areas, but the employment of such techniques helped increase agricultural efficiency and even the proportionate yield of the land. These developments served to free some members of the community for part, if not whole-time, specialization, and led ultimately to a definite polarization of society into those who controlled resources, such as land, manufacturing or trading enterprises, and those dependent on them. The viable size of settlements was directly related to these technological advances. Increased production resulted also in surpluses whose redistribution ultimately required special social and political institutions.

Although recent surveys undoubtedly demonstrate that *large-scale* integrated irrigation schemes did not appear in Babylonia until after the development of the political phenomenon known as the city state was well advanced, there can be little doubt that the necessity to control scarce water supplies amicably and distribute them equitably must have been a strong cohesive force within early village communities in Babylonia. Even the most modest irrigation schemes demand a level of local co-operation that would have been unnecessary under conditions of rain-fed agriculture, although co-operation is perhaps too strong a word in view of later textual evidence for the Babylonians' love of litigation, especially over water rights!

Thus in Sumer the growth of specialization, the internal differentiation of society and especially the development of community agencies to collect and distribute agricultural and manufactured surpluses, can be seen as clearly related to, though of course not an *inevitable* result of, the deficient environment. Such technological and economic developments were essential to local growth and prosperity; they also led to more effective methods for the acquisition of raw materials, not only for everyday necessities but also for a growing market in luxury items and the manufacture of products with which yet more imported goods could be acquired. At the same time, the open nature of the Babylonian terrain served both to discourage social isolation and to facilitate the rapid spread of new ideas, whether technical or political, while the lack of raw materials engendered an outward-looking attitude that was to influence political thinking and encourage expansion. This can be seen in the archaeological evidence already by 5000 BC and more specifically in the extensive trading networks of the late 4th and early 3rd millennia. In such ways the physical environment of Babylonia influenced and almost certainly served to accelerate the growth there of urban society, though we must remember that it was the people themselves who created their institutions within the limitations imposed on them by external forces.

1.2   Flannery, K.V. (1994) 'Childe the evolutionist' in D.R. Harris (ed.) *The Archaeology of V. Gordon Childe: contemporary perspectives*, London, UCL Press, pp.104–5

The 'Rank Revolution'

What happened next in the Andes and Mesoamerica was just as 'revolutionary' as the origins of food production: the leap from an egalitarian society to one with hereditary inequality, regional civic-ceremonial centers, loss of village autonomy, and lavish use of manpower in the construction of public buildings. I will call this 'the Rank Revolution', and suggest that it was out of focus for Childe because it involved changes in *ideology* and *social relationships* rather than the *means of production*. Childe made it very clear in his *Town Planning Review* article (Childe, 1950) that the 'technique of obtaining food [was used] to distinguish the consecutive stages termed savagery and barbarism' and that 'the density of population is determined by the food supply which in turn is limited by natural resources, the techniques for their exploitation, and the means of transport and food preservation available'. For Childe, 'social division of labor' was impossible at the stages of savagery and barbarism 'save those rudiments imposed by age and sex'.

But between 1800 BC and AD 300 in Peru, and between 1200 BC and AD 150 in Mexico, we see the rise of 'chiefdoms' or 'rank societies'. These new societies did not necessarily produce food differently from their Neolithic predecessors, but many used raiding or simple warfare to enforce tribute from outside their local resource sphere; they also supported craft specialists of high quality. Their great revolution lay not in a new way of producing food, but in a new ideology in which chiefly individuals and commoners had separate genealogical origins. That ideology justified the demands of greater output and productivity made by elite individuals and carried out by lower-status individuals. Such ideologies do not characterize egalitarian societies, who generally have leveling mechanisms to prevent society from being divided into elite and commoner. Among some Pueblo Indian societies of the south-western US, for example, families who were able to accumulate more luxury goods, store more surplus, and establish more ostentatious life styles often faced accusations of witchcraft and perhaps even expulsion from the community. The Rank Revolution overcame such leveling mechanisms by rationalizing the perquisites of the elite.

Surplus figured importantly in Childe's evolutionary framework, but he seemed to see it as the logical consequence of better technology, which in turn made inegalitarian society inevitable. Many of today's New World archaeologists, on the contrary, would see Neolithic people as 'satisfizers' for whom there was no incentive to produce surplus. The challenge, as Sahlins (1972) has pointed out, is 'to get people to work more, or more people to work'. Many New World archaeologists do not believe that this was likely to happen until there was an elite to command it. The Rank Revolution produced some of the most spectacular sites and public works in Nuclear America. Whether we are speaking of Complex A at La Venta in Mexico, with its earthen pyramids and buried serpentine pavements, or the stone masonry temple of Chavin de Huantar in Peru, with its Great Image and labyrinthine galleries, the scale of corvée labor and monumental construction exceeds anything seen before hereditary inequality arose.

It is at this stage of cultural evolution that we come face to face with the variable Childe left out of his scheme: warfare. Childe … was a pacifist. He saw warfare as a social ill – something that destroyed surplus, that crucial component of Childe's evolutionary model. Hence he could not see that warfare, long considered a destroyer of complex societies, was also one of the factors creating complex societies. But evidence for armed conflict, raiding, and terrorism is everywhere in Nuclear America at the rank society or 'chiefdom' level of evolution. From Cerro Sechin on the north coast of Peru, with its graphic carvings of trophy heads, enemies quartered and disemboweled, to the fortification walls and defensible locations of Late Formative sites in Mexico's Valley of Oaxaca, or the carved scenes of decapitation in Chiapas, it is clear that rank societies were violent societies.

1.3 Butzer, K.W. (1996) 'Irrigation, raised fields and state management: Wittfogel redux?', *Antiquity*, vol.70, no.267, pp.200–201, 203–4

The Wittfogel model, like Elvis, refuses to die. And like the impersonators of Elvis Presley who earn their keep by rocking around the clock, Karl Wittfogel's 'hydraulic hypothesis' (Wittfogel, 1938; 1957) continues to be repackaged in a variety of guises that assign a unique causal role to irrigation in the development of socio-political complexity. In analogy to the Industrial Revolution, V. Gordon Childe long ago propagated the concepts of Neolithic and Urban revolutions (see Harris, 1994). These were debated, but more importantly, they served to stimulate both archaeological and ethno-historical research of substantial importance. Thus, studies of urbanism revealed that the processes of urban evolution not only were incremental, but that the very nature of urbanism was to some degree unique to particular historical, cultural and ecological contexts. While the term 'Urban Revolution' has not been used for quite some time, the impact of Childe in channelling fresh investigations of historical urbanism has been substantial. Similarly, Childe's Neolithic Revolution set in train broadly conceived empirical research, first into the 'origins', then into the processes of plant and animal domestication. Again there was no universal model, but that no longer is disappointing: it is precisely the variety of alternative pathways to domestication and agricultural subsistence, and the many different social and ecological contexts of early agricultural transformation, that are interesting and informative. The 'Neolithic Revolution' is a term now only found in popularizing tracts. That does not discredit Childe, who was a major catalyst for archaeology and other disciplines interested in socio-economic evolution …

Wittfogel's thesis can best be overviewed in an early summation (Wittfogel, 1938), that is less complicated by tangential arguments and his anguish with the terrors of totalitarianism. He saw himself as an inductive student of social process, rather than a dogmatic, Marxist historian … he proposed a multilinear theory of cultural evolution, grounded in Marxist concepts, but elucidated by the empirical scholarship of his day. A 'Western' trajectory had led to industrial capitalism, primarily because authority there was traditionally decentralized. Another, 'eastern' trajectory saw the emergence of highly centralized societies, controlled by an autocratic sovereign and a powerful priestly-bureaucratic caste, that employed astronomy to predict nature's cycles. For the purposes of flood protection and a good water supply, they mobilized massive labour forces to build monumental dykes and canals, that assured high productivity as well as managerial control over their populations. As the empowered sought to perpetuate the economic and technological order that supported them … such societies tended to economic stagnation rather than innovation. For 1938, these were prescient and innovative ideas, if we are not sidetracked by the Marxist framework.

… during the 1960s, some social scientists pounced on the notion of irrigation as the 'engine' of history, resulting in a welcome spate of ethnographic and archaeological field studies. A long list of cross-cultural studies revealed that, despite some level of co-ordination, locally based socio-political organizations managed water-related problems in an un-centralized and non-hierarchical fashion. Furthermore, differences in managerial practices reflect particular cultural heritages at least as much as they do particular environmental contexts. The hang-up seems to be the tenacious assumption that early forms of intensification were a result of socio-hierarchical demands, rather than cumulative, small-scale, local decision-making. Obsessed with the vertical dimensions of socio-political complexity, we have neglected to give horizontal structures their proper due.

Yes, Wittfogel's method and his positions are dead, and they should be buried. But as in the case of Childe, we must acknowledge him as an innovative thinker in his day. He stimulated a suite of excellent studies on irrigation that eventually rendered his hypothesis obsolete … the underlying issues need reformulation, so that empirical research can be more sharply focused. If and when we can turn that corner, and stop rehashing Wittfogel, flowers will indeed be appropriate.

References

ADAMS, R. MCC. (1966) *The Evolution of Urban Society: Early Mesopotamia and Prehispanic Mexico*, London, Weidenfeld and Nicolson.

ALGAZE, G. (1993) *The Uruk World System: the dynamics of expansion of Early Mesopotamian civilization*, Chicago, University of Chicago Press.

ASHWORTH, G.J. (1991) *War and the City*, London, Routledge.

BIJKER, W.E., HUGHES, T.P. and PINCH, T.J. (eds) (1987) *The Social Construction of Technological Systems: new directions in the sociology and history of technology*, Cambridge, Mass., MIT Press.

BUTZER, K.W. (1996) 'Irrigation, raised fields and state management: Wittfogel redux?', *Antiquity*, vol.70, no.267, pp.200–204.

CARTER, H. (1983) *An Introduction to Urban Historical Geography*, London, Edward Arnold.

CASSON, L. (1974) *Travel in the Ancient World*, London, George Allen and Unwin.

CHANT, C. (ed.) (1999) *The Pre-industrial Cities and Technology Reader*, London, Routledge.

CHILDE, V.G. (1950) 'The Urban Revolution', *Town Planning Review*, vol.21, no.1, pp.3–17.

CHILDE, V.G. (1966) *Man Makes Himself*, London, Fontana (first published 1936).

CLARKE, S. and ENGELBACH, R. (1990) *Ancient Egyptian Construction and Architecture*, New York, Dover Publications (first published 1930).

COTTERELL, B. and KAMMINGA, J. (1990) *Mechanics of Pre-industrial Technology: an introduction to the mechanics of ancient and traditional material culture*, Cambridge, Cambridge University Press.

DAVID, A.R. (1986) *The Pyramid Builders of Ancient Egypt: a modern investigation of the Pharaoh's workforce*, London, Routledge and Kegan Paul.

EDWARDS, I.E.S. (1985, rev. edn) *The Pyramids of Egypt*, Harmondsworth, Penguin Books (first published 1947).

FLANNERY, K.V. (1994) 'Childe the evolutionist' in D.R. Harris (ed.) *The Archaeology of V. Gordon Childe: contemporary perspectives*, London, UCL Press, pp.101–19.

GAMBLE, C. (1995) *Timewalkers: the prehistory of global colonization*, Harmondsworth, Penguin Books (first published 1993).

GEORGE, A.R. (1993) 'Babylon revisited: archaeology and philology in harness', *Antiquity*, vol.67, pp.734–46.

HAMMOND, M. (1972) *The City in the Ancient World*, Cambridge, Mass., Harvard University Press.

HARRIS, D.R. (ed.) (1994) *The Archaeology of V. Gordon Childe: contemporary perspectives*, London, UCL Press.

HODGES, H. (1971) *Technology in the Ancient World*, Harmondsworth, Penguin Books.

HUGHES, T.P. (1994) 'Technological momentum' in M.R. Smith and L. Marx (eds) *Does Technology Drive History? The dilemma of technological determinism*, Cambridge, Mass., MIT Press, pp.101–13.

KEMP, B.J. (1977) 'The early development of towns in Egypt', *Antiquity*, vol.51, no.203, pp.185–200.

KEMP, B.J. (1989) *Ancient Egypt: anatomy of a civilization*, London, Routledge.

KENYON, K.M. (1979, 4th edn) *Archaeology in the Holy Land*, London, E. Benn.

KENYON, K.M. (1981) *Excavations at Jericho: the architecture and stratigraphy of the tell*, London, British School of Archaeology in Jerusalem.

KUHRT, A. (1995) *The Ancient Near East c.3000–330 BC*, London, Routledge, vol.1.

LAYARD, A.H. (1853) *Discoveries among the ruins of Nineveh and Babylon, with Travels in Armenia, Kurdistan and the Desert*, New York, Putnam.

MacKENZIE, D. (1984) 'Marx and the machine', *Technology and Culture*, vol.25, pp.473–502.

MELLAART, J. (1967) *Çatal Hüyük: a Neolithic town in Anatolia*, London, Thames and Hudson.

MISA, T.J. (1994) 'Retrieving sociotechnical change from technological determinism' in M.R. Smith and L. Marx (eds) *Does Technology Drive History? The dilemma of technological determinism*, Cambridge, Mass., MIT Press, pp.115–41.

MOOREY, P.R.S. (ed.) (1982) *Ur 'of the Chaldees': a revised and updated edition of Sir Leonard Woolley's 'Excavations at Ur'*, Ithaca, NY, Cornell University Press.

MOOREY, P.R.S. (1994) *Ancient Mesopotamian Materials and Industries*, Oxford, Clarendon Press.

MORGAN, L.H. (1877) *Ancient Society, or Researches in the Line of Human Progress from Savagery, through Barbarism to Civilization*, London, Macmillan.

MORRIS, A.E.J. (1979, 2nd edn) *History of Urban Form: before the Industrial Revolutions*, London, George Godwin (first published 1974).

MUMFORD, L. (1966) *The City in History: its origins, its transformations and its prospects*, Harmondsworth, Penguin Books (first published 1961).

NISSEN, H.J., DAMEROW, P. and ENGLUND, R.K. (1993) *Archaic Bookkeeping: early writing and techniques of economic administration in the Ancient Near East* (trans. P. Larsen), Chicago, University of Chicago Press.

OATES, J. (1986, rev. edn) *Babylon*, London, Thames and Hudson (first published 1979).

PIGGOTT, S. (1983) *The Earliest Wheeled Transport: from the Atlantic coast to the Caspian Sea*, London, Thames and Hudson.

POSTGATE, J.N. (1992) *Early Mesopotamia: society and economy at the dawn of history*, London, Routledge.

REDMAN, C.L. (1978) *The Rise of Civilization: from early farmers to urban society in the Ancient Near East*, San Francisco, W.H. Freeman.

ROAF, M. (1990) *Cultural Atlas of Mesopotamia and the Ancient Near East*, New York, Facts on File.

SAHLINS, M.D. (1972) *Stone Age Economics*, Chicago and New York, Aldine, Atherton.

SHERRATT, A. (1981) 'Plough and pastoralism: aspects of the secondary products revolution' in I. Hodder, G. Isaac and N. Hammond (eds) *Pattern of the Past: studies in honour of David Clarke*, Cambridge, Cambridge University Press, pp.261–305.

SHERRATT, A. (1996) 'Secondary Products Revolution' in B.M. Fagan (ed.) *The Oxford Companion to Archaeology*, Oxford, Oxford University Press, pp.632–4.

SINGER, C., HOLMYARD, E.J. and HALL, A.R. (eds); assisted by JAFFE, E., THOMSON, R.H.G and DONALDSON, J.M. (1954) *A History of Technology*, Oxford, Clarendon Press, vol.1.

SJOBERG, G. (1960) *The Pre-industrial City: past and present*, Glencoe, Ill., Free Press.

SMITH, M.R. and MARX, L. (eds) (1994) *Does Technology Drive History? The dilemma of technological determinism*, Cambridge, Mass., MIT Press.

STOCKS, D.A. (1989) 'Ancient factory mass-production techniques: indications of large-scale stone bead manufacture during the Egyptian New Kingdom Period', *Antiquity*, vol.63, no.240, pp.526–31.

UPHILL, E.P. (1988) *Egyptian Towns and Cities*, Princes Risborough, Shire Publications.

VAN DE MIEROOP, M. (1997) *The Ancient Mesopotamian City*, Oxford, Clarendon Press.

WEBER, M. (1958) *The City* (trans. D. Martindale and G. Neuwirth), Glencoe, Ill., Free Press.

WHITE, L.A. (1959) *The Evolution of Culture: the development of civilization to the fall of Rome*, New York, McGraw-Hill.

WHITE, L., Jr (1962) *Medieval Technology and Social Change*, Oxford, Clarendon Press.

WITTFOGEL, K. (1938) 'Die theorie der orientalischen Gesellschaft', *Zeitschrift für Sozialforschung*, vol.7, pp.90–122.

WITTFOGEL, K. (1957) *Oriental Despotism: a comparative study of total power*, New Haven, Yale University Press.

Chapter 2: GREECE

by Colin Chant

2.1 Urbanization in the Aegean region

A necessary condition of the emergence of cities in Europe was the diffusion of Neolithic agriculture and animal husbandry from the Near East, first of all to mainland Greece and Crete. From there, these technologies spread through the Danube Valley, and by about 3000 BCE had been adopted throughout Europe. How much the emergence of cities in Europe owed to direct contact with the urban civilizations of the Near East, rather than to the diffusion of agriculture, is more debatable. According to V. Gordon Childe's diffusionist model of the Urban Revolution, the interacting primary centres were Egypt, Sumer and the Indus Valley; the demand of the primary centres for raw materials drew secondary centres, including the Aegean region, the subject of this chapter, into economic specialization and external trade. The preconditions for the transition from self-sufficient agricultural village to city were thereby spread (Childe, 1966, pp.169–70). In particular, after the discovery of bronze's advantages over copper, the quest for tin ores was stepped up; this would take Phoenician traders from the Levant as far as southern Spain during the first millennium BCE. The existence of desirable raw materials in a locality was insufficient to bring about urbanization; there also had to be fertile land close by, to generate the agricultural surplus necessary to feed specialists in the production and processing of industrial raw materials. Accordingly, there was no urbanization around the copper and turquoise mines of the desert region of Sinai, and it was limited in other mineral-rich, mountainous areas.

Childe's economic model at least has some explanatory power, in which respect it is preferable to older diffusionist accounts. According to these, the 'idea' of the city was invented in the Near East; it diffused to the east, to the Indus Valley and south-east Asia, and to the west, to Crete, Greece and Rome, and thence to Europe as a whole (Carter, 1983, p.10). Such accounts suggest that villagers were persuaded of the advantages of building a city, in much the same way that traders might have demonstrated to them the superiority of copper tools over stone. The developmental process through which a society comes to channel substantial resources into building cities is surely more complex than this. Archaeologists now find even Childe's version of diffusionism an over-simplification, and insist that the causes of urbanization in the Aegean cannot be reduced to 'the irradiation of European barbarism by Oriental civilization', as Childe put it (quoted in Renfrew, 1972, p.xxv); the emergence of cities was also the result of interacting, indigenous developments in technology, culture and social organization. This modern interpretation complements one of the main aims of this chapter, which is to show that the early history of cities in the Mediterranean basin reinforces conclusions already drawn in Chapter 1, from a comparison between Mesopotamia and Egypt: that urbanization is no simple, unilinear process, and that the technological activity of city-building is embedded in social, political, religious and environmental contexts peculiar to the region.

The author of this chapter is grateful for detailed advice and corrections from E.J. Owens, and for the comments of colleagues at The Open University, especially Lisa Nevett. The overall approach adopted, however, is the author's own.

The first European Neolithic communities were established around 7000 BCE in Greece and the Balkans; their agriculture was based on crops (emmer wheat) and animals (sheep and goats), almost certainly brought over from western Anatolia. Copper metallurgy was practised in the region during the late Neolithic; it may have diffused from the Balkans or Anatolia, though a local origin cannot be ruled out (Renfrew, 1972, p.312). A sudden expansion in metalworking, associated with the use of bronze and other copper alloys, dates from about 3000, and was accompanied by rapid population growth, the domestication of the grape and other tree fruits, and the development of a more ranked or hierarchical society.

Although agriculture and bronze metallurgy were soon widespread in Europe, urbanization was not their inevitable consequence. Cities first developed as a result of intensive trading in the eastern Mediterranean region, including the Aegean Sea (see Figure 2.1). Although Childe was right to emphasize trade in raw materials as a stimulus to urbanization, it would be wrong to see this trade as flowing in one direction only, from the Aegean to the Near East. As bronze became more widely used, the first Aegean civilizations themselves looked further afield for sources of copper and, especially, of tin (its constituent metals), as local deposits were exhausted. There are a number of other reasons for this intense trading activity. Apart from the accessibility of the region to the urban civilizations of the Near East, its geography encouraged the expansion of sea-borne trade. Considerable mountain ranges to the north and west of the Aegean exacerbated the difficulties of overland transport; pack-animals were the main means of carrying goods over the broken terrain. The prevailing geology of porous limestone discouraged the formation of major river systems, and of the

Figure 2.1 The eastern Mediterranean

Table 2.1   Periodization of ancient Aegean history

Dates (BCE)	Civilization (Period)
2600–1400	Minoan
1400–1200	Mycenaean
1100–750	Greek (Dark Age; also subdivided into Protogeometric and Geometric, after distinctive pottery styles)
750–500	Greek (Archaic)
500–338	Greek (Classical)
338–31	Greek (Hellenistic)

extensive alluvia found in the Near East. Aegean cities were usually located on small coastal plains, or a little way up the valley of one of mainland Greece's relatively small rivers. There was, however, abundant building stone, clay for pottery and good-quality timber from northern Greece; iron ores were available in Greece itself, and copper in Cyprus. The situation was the reverse of that in Mesopotamia, which had abundant agricultural land and limited raw materials, but the effect was the same: a resource imbalance that stimulated trade.

Favouring marine trading was an indented mainland coastline full of natural harbours, and a multitude of accessible islands: the northern Sporades group extending from Thessaly, the Cyclades to the south of Euboea and Attica, the southern Sporades off the coast of Anatolia, and a southernmost stretch of islands from the Peloponnese through Crete and Carpathos to Rhodes, near the mainland of Anatolia (see Figure 2.1). In this last chain there developed the first European civilization with urban characteristics. A simplified historical framework within which to set urban developments in the region is given in Table 2.1.

The first European cities

The first Bronze Age civilization was called Minoan, after Minos, the legendary ruler of Crete. The name was chosen by Sir Arthur Evans, who began to excavate the palace at Knossos at the very start of the twentieth century. Dating is controversial, but Minoan civilization seems to have begun around 2600 and lasted until about 1400 BCE. The Minoans were skilled in the making of pottery and faience, metalworking, weaving and dyeing, and building in local stone, especially limestone, gypsum and alabaster. At its height, from about 2000, Minoan civilization was dominated by independent palaces of monumental stone, and a good deal of Minoan economic activity took the form of a carefully calculated exchange of gifts between palace rulers. The sea being the primary medium of this activity, there must have been specialist production of wooden sailing ships; however, their design can only be inferred from a very few shipwrecks of contemporary craft. Land communications were not neglected: a two-lane highway, strengthened in places by stone terraces, connected the main southern port with the palace of Knossos, in the north of the island. The final stretch of road leading to the palace was carried on a massive stone viaduct (Casson, 1974, p.27).

Each Minoan palace, typically an assortment of rooms built around a large court, controlled a network of villas in the surrounding countryside. The palaces and villas had numerous storage rooms for the agricultural surplus, which was recorded by scribes in the indigenous script that Evans called 'Linear A'. This surplus supported the full-time craft specialists (potters, weavers and metalworkers) who worked in the palaces. With the exception of a palace at Gournia, in eastern Crete, there is no evidence that these unfortified Cretan palaces were surrounded by the communities that would have made them urban in form. Even then, Gournia was only a small town, in which domestic

and artisan quarters ranged along winding alleyways around the palace,
'like a small medieval town clinging to the skirts of some baronial castle'
(Ward-Perkins, 1974, p.10).

   After the first palaces were all destroyed in about 1700 BCE, apparently
because of an earthquake, their successors became more elaborate in design.
The new palace at Knossos had running water and a flush drainage system, and
rose more than two storeys up a hillside; the first storey was built from mortared
rubble reinforced with wood, and the second was of brick (see Figure 2.2). The
palaces contained shrines and religious imagery; no separate temple remains
have been found in the region. Apart from the palaces, there were some

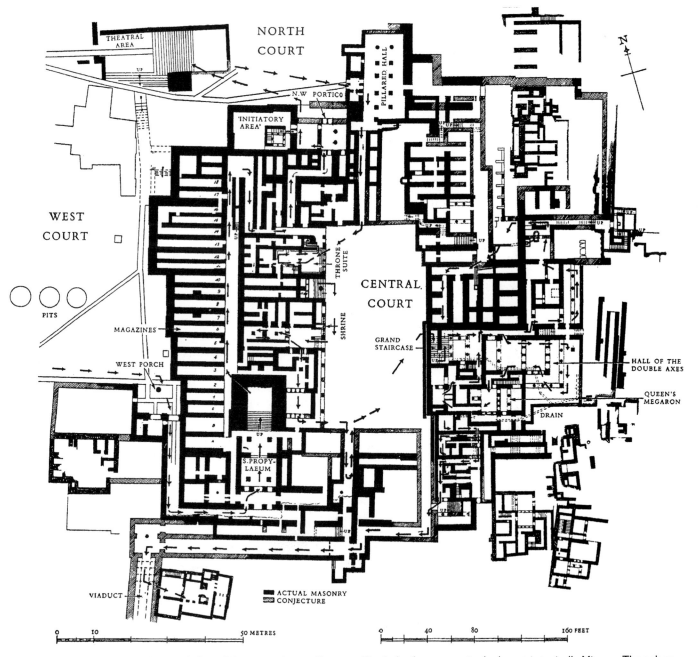

Figure 2.2 Restored ground-plan of the new palace at Knossos. The lack of symmetry in the layout is typically Minoan. The palace
was approached by a paved road, carried over a ravine in front of the palace by a massive viaduct, leading to a stepped portico at the
south-west corner. The numerous storage magazines on the west side reflect the economic, as well as religious and ceremonial,
functions of these palaces. The drains in the south-east domestic quarter connected with a latrine. The 'Queen's Megaron' is an
excavator's misnomer: there is no evidence that the space was for women only, nor is it a great hall, like a true Mycenaean megaron
(Lawrence, 1973, pp.32–3; by permission of Yale University Press)

settlements now recognized as towns, such as Palaikastro and Mallia, with maximum areas of twenty-three and thirty-six hectares respectively. These towns were typical of settlements throughout antiquity, in that most of the craft specialists, priests and administrators who lived in them owned and/or worked on the land (Dickinson, 1994, pp.64, 69). Given the Minoans' domination of trade in the eastern Mediterranean, and their craft skills, some degree of urbanization is to be expected, even if the royal palaces themselves seem more like manorial centres of agricultural production than nodes of urban growth.

Minoan civilization went into rapid decline from about 1400 BCE. Scholars disagree about whether this resulted from volcanic eruptions and associated earthquakes, or from invasion. In either event, the domination of the Aegean region passed from this time to mainland Greece, and to a Bronze Age civilization called Mycenaean, because its culture was first detected through excavation of the city of Mycenae. The Mycenaeans are of uncertain origin: they may have crossed over from Asia Minor, or come down from the Balkans. Having conquered Crete, they assimilated Minoan habits of sea-borne commerce, trading in textiles, olive oil and metals, and of palace-building. Mycenaean palaces were distinctive in being built around a massive rectangular hall, or megaron, overlooking a courtyard. The Mycenaeans adapted the Minoan script to their own language in a form Evans called 'Linear B'. Their fondness for chariot-riding was satisfied by the construction of substantial roads around the main settlements, sometimes paved in their immediate vicinity, and kept passable in the rainy season by bridges and culverts (Casson, 1974, p.27).

Great palaces were built at a number of sites, including Athens, though very little from the Bronze Age remains in the Greek capital. A palace of similar design and period has also been found in one of the levels of the ancient site of Troy, across the Aegean Sea, overlooking the Hellespont. The most fully excavated are those at Mycenae and nearby Tiryns, and at Pylos on the westernmost prong of the Peloponnese. Unlike their Minoan predecessors, the palaces at Mycenae and Tiryns had stone defensive walls. The building of an additional long wall in about 1250 BCE, across the Isthmus of Corinth, suggests that the threat of further invasion from the north was growing. At Mycenae, the remains of houses on both sides of the walls indicate a settlement of urban complexity. Like the Minoans, the Mycenaeans were skilled in the technologies of city-building. In the thirteenth century BCE, the walls of Mycenae were extended to protect the city's supply of water; a stepped passage from inside the extension led to an underground cistern outside the walls, fed by an aqueduct from a nearby mountain spring (see Figure 2.3). Again, this bloom of urbanization was relatively brief: most of these palaces, including the now reconstructed palace at Knossos and the fortress at Troy, were destroyed in

Figure 2.3 Plan of Mycenae. The site occupies an area of four to five hectares only, but had massive fortifications made of roughly shaped limestone boulders, with gateways made from limestone slabs, each several tonnes in weight. The walls were between six and eight metres thick, and were pierced by a chariot-road leading to the palace. Like their Minoan predecessors, Mycenaean palaces had bathrooms and drains (Morkot, 1996, p.27; map: copyright © Swanston Publishing, Derby)

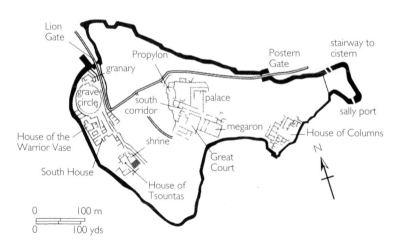

about 1200 BCE. Whether this was because of earthquake, drought, invasion or the kind of inter-city warfare described in Homer's *Iliad* is another of the great unresolved issues of ancient history.

The degree of urban development, and the extent to which the Mycenaeans controlled Aegean commerce, are also matters of historical conjecture. So too is any connection between their palace-dominated urban forms and the cities of the Near East. How influential were the first urban civilizations in the Aegean? Chapter 1 followed the development of Mesopotamian urban civilization to the first millennium BCE, during which ascendancy in the region passed from Assyria, in the north, to the New Babylonian Kingdom. Both these Mesopotamian empires at one time extended westward to Canaan (Roman Palestine and Syria), a volatile land of great cultural and ethnic diversity, which had already been subject to Egyptian rule in the second millennium. But a Canaanite urban civilization, based on agriculture and commerce, had emerged well before these annexations. As far back as the third millennium, the port of Byblos had been an important centre of trade, especially in cedarwood, with Egypt and Mesopotamia. During the second millennium, Ugarit, further up the coast, became the main Canaanite port until its destruction in about 1200 BCE. Some scholars believe that it directly influenced the form of Minoan settlements, but, although there are some similarities between Aegean and Canaanite cities, the consensus is now that the Aegean building tradition was essentially indigenous. By this account, it emerged independently out of the diffused technological and social conditions from which the Mesopotamian and Egyptian urban civilizations, and then the Canaanite, had themselves arisen.

From bronze to iron

In their attempts to reconstruct the later Aegean Bronze Age and the early Greek Iron Age, historians have often introduced the migrations of the so-called Indo-European peoples, among whose languages were those that would become central to the European identity: notably Greek, Latin and its derivative Romance languages, and the Germanic and Slavic tongues. Some have included the Mycenaeans among these migrants. This is a controversial topic; first, there is scant archaeological evidence for the Indo-Europeans' origins and supposed migrations, and much of their culture and technology is inferred from linguistics. Second, archaeologists are increasingly reluctant to identify cultures and technologies with ethnic groups; the spread of technologies and languages is seen as much more piecemeal and complex than the simple supplanting of one people by another (Renfrew, 1987). It seems clear at least that at the time of the urban civilizations of the Aegean, there were agriculturists to the European north speaking Indo-European languages; and, to the north-east, other such peoples were engaged in nomadic pastoralism on the Asiatic steppes. It would be easy to dismiss these cultures as technologically less advanced than those of their urban contemporaries. But they offer a salutary corrective to any notion that history presents an orderly sequence of technological progress, and also to any view of historical progress that places urban civilization on a necessarily higher plane than an agricultural or a nomadic way of life.

There is no reason to believe that nomadic encampments of covered wagons and tents were less well equipped than many contemporary urban settlements. The main difference between the material culture of nomadic and that of sedentary societies is obvious but worth stating. Nomads developed technologies suited to their mobile way of life, rather than those presupposing commitment to place. Among the latter are building technology and manufacturing processes that employ complex plant and equipment, such as glass-making and pottery of any intricacy. Innovations helping mobility and

Figure 2.4 Chariot with spoked wheels, from a Hittite relief, *c.*1200 (Hodges, 1971, p.140; by permission of Hirmer, Munich)

hunting were another matter. Some historians have attributed to these steppe peoples the invention of wheeled carts and wagons, and subsequently of the military chariot, introduced around 1600 BCE (see Figure 2.4). Debate over the invention of the chariot continues; Stuart Piggott argued in favour of its steppe origins, but a case has recently been made for its development from Mesopotamian wagons and carts (Littauer and Crouwel, 1996; Piggott, 1983, pp.103–4). Less controversially, the steppe peoples are accorded priority in domesticating the horse, and in developing the bridle and bit needed to control it. These technologies, moreover, had obvious military applications. Wherever it was actually invented, the light chariot, with its pair of yoked horses, light frame and spoked wheels, represented a marriage of woodworking skills, developed by the Mesopotamians, and the horse's speed and mobility, harnessed by the nomadic steppe peoples.

The northern agriculturists and pastoralists have sometimes, though without much hard evidence, been credited with the first steps in iron metallurgy. Iron was known to the leading Bronze Age civilizations from the third millennium, but little used; it was not until the tenth century BCE in Greece that it became widely preferred for weapons and tools. Ores of iron are more common than those of copper and tin, but iron's melting temperature (1,530°C) is much higher than that of copper (1,083°C); obtaining useful metal, and devising ways to work it, thus posed a much greater technological challenge. Indeed, unlike the other metals known in antiquity, iron could not be cast until the advent of the blast furnace in medieval Europe; all ancient forms of iron, and of steel (iron with carbon added), were wrought – that is, forged from a solid 'bloom' of metal smelted in a simple bloomery hearth (see Figure 2.5).

Iron metallurgy was the final technology of the 'three-age system' (stone, bronze and iron), proposed by the Danish museum director Christian Jürgensen Thomsen in the early nineteenth century and still applied to Old World prehistory. Nowadays, the Iron Age is seen as a convenient umbrella term for a complex of technological and cultural changes. But, in the past, the development of iron-smelting has been invested with great significance in world history, not least by the US pioneer of anthropology Lewis H. Morgan. As noted in Chapter 1, Morgan saw the history of human culture as a progression from savagery, through barbarism, to civilization. He identified 'four events of pre-eminent importance' in the transition from barbarism to civilization: the domestication of animals; the discovery of cereals; the use of stone in architecture; and, finally, the invention of metallurgy, through which 'nine-tenths of the battle for civilization was gained'. The element of technological determinism in his scheme is at its most explicit in this case:

Figure 2.5 Bloomery hearth for the production of wrought iron. Alternate layers of charcoal and ore were heaped on a circular stone platform and covered with clay. Bellows inserted in the dome were kept on blast for hours, and the iron concentrated in a spongy mass at the bottom (Singer *et al.*, 1954, p.123; by permission of Oxford University Press)

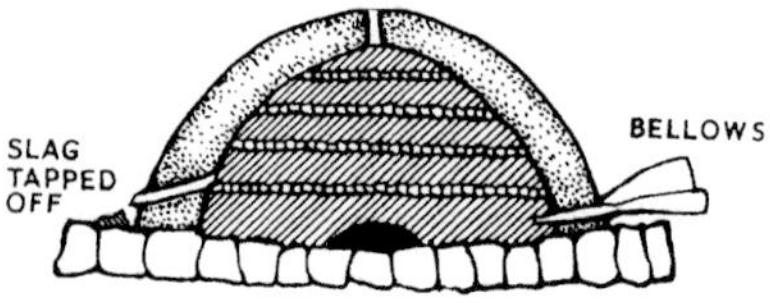

SLAG TAPPED OFF   BELLOWS

> The production of iron was the event of events in human experience, without a parallel, and without an equal, beside which all other inventions and discoveries were inconsiderable, or at least subordinate. Out of it came the metallic hammer and anvil, the axe and the chisel, the plow with an iron point, the iron sword; in fine, the basis of civilization, which may be said to rest upon this metal.
>
> (Morgan, 1877, p.43)

It is much less obvious than in the case of the domestication of the horse why the northern nomads and farmers should have taken the lead in iron metallurgy, if indeed they did. One suggestion is that the improved bellows necessary for the smelting of iron ores was an offshoot of the established nomadic tradition of working in hides and leather (Hodges, 1971, p.233). Another interpretation takes a wider perspective. The Minoan and Mycenaean palace economies had been based on the centralized control of trade in copper and tin, the constituent metals of bronze. The democratic, hard-working, self-sufficient communities that succeeded them turned to iron metallurgy because iron ores were much closer to hand than those of copper or tin, and they were prepared to put in the greater effort needed to convert them into high-quality materials. Paradoxically, the strength of iron was eventually to bolster the power of later empires (Snodgrass, 1989). However they originated, iron tools and implements, notably the iron ploughshare, were vital in extending agriculture to areas of forest and clay soils, where stone tools would have been inadequate, and bronze too rare and expensive. In this way, iron metallurgy was one of the technological conditions for the spread of urbanization into Europe.

The Bronze Age urban civilizations of southern Asia, as well as of the Aegean, proved vulnerable during the second millennium BCE to the military prowess of their northern neighbours: one grouping of these peoples overran the first Harappan urban civilization of the Indus Valley, and became the rulers of India. Another, the Hittites, took over Anatolia, establishing a network of packed-dirt roads for their chariots. In a way that was often to be repeated in urban history, these invaders were assimilated into the urban cultures they overran. And, in their turn, they were subject to further pressure from the northern tribes. Another innovation in steppe technology, the riding of horses, along with the invention of the saddle, posed another military challenge – that of cavalry – to the agricultural and trading civilizations of the Near East and the Mediterranean basin.

During the first millennium BCE, the Medes and Persians, whose languages were also part of the Indo-European family, moved into Iran and conquered Babylonia. Their vast empire contrasted with the Greek maritime civilization with which it would compete over the next few centuries. It was held together by a system of well-made, often paved, roads, which facilitated communication by relays of expert horsemen: an example that impressed the Greeks was the 'Royal Road' from Sardis across Anatolia to the Persian capital of Susa. It seems that it was others of these northern peoples who precipitated the decline of the Hittite Kingdom of Anatolia, and the Mycenaean civilization of the Aegean. The Greeks themselves attributed the fall of the latter to an invasion by the Dorians, though this is probably an over-simplification of the migration patterns in the region. At any rate, there followed a period controversially known as the Greek Dark Age, comparable to the much later period of European history following the collapse of the Roman Empire in the West. The term is controversial, because the period saw the introduction of iron technology to Greece, and certain distinctive pottery styles. A question debated by historians is whether there was any continuity of the culture of the Mycenaeans with that of the early Greeks between the approximate dates of 1200 and 1100 BCE (see Table 2.1). One possibility is that Mycenaean civilization was preserved in the Ionian settlements across the Aegean Sea, established by those fleeing the northern invaders.

If there was discontinuity, how else might the Near East have directly influenced the development of an urban civilization in Greece? The Greeks had close contacts with Sardis, the capital of Lydia in western Anatolia, situated on the main overland trade route to Syria and Mesopotamia. Tradition has it that the Greeks took over the idea of coinage from the Lydians. Metals such as silver had long been used in Mesopotamia as a means of exchange and payment, their value generally being reckoned by weight. The Lydians' invention in the early seventh century BCE of coinage, based on the concept of face value, was a further stimulus to trade, and thence indirectly to urbanization. The use of currency also had a direct bearing on the built environment, as the rich and powerful had less need to store their wealth in the form of agricultural surplus or manufactured goods.

But the Greeks' main experience of cities was through sea-borne trade. Sidon, to the south of the Canaanite port of Byblos, became the main centre of Phoenician traders in the first half of the first millennium BCE. The Semitic Phoenician merchants, who exchanged luxury goods from the East for grain, timber, metals and wool, stepped into the Mediterranean vacuum left by the decline of the Minoans and Mycenaeans. They developed biremes, or ships with two banks of slave oarsmen, and established trading colonies throughout the Mediterranean in the ninth and eighth centuries: in Sardinia, western Sicily, Spain and North Africa, where the Phoenician colony of Carthage became a great commercial city in its own right. Ports such as Carthage, Byblos and Sidon (and, indeed, inland caravan cities such as Damascus, situated by an oasis to the east of the mountain ranges of Lebanon) represent a different kind of city from the Mesopotamian and Egyptian examples discussed in Chapter 1. They were primarily based on trade, rather than on imperial power, or on the control of an agricultural surplus; and the cities of the ancient Greeks were to follow in this tradition.

2.2 Greece

The emergence of an ancient Greek urban culture dates from about 900 BCE. Its first distinctive feature was a geometric style of pottery decoration, the technological basis of which was an improved potter's wheel. Pottery was to become one of the main trading commodities of ancient Greece, along with wine and olive oil, two of the definitive products of Mediterranean agriculture.

The Mediterranean trading and transport system powerfully influenced the pattern of Greek urbanization. The Minoan and Mycenaean elites had invested in good roads to satisfy their enthusiasm for chariot-riding, but these were neglected by the Greeks. There were some rocky stretches with ruts cut into them for wheeled vehicles (see Figure 2.6), but otherwise overland routes were unsurfaced and undrained tracks, impassable in winter: 'Greece in the thirteenth century [BCE] probably had a better system of roads than it did in the third' (Casson, 1974, p.27).

Like the Minoans and the Mycenaeans, the Greeks conducted their trade mostly by sea. Increased trading, along with intensifying military competition in the Mediterranean, stimulated some developments in shipbuilding. Few wooden ships have been recovered, but the evidence overall suggests that the Greeks followed the lead of the Phoenicians. They built ever larger oared warships: first, biremes with two banks of oars, then triremes with three, with bronze beaks intended for ramming enemy ships and holing them below the water-line. Cargo ships, for reasons of economy and space, relied mainly on sail, using simple square-rig arrangements that could operate only with a following wind.

Figure 2.6 Greek rut-road, showing a siding branching from the main track (Singer *et al.*, 1956, p.499; by permission of Oxford University Press)

The rise of the city state

The system of cities was also shaped by fundamental political changes. The rulers of Greece during the Dark Age were tribal chieftains. Their power was weakened by the rise of agricultural aristocracies, consisting of the most successful farmers. Between the eighth and sixth centuries BCE, the power of the aristocracies was itself reduced; one explanation for this hinges on changes in military technology and strategy (Snodgrass, 1980, pp.97–107). Formerly, it is argued, warfare was dominated by the aristocratic warrior, who rode into battle on horseback, though it is likely that he dismounted to fight (chariots were probably used only for racing and processions). Between 700 and 650 BCE, the aristocrat gave way to the tightly formed phalanx of foot soldiers, or hoplites, armed with much-improved bronze helmets, breastplates, shields and spears (Greek *hoplon* means 'spear'). This transformation was associated with the rise of a separate class of soldiers, whose interests were not necessarily those of the aristocracy. It has also been seen as a more egalitarian mode of warfare, prefiguring the rise of the *polis*, or democratic city state of Classical Greece.

Some scholars have questioned whether this fundamental shift from aristocratic to hoplite warfare actually occurred (Morris, 1987, pp.196–201); but, assuming that it did, it is hardly sufficient to explain the rise of the democratic *polis*. The economic position of the aristocracy had been undermined more profoundly by city-based merchants and manufacturers, as Greek participation in Mediterranean trade developed. The migration of Greek settlers to the Mediterranean and Black Sea coasts of Anatolia, and to southern Russia, began in the ninth century BCE, and, from the late eighth century, they struck out across the Mediterranean to southern Italy and Sicily, southern Gaul, north-eastern Spain and Libya. These settlers were in part the surplus population of established cities seeking agricultural land elsewhere; but the movement also betokened a great expansion of trade, in which pottery, metal goods, wine, oil and fabrics were exchanged for primary materials such as grain, timber and minerals. This growth in trade and industry led to an expansion of the population of mainland cities well positioned to exploit sea-borne trade, such as Athens and Corinth, and the Anatolian colonies Ephesus, Miletus and Smyrna. 'Colonies' is a disputed word to use, as many of these planted cities were legally independent of their mother cities on the Greek mainland. This seeding of cities by parent settlements that placed a conscious limit on their own size was a distinctively Greek means of spreading an urban way of life.

In due course, these Greek city states developed forms of government quite unlike those of the Near Eastern empires. Instead of monarchical rule, political authority came to reside in an assembly of citizens who elected their leaders. Although the development of participatory modes of government might well be seen as a product of enlightened Greek culture, the influence of the Aegean environment must also be considered. The limited availability of agricultural land placed real practical limits on settlement size, and predisposed cities to sea-borne trade and the establishment of colonies. There were, it should be noted, some Greek innovations in food-processing, such as the animal-powered grain mill, and the screw-press for the manufacture of wine and olive oil, but not such as to alter fundamentally agricultural productivity. The natural environment therefore strongly influenced the Greek pattern of urbanization.

The distinction between *polis* and *astu* is fundamental to any appreciation of the distinctive nature of urbanization in the Greek Classical Period. The concept of the city state (*polis*) transcended that of the physical city (*astu*). The city state of Athens, through the federative process known as synoecism, came to include male residents of the neighbouring towns and villages of Attica; Athens effectively controlled a territory of some 2,500 square kilometres. Apart from its evident political connotations, the concept of the city state also reflects the blurring of divisions between town and country; many city-dwellers retained strong connections with agriculture. Overall, probably only two-fifths of the population of the Athenian *polis* actually lived in the city during the fifth century BCE (Hammond, 1972, p.181). There were also many rural features within the area enclosed by the substantial stone fortifications in which most Greek cities invested.

In the city state, pre-urban notions of tribal identity and kinship were preserved in the way the population was grouped for social, political and military purposes. Beneath this traditional veneer, however, a profound change was taking place: the political basis of Greek urban life became the concept of citizenship, and its socio-economic basis a hierarchy of social class determined by wealth. Another departure was that legislation through assemblies of citizens was clearly a human activity, rationalized through abstract concepts such as 'justice'. Although Greek civic life was conducted within a religious context, laws and political decisions were no longer presented as unchallengeable divine commands, as they were in Egypt and Mesopotamia. An obvious inference is that this greater responsiveness on the part of governments to the requirements of the populace would be reflected in the urban built environment, not least in the attention given to public amenities. This perception certainly needs qualification: tyrants and emperors were among the most active sponsors of water supply and sanitation schemes (Camp, 1986, p.40).

The *polis* of Athens was, in theory, a democracy, so far as its male citizens were concerned. In 432 BCE, these numbered some 43,000 out of a total population of around 330,000, including women, children, slaves and metics (foreign residents). The metics were prevented from participating in political life, but, along with slaves, were very active in commerce and industry, which was scorned by the governing class. Although the constitution of Athens looks less than democratic in the modern sense, it was a far more inclusive form of government than that of other sizeable cities on the Greek mainland, such as Corinth and Thebes, Syracuse in Sicily, or Smyrna, Miletus and Rhodes across the Aegean Sea. Most of the city constitutions of the Classical Period were Athenian-style democracies, or variations on a theme of oligarchy, with interludes of tyranny, especially in the seventh and sixth centuries. As will be seen, this political diversity had some bearing on the practices of city-building.

The Hellenistic city

The Classical Greek era of urban civilization was brought to an end, yet again, by a supposedly technologically inferior northern people. The Hellenistic Age may be dated from the conquest of Greece by Philip of Macedon in 338 BCE until Augustus' defeat of Antony and Cleopatra at Actium in 31 BCE. Macedonia was an agricultural, tribal state to the north of Greece, which used its grain, timber and gold resources to support a strong army. Through their military prowess, these farmer-soldiers acquired a territorial empire of unprecedented extent – one, indeed, that briefly swallowed up all the older empires of the Near East. Through an audacious campaign conducted between 334 and his death in 323 BCE, Philip's son Alexander the Great occupied Anatolia, Syria, Palestine, Egypt, Mesopotamia, Persia and the Indus Valley. This empire marked a return to territorial monarchy, and away from independent city states, thus mirroring developments in Mesopotamia. Almost fittingly, Alexander died in Babylon, shortly after the end of his eastern campaign. According to Plutarch, Alexander founded seventy cities; through them he exercised political and military control over his vast territory.

In fact, there were many political shades between territorial empire and independent city state. The Athenians in the fifth century created a kind of non-territorial empire out of the Delian League of city states, which centred on the shrine of Apollo on Delos, and they used the Delian treasury to finance their military campaigns and monumental public buildings. Other such leagues or federations of cities included those dominated by Thebes and Sparta. Part of the motivation for these alliances was defence against rival leagues, or against the threat of the Persian and Macedonian territorial empires. At a deeper level, they showed that the expansion of commerce, while stimulating the rise of independent city states, brought in its train competition for political and military control of the main trade routes.

Three Hellenistic territorial empires emerged in the aftermath of Alexander's conquests, under dynasties founded by Alexander's generals or their descendants: Macedonia under the Antigonids, Syria under the Seleucids, and Egypt under the Ptolemies. Macedonia remained an agricultural society, and maintained its control of Greece not through imperial cities, but through three strategically placed fortresses, including the Acrocorinthus, or Citadel of Corinth. The Macedonian capital of Pella, though probably quite sizeable, was essentially a settlement based around a royal palace. The empire of the Seleucids was the rump of the Alexandrian conquests in the Near East and Indus Valley, corresponding to the ancient region of Canaan. An important feature of the Seleucids' rule of the area, like that of Alexander, was the founding of cities of Greek and Macedonian settlers. Around 312 BCE, the first Seleucus built his capital, Seleucia, on the Tigris, to the north of Babylon. According to the Roman writer Pliny the Elder, it grew to a population of 600,000. In Syria, Seleucus founded Antioch on the River Orontes, on one of the main trading routes between the Mediterranean and Mesopotamia. Antioch became the capital of the contracting empire of his successors, and the most important Near Eastern city after Alexandria.

Ptolemy, the general who founded the Egyptian dynasty, reputedly seized Alexander's body in Babylon, and entombed it in the new city of Alexandria, which Alexander had planned and Ptolemy completed. Ptolemy's city included a great palace; attached to it was the renowned library and museum, which would attract scholars from all over the Hellenistic world. An important entrepôt of trade, and centre for the manufacture of glassware and jewellery, Alexandria became the pre-eminent Hellenistic city. Its built environment, laid out to a grid plan by the architect Dinocrates, was one of the most advanced of its time. The administration of its public services offered Augustus, the first Roman emperor, a model for his own imperial capital. Construction was generally of stone, with vaulted roofing; private houses had cisterns and sewerage arrangements. Ptolemy's son was responsible for the famous lighthouse on the island of Pharos, designed by Sostratos of Cnidos, along with the harbour works. But, apart from Ptolemy's own city of Ptolemais, Alexandria was the only new city founded by his dynasty in Egypt, and very much apart from it; Egypt proper still did not seem to provide the economic and geographical environment for vigorous urban growth. Moreover, both the Seleucids and the Ptolemies took on the divine trappings of the Near Eastern monarchs they succeeded, and, although the Near Eastern regions over which they ruled were culturally and linguistically Hellenized, the fundamental political and economic characteristics of these areas persisted.

What were the implications for cities of the transition to imperial rule? The loss of independence meant that many former city states no longer supported their own military and naval establishments, nor did they invest in fortifications. There was also a trend, within the wider setting of imperial rule, from democracy to oligarchy in their internal politics. The eclipse of the city states has often been seen as a decline in Greek civilization, but this does less than justice to the cultural, architectural and technological achievements of the Hellenistic Age; at the very least, the great prosperity that resulted from the Hellenistic conquest of the Near East brought about a general rise in the material quality of urban life. Another consequence of imperial rule was greater uniformity in the urban built environment. The same Hellenistic styles, building methods and urban design spread throughout the Mediterranean and into the Near East. Each Hellenistic city was similar in its temples, assembly halls, stoas (colonnaded public buildings), market-places, roads, gates and public fountains.

2.3 Greek urban planning and morphology

As in the Near East, developments in military technology had both direct and indirect implications for the Greek system of cities and for Greek urban morphology. Defence was a prime consideration in the foundation of Greek cities: the earliest usually centred on a rocky hill, or acropolis. Security from piracy and naval raids was essential in the Aegean region: as noted above, the early cities were generally set back from the sea for this very reason, and heavily fortified; the building of defensive walls was 'the most laborious and expensive task' faced by a city, and 'as feats of engineering they are not surpassed in the whole range of Greek architecture' (Wycherley, 1976, p.38; see Figure 2.7). Inland locations, however, were ill-suited to marine trade, and, in a period of continuous warfare, weapons of defence took their place in the repertoire of urban technologies in the newer port cities: the Greeks were responsible for improvements in the design of catapults, and in devices for dropping weights on ships involved in siege warfare.

The apparently spontaneous development of cities in Greece, from about 900 BCE, contrasts with both the planned rebuilding of cities after their devastation by the Persians in the fifth century, and the planned laying out of

Figure 2.7 Restoration of fortifications enclosing the planned colony of Heraclea, in Ionia. Greek defensive walls and towers were noted for their fine ashlar (squared) masonry, which had to be strong enough to withstand missiles from siege-engines. The walls were provided with parapets and battlements, and the towers with loopholes or archer-slots (Wycherley, 1976, p.45)

colonies in the Hellenistic Period. The first phase of rebuilding inspired the formulation of a theory of town-planning, attributed in written sources, such as Aristotle's *Politics* (Book II, Chapter 8), to Hippodamus in the fifth century BCE. Its definitive features were a gridiron, or orthogonal layout of streets, centring on an agora, an open space surrounded by the main public buildings. But Hippodamus was not the inventor of the grid layout. Archaeological evidence shows that earlier Greek cities in both Asia Minor and the western Mediterranean were designed in this regular fashion: Smyrna, in the seventh century BCE, was rebuilt after a fire to a simple plan with an agora, a temple and parallel streets running north to south. The essentially planned allocation to Greek colonists of building plots within the city, as well as farming strips outside the city walls, probably dates from as early as the seventh century BCE; a rare survival of this form of land division is at Metapontum, in southern Italy (Carter, 1980).

One of the best-known examples of Greek town-planning is Miletus, rebuilt following its sacking by the Persians in 494 BCE (see Figure 2.8). A key city for Black Sea trade, Miletus was from the outset built around its two natural harbours, unlike the older inland cities of the Greek mainland. The new plan of Miletus, traditionally attributed to Hippodamus, a native of the city, was based on a rectilinear grid of streets enclosing oblong blocks of buildings. Miletus had two distinct residential areas, which were laid out to a highly repetitive grid plan and separated by an area of public spaces and buildings, oriented towards the two harbours. There were two agoras, defined by stoas; the stoa, a distinctively Greek building type, was essentially a free-standing, covered colonnade, open

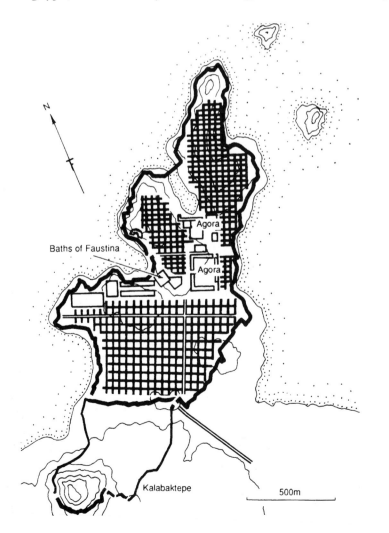

Figure 2.8 Plan of Miletus. The occupied area was about ninety hectares. After rebuilding, from 479 BCE, the walls included the original acropolis; this was later excluded by a new wall across the peninsula. The Baths of Faustina and the southern residential area, with its larger housing blocks, date from the later Roman period (Owens, 1991, p.53)

on one side, and used for a variety of civic and commercial purposes. From the second century BCE, the two agoras were separated by a bouleuterion (the meeting-house of the legislative council). Among its numerous roads, the plan included fewer wide, arterial avenues than most Greek cities had. The whole settlement was enclosed by a wall, the line of which, as in most Greek fortifications, was fixed by the site's contours, rather than the regular street plan.

The grid was no merely neutral device: it embodied notions of land-use and occupational separation, and presupposed central organization, and co-operation in the building of housing blocks with party walls and shared drainage. It has been argued that it is an inherently more democratic form of plan, involving a more equal division of land, when compared with early royal cities, in which a jumble of poor dwellings on the periphery contrasted with substantial, higher-class dwellings nearer the centre. Certainly, in some Greek colonial cities, street grids were accompanied by housing blocks of identical dimensions, though the reasons for this may have been less political than military or utilitarian, in that such uniformity makes for speed of construction. Another important feature of the grid, which made it a practical choice for new settlements, was that it allowed for gradual filling in and, depending on the site, easy extension as a city grew. One objection to the regular grid layout, however, was made by Aristotle in his *Politics* (see Extract 2.1 at the end of this chapter). He argued that the old-fashioned irregular siting of houses was harder for attackers to penetrate, and that cities should have a mixture of the two arrangements.

These practical considerations need to be weighed against the ideological influences on city-building. The French historian François de Polignac has recently reasserted the case for the religious origins of the city. His argument is that the blossoming of cults in the ninth and eighth centuries BCE, and the commitment to building sanctuaries for the deities, ultimately caused the revival of city-building in Greece (Polignac, 1995, pp.152–3). This is bound to remain a controversial issue in the case of the earliest Greek cities, for which evidence is so sparse. As for the founding of subsequent Greek colonies, J.B. Ward-Perkins insisted that practical concerns were more important:

> The archaeological record emphatically confirms the explicit testimony of classical writers from Aristotle onwards that the criteria followed in establishing the orientation of a new town were primarily topographical and hygienic, not religious. The direction of the prevailing winds figures prominently. Other common factors were slope (facilitating drainage), the passage of a major road, or a sea frontage. At the level of individual buildings care was taken to give a southern exposure to the living rooms of houses, for penetration by, and exclusion of, the sun in winter and summer respectively. [1] Bath buildings were sited so that the hot rooms faced southwards or south-westwards. Temples on the other hand regularly, though not invariably, faced east. In short, orientation was an important item in the city planner's calculations, but except for religious buildings the reasons once again were severely practical.
>
> (Ward-Perkins, 1974, p.40)

How was the foundation of these colonies organized? Written evidence suggests that a standard procedure became established, involving the setting up of a supervisory commission. There was assuredly a religious dimension: an oracle would be consulted, and sacrifices and prayers would precede the actual laying out. But religious and practical considerations went hand in hand throughout the foundation procedure. Overall control was exercised by a single official, the *oikistes*; the establishing of boundaries, or *horoi*, was the responsibility of the *horistes*, who would decide where the lines of defence would go, and plan the layout of the street system and the building plots. The plan set aside land for the public spaces and buildings, and the remainder was put up for sale.

[1 The summer sun was, of course, much higher in the sky, and was excluded by a portico in front of the main rooms.]

The religious dimension of the founding of a city, and the increasingly religious function of the acropolis (the rocky height that was the defensive focus of many of the first Greek cities), have to be set against the diversification of the urban built environment, as populations grew, and trade and occupational specialization increased. Political and economic functions were transferred from the acropolis to the agora, originally an open space serving as a market-place, and as a place for public assemblies and spectacles. Next, specific buildings attached to the agora were assigned for these functions; these might include a bouleuterion, temples, lawcourts and stoas (see Figure 2.9). Not that religion was marginalized in the Greek city: far from it. All Greek public buildings were dedicated to the gods, and contained shrines (Wycherley, 1976, p.88). But, apart from temples, these were buildings defined by their secular functions, set against a religious background; in this respect, the ancient Greek city differs from the Sumerian city, dominated by its ziggurat and shrine, or the Egyptian city, dedicated to the divine Pharaoh's afterlife.

The shift of functions from the acropolis to the agora was followed by development beyond the agora, notably of recreational and commercial structures, such as theatres, gymnasia, stadia, colonnaded public markets and enclosed shops, and commercial and manufacturing quarters. In time, the

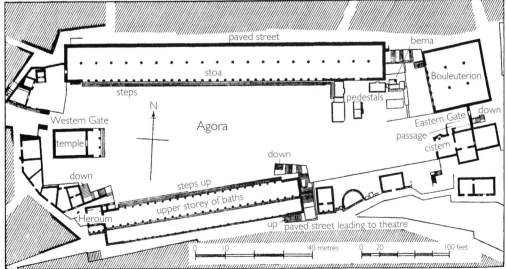

Figure 2.9 Agora at Assos, in north-west Anatolia, second century BCE. This agora was built into a slope, and demonstrates Hellenistic skill in terracing. The stoa occupying most of the southern side of the agora was multipurpose, functioning not only as a baths building but also as a bazaar, which was accessed from the rear of the building, facing away from the agora. The plan shows the need for internal columns in the larger buildings (Lawrence, 1973, p.266; by permission of Yale University Press)

physical structures of the Greek city, rather than its political constitution, would come to be seen as definitive: the Greek travel writer Pausanias was reluctant in the second century CE to count as a city a small settlement called Panopeus, in Phocis, as it had 'no government offices or gymnasium, no theatre or agora or water flowing down to a fountain' (quoted in Wycherley, 1976, p.xix).

The splendid public buildings of the older unplanned cities made a stark contrast with the modest, and often squalid, houses, squeezed within the city walls along narrow, irregular streets. According to the orator Demosthenes, the houses of leading Athenian citizens were indistinguishable from the rest (Wycherley, 1976, p.186). This may have been rhetoric, but there was certainly little segregation of the rich and the poor in Greek cities. The patchy archaeological record makes generalizations about Greek housing difficult. Most of the houses in unplanned cities were built of sun-dried brick on a stone foundation, and consisted of a few rooms only, around an open courtyard. In planned colonies, some uniformity in the size of housing blocks was a corollary of the grid layout, though the blocks usually accommodated a variety of internal designs. A distinctive example is the planned residential quarter of the hilltop city of Olynthus, in northern Greece (see Figure 2.10). Dating from the fifth and

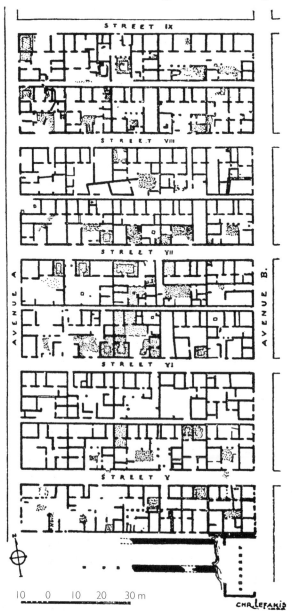

Figure 2.10 Plan of housing blocks, Olynthus, northern Greece, fifth and fourth centuries BCE. One of the blocks was apparently built in two stages: the outhouses of the first row were incorporated into the second, when that was added. The courts, usually cobbled and sometimes with a water cistern, had a portico on the north side at least, with a tiled roof carried by wooden posts or stone pillars. The houses had drains, emptying on to the street, but there is no indication of any latrines (Wycherley, 1976, p.24)

fourth centuries BCE, it consisted of long, rectangular blocks of ten two-storey houses, in two rows of five, divided by an alleyway. The houses, built of unbaked brick on a rubble foundation, were somewhat different from each other in internal layout, but similar in having their main rooms facing south across a courtyard. Greek houses always turned in on their courtyard and away from the street, a configuration that reflected the importance of the private household (*oikos*) as the social and economic unit of Greek society. Interior courtyards were also practical in the Mediterranean climate, providing cool and shade, and also light, windows being few and small (Jameson, 1990). Houses also generally faced south, to catch the winter sun, and for protection from the northerly winds. Hellenistic houses, some of which have been excavated at Priene and Delos, were built to higher standards; more often than in the Classical Period, they featured peristyle courts (courts with colonnades on all four sides).

How important were the various Greek political and social structures to the physical structure of their cities? A quite distinctive city, politically and socially, was Athens' great rival, Sparta. Sparta was the creation of an agriculture-based, closed, militaristic society: citizenship was limited to a small, closed group of Spartiates, which shrank as they failed to reproduce themselves. Instead of slaves, there was a special class of state-owned serfs, or helots, who were rarely freed. There was relatively little commerce or industry in the city itself, and it lacked the imposing public buildings of Athens. At its fifth-century height, Sparta apparently had no defensive wall; the Spartiates preferred to rely on military prowess, an attitude which inspired the indignation of Aristotle (see Extract 2.1 at the end of this chapter).

Many of the early Greek cities were ruled by tyrants, essentially non-hereditary monarchs, who drew their support from coalitions of the disaffected. These tyrants, rather like the Near Eastern *lugals* and Pharaohs before them, sometimes had a great impact on the physical fabric of the city. By sponsoring building projects, such as temples and water-supply systems, they intended to impress and enlist the support of the urban poor. In the Classical city state, city-building was more regulated: magistrates, known as *astunomoi*, were responsible for public services such as streets, water supply and drainage; markets and other commercial activities were administered by *agoranomoi*. Private property rights were recognized, though there was a right of expropriation in the public interest. Laws of expropriation are, as this series will demonstrate, crucial to the responsiveness of the urban fabric to the potentials of new technologies.

2.4 Greek technologies and city-building

How were Greek practices of city-building influenced by prevailing technologies? It would be difficult to argue that the grid layout is a function of any particular set of technologies, but there are some connections worth exploring. Urban planning should, perhaps, be considered more as a shaper of the technologies of city-building than as a technology itself. Nevertheless, the idea that suitable environments should, as far as possible, be made to conform to a human plan is consonant with a shift from the technologies through which prehistoric humans adapted to the natural environment, to the more ambitious, environmentally aggressive projects of antiquity. It also indicates a notable transition from the kinds of cities discussed in Chapter 1: in these, a prodigious proportion of the building effort went into the glorification of the monarch and the preparation of the monarch's after-life. In Greek cities, there begins to be an emphasis on buildings for public use (theatres, gymnasia, stadia, baths) and public works and services (artificial harbours, aqueducts, fountain buildings). In other words, the planned city can be seen as an engine of technological innovation; indeed, the planned Greek colonies were often architecturally and technologically more advanced than their mother cities in Greece proper. Another important point about the grid is that its realization at a given place and time involves assumptions about building materials and techniques, and modes of transport, among other technological considerations. Even in ancient cities, where most people got about by walking, and goods were carried by human porters or pack-animals, the main avenues would have to cope with ox-drawn carts, unless, as apparently in the case of Ur, these were excluded. Certainly, as this series will make clear, changes in these technological parameters were reflected in the dimensions of grid components, such as street widths, and block height and depth. And, last but not least, the grid layout presupposed the technical skills of the surveyor.

Building construction

The Greeks have been portrayed as conservative in their building styles, techniques and materials. Greek public buildings of the Classical and Hellenistic Periods relied on dry-stone building techniques (laying masonry without mortar) and on 'post-and-lintel', or trabeated, construction: a combination of vertical columns and horizontal beams, already seen in the stone temples of Thebes.

Figure 2.11 Sectional view of the front of the Parthenon, on the Acropolis, Athens. The Parthenon ('the Virgin') was a Doric temple dedicated to Athena, and was built entirely of marble, in the middle years of the fifth century BCE. The portico columns support the standard arrangement of architrave, frieze and pediment (adapted from Lawrence, 1973, p.160; copyright © The Trustees of the British Museum)

Arches and vaults are seldom found; the Greeks' architectural predilections partly reflected the abundance of good roofing timber, which allowed spans as wide as ten metres, and of local building stone, particularly the various types of limestone and marble, both of which yielded beams of some four metres. Since the Romans are often portrayed as excessively deferential to Greek architectural practice, the Greeks' own debt to the Near East is worth emphasizing:

> The 'Aeolic' capitals of the earliest Greek architecture in stone echoed a type that was widespread in Phoenicia and Palestine, while more generally the whole Greek architectural system of column, capital, and entablature represents the impact of the masonry architecture of the ancient East, and in particular of Egypt, upon the primitive timber and mud-brick construction of Aegean Greece.
>
> (Ward-Perkins, 1974, p.11)

Despite the limitations on dimensions and design associated with these methods, or perhaps because of them, buildings of stunning simplicity and symmetry resulted. Their sturdy structures had the signal advantage of resistance to the earthquake tremors of the Aegean region.

Of the main Greek building types, temples represent the most successful union of Greek building methods, aesthetics and social function. The various governments of Greek cities undoubtedly put a disproportionate amount of their limited resources into such structures, and the rest of the urban built environment generally suffered by comparison. Again, there may be continuities with earlier building traditions: comparisons have been drawn between the Mycenaean megaron and the earliest Greek temples (Ward-Perkins, 1974, p.10). The first temples had mud-brick walls, with wooden columns and beams, and thatched roofs; even though stone became the main material, many have argued that the design of the Greek temple reflected its wooden origins. In its mature form, the temple centred on a rectangular inner shrine with masonry walls and a doorway of two vertical jambs and a horizontal lintel. The shrine was surrounded by a continuous outer colonnade, and covered with a tiled pitched roof. As the temples grew in size, and terracotta roof tiles and decorated pediments at each end became standard, their great roof structures were carried by increasingly large and heavy longitudinal and transverse beams of timber or stone resting on rows of stone columns (see Figure 2.11). The profusion of internal and external columns that characterized Greek temples was both structural and decorative in function; it was acceptable in buildings that were intended to house the image of a god, not a congregation of worshippers.

In more utilitarian structures, the limitations of Greek building technology are more evident. In Athens, the custom was for large assemblies to take place in the open air, and the Mediterranean climate allowed Greek recreational building types such as theatres to be uncovered. If desired, however, a small public building could be roofed easily enough with timber. From the fourth century BCE, there was a political need in many smaller cities for roofed structures in which the entire assembly of citizens could meet. Again, this meant a proliferation of internal columns, which inevitably blocked the view of many in the audience. An unusual attempt to get round this difficulty was the design of the grandiose Thersilion at Megalopolis, 'an exceptional *tour de force*' in Greek building (Wycherley, 1976, p.120). Serving as the assembly hall for the entire Arcadian League of city states, the Thersilion could seat 6,000; its floor was slightly raked, and the columns were set in lines radiating from the speaker's rostrum (see Figure 2.12). Extensive colonnades (stoas) were better adapted to commercial buildings, such as markets, in which simple repeated spaces were adequate for small-scale retailing and manufacturing.

Figure 2.12 Restored plan of the Thersilion, Megalopolis, fourth century BCE. The building was fifty by sixty-three metres, with a floor space of 3,000 square metres. The sixty-seven columns supporting the roof were arranged in five concentric rectangles and also along lines radiating from the rostrum (White, 1984, p.77; by permission of Professor K.D. White)

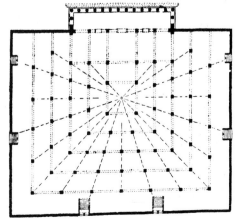

A relatively late development was the gymnasium, a building type based on the three great philosophical schools at Athens: the Academy, the Lyceum and the Cynosarges. Gymnasia became a fixture of Greek urban life. From their original brief of providing young citizens with a combination of physical training and cultural stimulation, they expanded during the Hellenistic Period into cultural centres with libraries and auditoria. From early in the design history of the gymnasium, associated with the exercise area (*palaestra*) there seems to have been a *loutron,* or area in which covered bathing arrangements were provided. Basins and running water were available, though there is no evidence that the water was heated. Hot bathing was nevertheless available in the public baths (*balaneia*), which typically had a circular room, or *tholos,* for this purpose, the circular shape helping to conserve heat. In many baths the source of heat was a simple charcoal brazier, though combination furnaces and boilers were sometimes installed, as in the baths dating from the mid-fourth century BCE at Olympia (see Figure 2.13). At Gela, in Sicily, another early hot bath has been excavated, dating from the fourth or early third century:

> If it is from the fourth, it is early for a bath with several rooms each with banks of tubs, and with a below-floor heating room … Gradually we are becoming aware that public baths of several rooms, with provision for heated water and oriented to the south for solar heating of the spaces, are a standard feature of Greek water management from the fifth century on, rather than being a development of the Romans.

(Crouch, 1993, pp.320–21)

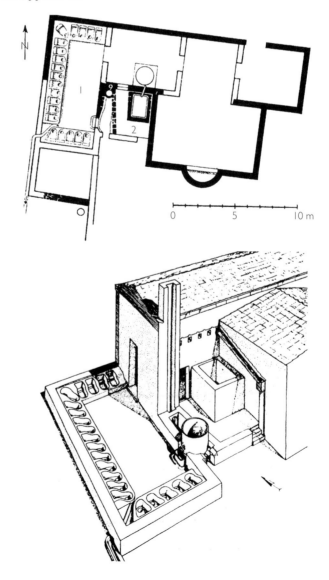

Figure 2.13 Baths, Olympia, mid-fourth century BCE.
Top plan of the bathhouse, with its hip-baths (1), which was served from a furnace–boiler combination in an outside service yard (2).
Bottom restored cutaway view, showing the furnace flue and, embedded in the wall, the bronze boiler that supplied water directly into the bathing unit. Athletes at the Olympic Games must have used this arrangement for some two centuries (adapted from Yegül, 1995, p.378; by permission of the Deutsches Archäologisches Institut, Berlin)

Several technological systems were involved in the construction of a building such as a temple, from quarrying and transporting building materials (usually local in origin), to the organization of labour and methods of construction on site. Stone was quarried as in Egypt, though there is evidence that less labour-intensive means of transporting the blocks were employed, such as heavy wagons drawn by teams of oxen. According to Vitruvius, column drums were sometimes moved by ox-drawn cradles with large wooden wheels (see Figure 2.14). On site, the geometrical perfection of the structures was achieved with simple, traditional instruments: the cord, plumb-line, level and square. The drums of columns were centred using countersunk wooden plugs and pins, and various means were used to secure the component stones of walls, including rebating the pieces, and bonding them with iron clamps. Iron was used sparingly in construction, though two-metre iron beams supplemented those of marble in the Propylaea, the monumental entrance to the Athenian Acropolis. Building continued to be labour-intensive, though there was some saving of effort through the hollowing of stone beams, and the introduction of

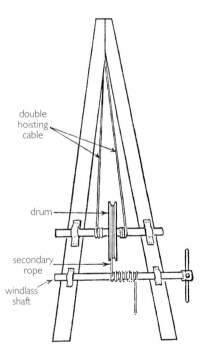

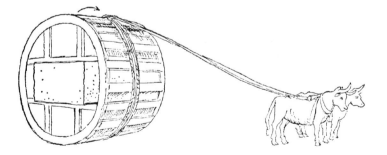

Figure 2.14 Transport of column drums and stone blocks, based on descriptions by Vitruvius. *Top* in this method a column shaft or drum was used as a roller, and was attached to a wooden frame by iron spigots. *Middle* an architrave block, which could not be rolled because of its square or rectangular cross-section, could be enclosed in suitably large wheels, as shown here, as well as being attached by spigots to a frame. In both methods the quarry had to be relatively near to the site, and the ground in between had to be flat and level. *Bottom* this method was a failure, and, according to Vitruvius, caused its inventor's bankruptcy. The use of an uncoiling rope, instead of a frame through which the oxen could steer the wheels, meant that the structure kept swerving off the track (Landels, 1978, p.184; by permission of University of California Press)

double hoisting cable

drum

secondary rope

windlass shaft

Figure 2.15 Crane, based on a description by Vitruvius. Heavy blocks were lifted by means of a system of ropes and pulleys powered by a windlass. For especially heavy blocks, the hoisting rope was doubled and wound on to an additional shaft with a drum at its centre (a reduction gear, in modern terminology). The drum was turned by a secondary rope winding on to the windlass shaft (Landels, 1978, p.86; by permission of University of California Press)

Figure 2.16 Artist's impression of Priene, a good example of the form of a small planned and terraced Hellenistic city. Much of the grid, especially the north–south components, consisted of stepped footpaths. Surface and underground drains were needed to cope with the problem of storm run-off on such a steeply sloping site

Key A: agora, with the Sacred Stoa prominent on the north side; B: Temple of Zeus; C: gymnasia (there are two); D: theatre; E: Temple of Athena; F: stadium; G: main entrance to city (Morris, 1979, p.29; by permission of Addison Wesley Longman)

cranes for lifting them. These cranes were based on two of the main mechanical devices of antiquity for augmenting human muscle power, both developments of the wheel: the pulley, first depicted on Assyrian monuments early in the first millennium BCE, and the windlass, a wheel-and-axle mechanism that was rotated about a horizontal axis (Coulton, 1974; see Figure 2.15).[2]

Although Greek building methods are often seen as conservative, there were notable developments in practical construction, demanded by both the rocky terrain on which many cities were built and the Greeks' desire to adapt it to an orthogonal layout. The commitment to Hippodamean planning is striking, and could result in 'an uneasy, not to say impractical, union of the gridded plan and the terrain' (Owens, 1991, p.7); it was retained even where some of the lesser streets had actually to be steep flights of stairs. Greek architects and builders became adept at constructing platforms, terracing and landscaping, and deploying multi-storey porticoes to connect different levels. A fine Hellenistic instance of these skills is the small Anatolian hilltop city of Priene (see Figure 2.16).

As might be expected from their investment in maritime trade, the Greeks became active harbour-builders. Natural harbours were frequently chosen as the sites of colonies, and in the tideless Mediterranean small ships could be loaded and unloaded in the surf at open roadsteads (stretches of water near the shore where ships can ride at anchor); the roadstead at Phaleron served

---

[2] Extracts from J.J. Coulton's article are included in the Reader associated with this volume (Chant, 1999).

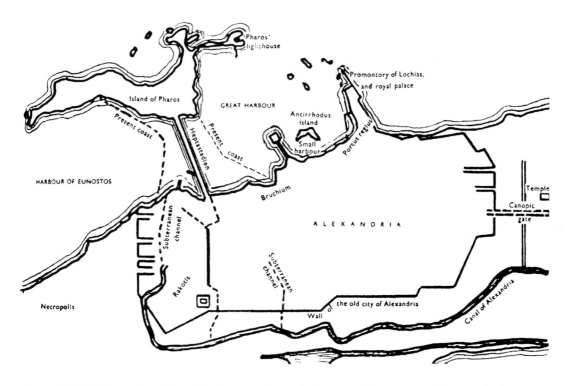

Figure 2.17 Harbour plan, Alexandria, showing the two harbours and the causeway (Heptastadion) joining the mainland to the island of Pharos (White, 1984, p.105; by permission of Professor K.D. White)

Athens' early needs. But, as early as the eighth century BCE, an artificial harbour was created by a long mole at the Cycladean island of Delos; the whole was somehow constructed from blocks of local granite more than ten tonnes in weight. Substantial artificial facilities were developed at Lechaeum, one of the ports of Corinth, at the Piraeus, the main port of Athens, and at Syracuse in Sicily. The most ambitious undertaking was the installation at the Hellenistic city of Alexandria, where two harbours were formed by the construction of a massive stone mole, the Heptastadion. Over 200 metres wide and about one and a half kilometres long, it connected the mainland to the island of Pharos (see Figure 2.17). The famous multi-tier lighthouse, designed by Sostratos of Cnidos, was built later, and was intended to guide ships away from reefs near the east harbour entrance.

Water supply and sanitation

A reliable supply of water other than rainfall or river water was essential for urbanization in ancient Greece: the first step in founding a city, apart from consulting an oracle, was locating a spring. This and other considerations have led Dora Crouch to assert for Greece something akin to Wittfogel's insistence on the fundamental importance of water supply and management in shaping oriental civilization. Her emphasis, however, shifts from irrigation to the prevailing karst geology of the Aegean region. Karst, which Crouch calls 'the hydrogeological basis of civilization', is a geological area, usually of limestone, which 'interacts with water to form characteristic surface features (sinks, ravines, etc.) and underground water channels' (Crouch, 1993, p.67; see Figure 2.18 overleaf). The channels form because the rock is soluble; a variety of springs and water sources, often intermittent or short-lived, result from this process. Crouch argues that the Greeks, having learned to manage water in the karst geological formations of the Greek mainland and islands, looked for similar rock formations for their colonies; indeed, karst phenomena are also common in Anatolia, Sicily and the heel of Italy.

Figure 2.18 General scheme of interaction between water and limestone in karst formations. Retreating escarpments like this are visible along many Greek coasts, as well as inland

Key 1: impermeable rock strata; 2: limestone; 3: place where water enters the limestone, continually enlarging the original fissure; 4: shaft formation, resulting in spaces like the domed chapel and spring of the Askeplion at Athens; 5: collapse sink, one of a series; 6: resurgence, typical of a Greek spring (based on diagrams by E.R. Pohl, 1955, p.23, figure 5; by permission of the National Speleological Society; and Nicholas Crawford, 1984)

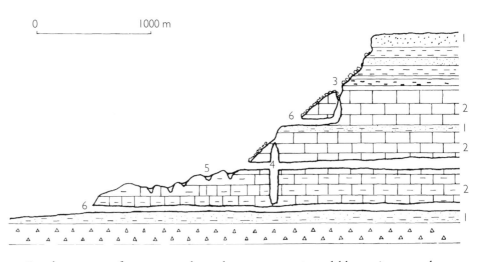

Greek systems of water supply and management could be quite complex, involving aqueducts, closed-pipe systems, settling tanks, cisterns and fountain-houses (see Figure 2.19). A striking early example of hydraulic engineering is an aqueduct carried in a tunnel through the base of a mountain on the island of Samos (see Figure 2.20). Constructed by Eupalinos of Megara in about 530 BCE, the one-kilometre tunnel itself was horizontal; but, to facilitate the flow of water, the aqueduct was carried in a sloping cutting in the tunnel's east wall, a cutting that ended up over eight metres below the tunnel floor:

> This method of construction was probably employed because it was extremely difficult with the tools of the time to hold a constant shallow slope, whereas there were adequate tools to hold the horizontal.
>
> (Rihll and Tucker, 1995, p.420)

This design of tunnel was imitated by other Greek cities, including Acragas and Syracuse in Sicily, and Athens. In most cities, though, arrangements were quite simple, consisting of a large public fountain fed by a local spring, perhaps

Figure 2.19 Artist's reconstruction of a fountain-house at Megara on the Greek mainland, dated between the seventh and fifth centuries BCE. The user stands on the paved porch as she pulls her full amphora (storage jar) out of the draw basin. Constant use has worn U-shaped depressions in the outer wall. Behind the columns on the inner wall of the draw basin are the octagonal piers of the reservoir. Meeting at fountain-houses was an important part of the social life of women and girls (Crouch, 1993, p.286; drawing: copyright © 1993 Dora P. Crouch; used by permission of Oxford University Press Inc.)

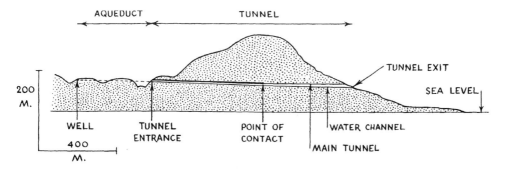
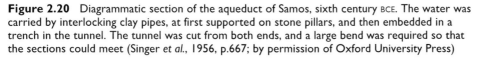

Figure 2.20 Diagrammatic section of the aqueduct of Samos, sixth century BCE. The water was carried by interlocking clay pipes, at first supported on stone pillars, and then embedded in a trench in the tunnel. The tunnel was cut from both ends, and a large bend was required so that the sections could meet (Singer *et al.*, 1956, p.667; by permission of Oxford University Press)

supplemented by private wells for drinking-water, and rainwater cisterns for other domestic uses. The layout of Corinth within its town walls, so much more open and scattered than in Athens, owes something to the concentration of groups of dwellings around certain springs. In the Classical Period, the Greeks were prepared to pipe water from any distance only if there was no alternative. One instance is Olynthus, in northern Greece, which was supplied by a spring eleven kilometres away, by way of a terracotta pipeline in an underground tunnel; another is Syracuse, in Sicily, one of the largest cities in the Greek world, with a network of aqueducts, one of which was seventeen kilometres long (Burns, 1974, p.392; Owens, 1991, p.158). In the Hellenistic Period, more ambitious projects were undertaken. Pergamon, in Anatolia, was supplied from the third century BCE by a system utilizing the 'inverted siphon' principle; water was pressurized in closed pipes so that it could be made to cross two valleys, and travel up the city's terracing to a point only forty metres below the springs (Crouch, 1993, p.119; White, 1984, pp.161–2).

The planned street layout of Rhodes, founded in 408–407 BCE, incorporates one of the earliest-known Greek examples of a system of underground drains, some of which may well have been there from the start. There is literary evidence that the large planned Sicilian colony of Acragas had drainage tunnels in the early fifth century (Burns, 1974, p.400). Extensive systems of underground drainage were a feature of many planned cities, but in fifth-century Athens and most other unplanned cities at that time, sanitation in houses was virtually non-existent; domestic drains discharged into open gutters, and much waste went out of the window, or into private cesspools (*koprones*). There were no public latrines in Athens before the Roman period, though from the fourth century cesspools were abolished and latrines in private houses connected to the main system of drainage tunnels beneath the streets (Crouch, 1993, p.304). The Classical Greeks were among the first to be concerned about the effects on public health of urban conditions, and laws were passed controlling the dumping of waste, and the digging of open drains and cesspools in public places. The great philosopher Plato objected to city walls partly because he thought them less than conducive to health (*Laws*, VI.778d). In Hippocrates' collection of medical writings, there is a treatise entitled 'About airs, waters and locations', which gives advice on the healthiest street alignments. Aristotle also advised on the most healthy locations and orientations in his *Politics* (see Extract 2.1 at the end of this chapter).

Urban industry

The Greeks were responsible for a number of craft innovations that affected the labour process and, to a limited extent, the industrial premises of cities, and their equipment. There were notable developments in pottery, especially

Figure 2.21 Black-figure painting on an Attic vase, showing the forge of Hephaistos, sixth century BCE. Hephaistos was the Greek god of fire, whose mythological workshop at Olympus housed twenty pairs of bellows. The cult of Hephaistos as divine smith and master craftsman was especially popular among Athens' large industrial population (Singer *et al.*, 1956, p.59; by permission of Oxford University Press)

the wheel, which came to take the form of a kick-wheel, powered by the potter himself rather than by an assistant. The lathe was introduced into woodworking, and tools, such as chisels and punches, were made of iron, rather than of copper or bronze, as iron gives a better edge. For iron-smelting, the Greeks replaced the simple domed clay bloomery hearth with a shaft-furnace into which air was blown by bellows. Larger forges, with improved bellows, were developed to work the smelted iron into the desired shapes, though iron still could not be melted and cast with the pyrotechnology of the time (see Figure 2.21). The changes in the industrial premises of cities would have been subtle, involving some specialization of workspaces. An example is the fullery, with its vats in which clothes were cleaned, using fuller's earth, and cloth was dyed (White, 1984, pp.39–41).

Any discussion of the relationship between technological change in manufacturing and city-building in ancient Greece needs to be set in the context of the general derogation of commerce, industry and manual labour by political elites, who viewed land as the only respectable form of property. But this generalization needs to be qualified. The legend of Prometheus, who stole the gift of fire from heaven, shows some reverence for practical invention. Some of the early Greek philosophers were reputedly as interested in practical problems as in theories about the nature of the universe. Thales of Miletus (*fl.* sixth century) is said to have studied Egyptian techniques of land-surveying and astronomy, and applied them to navigation. Early inventors were held in high esteem: Glaukos of Chios, for example, was credited with the seventh-century innovation of welding separate pieces of iron by hammering them together when red-hot: 'for a few centuries at least, the status of the technician amongst the Greeks was raised far above that of his fellow technicians in the Asiatic countries' (Hodges, 1971, p.163). But, as the elite grew more distant from industry and commerce, craft production became the exclusive preserve of slaves (whom Aristotle likened to 'living tools' in his *Politics*, Book I, Chapter 4) and of foreigners. Classical Athens, however, as a centre of international trade, did more to encourage technical skills, requiring its citizens to teach their sons a trade (White, 1984, p.25).

The political and intellectual elite had some interest in mechanical devices, but not for their economic advantage. Among the scholars attracted to the library and museum at Alexandria, some invented machines, based on mechanical principles such as the pulley, the screw, the lever, the spring, cogs and valves, and, potentially the most powerful of all, the expansion of air when heated. These included Hero, the inventor in the first century BCE of the world's first steam turbine; Ctesibius (*c*.300–230 BCE), a native Alexandrian, who invented a suction pump, sometimes used for fire-fighting; and Philo of Byzantium, who in the second century BCE wrote works of direct relevance to city-building, on the building of harbours and the defence and siege of towns. Some of these mechanical devices were, like ancient building technology, put to the purpose of inspiring religious wonder: for example, a system of gears and pulleys that was operated by expanded air from a fire on the altar was used to open the large doors of a temple.

This selective attitude to technology on the part of the later Greek intellectuals has been attributed to the great third-century Greek mathematician Archimedes, the inventor of the screw, one of the fundamental principles of mechanical engineering. His other innovations included devices for the defence of his native Syracuse, such as his projected use of huge concave mirrors to burn enemy ships by focusing the rays of the sun on them. Perhaps the idea was more valued than its chances of being realized. According to the historian Plutarch, in his life of Marcellus, Archimedes was ashamed of this kind of work; he regarded engineering and every art that ministers to the material needs of life as ignoble and sordid. Care, however, must be exercised over this much-quoted verdict; it may convey less about the attitude of Archimedes than that of Plutarch himself, a well-born Greek citizen writing during the early Roman Empire, some three centuries after Archimedes' death (Rihll and Tucker, 1995, p.425).

2.5 *Athens*

The city of Athens enjoyed political, economic and military ascendancy in the Aegean region for most of the fifth century, following the defeat of the Persians in 479 BCE. Its location partly explains the Athenians' ability to dominate the eastern Mediterranean trading system. They were able to assemble the various materials (notably timber, bronze and hemp) necessary for a strong fleet, through which they could levy their wealth from other cities. With the population rising to about 330,000, quite unprecedented for a Greek city, food supply had become a problem in the fifth century, and Athens looked to the Black Sea region for additional grain. The Athenian fleet secured this supply by dominating the straits between the Black Sea and the Aegean, and forcing grain ships to stop first at the port of Athens. Another important source of the Athenians' economic strength was their control of the nearby Laurion silver mines, an operation largely dependent on the labour of criminals and slaves, who wielded iron picks, chisels and wedges. This was an operation of some technological note: by the fourth century, there were shafts 100 metres deep, additional shafts for ventilation, and a system of recycling water for washing and grading the ores (White, 1984, pp.114–16, 120–21).

Apart from its exceptional size, Athens offers a stark contrast with the colony cities, in its largely unplanned, organic growth around the Acropolis, one of a group of five hills on the Attic plain enclosed by the city walls. Unusually too, there was continuity of development from the palace of a Mycenaean ruler. The changing layout of the city reflects a trend in the ancient Greek world: the development from a concentration of crafts and trade around the king's palace, to a more complex built environment as government passed to a college of magistrates. The Acropolis became a separate realm of the gods, conferring legitimacy on the magistrates following the weakening of the kingship (Hölscher, 1991, pp.360–61). The Agora, connected to the Acropolis by the Panathenaic Way, became the new centre of communal life, once the land was drained and water was supplied to a public fountain. Among the first buildings added to the Agora was a bouleuterion, where Solon's new elected council met in the early sixth century, as well as lawcourts and a stoa (Camp, 1986; see Figure 2.22 overleaf).

Buildings of the Agora

1 Stoa of Zeus Eleutherios
2 Temple of Apollo Patroos
3 Metroon
4 Bouleuterion
5 Tholos
6 Greek building
7 Enneakrounos
8 Heliaia (lawcourt?)
9 South Stoa
10 Library of Pantainos
11 Stoa of Attalos
12 Painted Stoa
13 Royal Stoa
14 Altar of the Twelve Gods
15 Temple of Ares
16 Odeion
17 Middle Stoa
18 Hephaisteion (Theseion)
19 Hellenistic building
20 Temple of Aphrodite Ourania
21 Sanctuary of Demos and Graces
22 Early Roman colonnade
23 Poros building
24 archaic cemetery

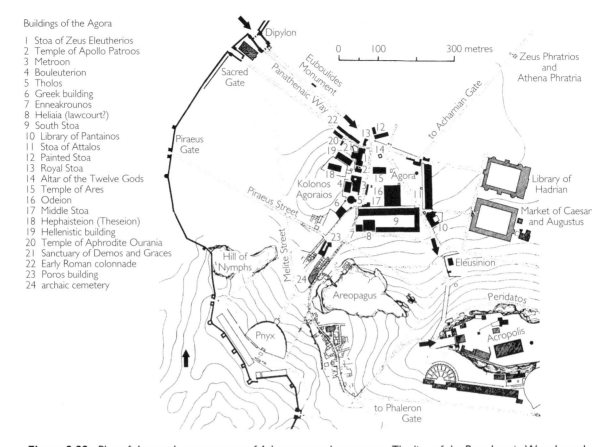

Figure 2.22 Plan of the north-western part of Athens, second century CE. The line of the Panathenaic Way through the Agora to the Acropolis is shown by a sequence of arrows. The Agora has a mix of Classical, Hellenistic and Roman buildings; the main buildings of the later fifth century, other than the dominating Hephaisteion Temple (18), were the Stoa of Zeus Eleutherios (1), the new Bouleuterion (4) and the South Stoa (9). The building named Enneakrounos ('nine-spouted') (7) was a fountain-house. A typical linear group of organic-growth housing is shown south of the hill named the Areopagus (adapted from Morris, 1979, p.31; by permission of Addison Wesley Longman)

Figure 2.23 The Acropolis, Athens. The structure at the top left is the Propylaea, the monumental entrance to the Acropolis; dominating the middle is the Parthenon; and visible to the right, beneath the distant mountain, are some of the massive retaining walls of the site (Mumford, 1966, plate 9; copyright © Ewing Galloway, Inc., all rights reserved)

The destruction of many buildings and fortifications by the Persians in 480 BCE led eventually to a comprehensive rebuilding programme in the second half of the fifth century, initiated by Pericles and financed by the Delian League of city states. Four major marble structures were erected on the Acropolis: the Propylaea, a monumental gateway, and three magnificent temples, the Temple of Athena Nike, the Erechtheum and, dominating the whole city, the Parthenon (see Figure 2.23). Several new buildings were added to the Agora, mostly towards the end of the fifth century (see Figure 2.22). The Persian sack also led to the redesign of the Piraeus, a peninsula with three harbours which became the main port of Athens. The plan, attributed by Aristotle to Hippodamus, incorporated a gridiron street layout, zones marked by boundary stones, such as the military and commercial ports, and two agoras.

There was a tension between pressures for outward growth and the need for defensive walls, resulting in a discontinuous, ratchet-like process of expansion. Walls enclosed both the Acropolis and the city as a whole from the beginning of the fifth century. If Hippodamus was indeed involved in the planning of the Piraeus, it may have been at the time that it was fortified and connected to the city by the Long Walls, built probably in the middle of the fifth century. The fortifications also enclosed Athens' first port, the open roadstead of Phaleron (see Figure 2.24). In this way, Athens reconciled the early military advantages of its inland setting with the commercial necessity of immediate access to the sea. At a time of increasing inter-city rivalry and warfare, some other older cities reached out to their ports with a similar stony embrace; for example, Corinth built long walls to Lechaeum, its port on the Gulf of Corinth.

As well as food supply, water supply was clearly a pressing concern for a city the size of Athens. The Acropolis at Athens was particularly well watered; according to Crouch, this hydrogeological endowment was as important as the hill's defensive advantages in establishing Athens' domination of its region. There is evidence of a Mycenaean spring and fountain dating from the thirteenth century BCE. As early as the sixth century, Peisistratos constructed an underground aqueduct from Mount Pentelikos to the north-east of the city: a good example of a tyrant providing an urban amenity in order to foster popular support. Water from the aqueduct was eventually distributed throughout the city by terracotta pipes. In the late sixth century large cisterns were cut into the rock, and were fed by rainwater by way of cut channels. Over the centuries of Athens' ancient history, springs came and went, a reflection of the prevailing karst

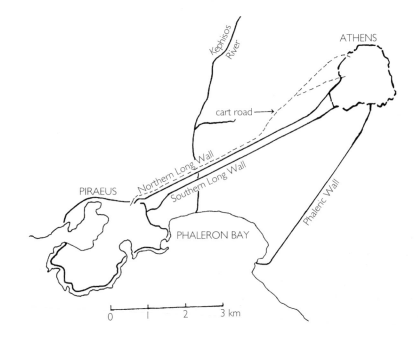

Figure 2.24 Plan showing the walls of Athens and its ports. The Long Walls connecting Athens with the Piraeus formed a protected corridor six kilometres long, averaging 165 metres in width (Garland, 1971, p.23, figure 7; drawing: based on Travlos, 1971, figure 213)

geology. Each was exploited by a system of wells, fountains and cisterns. Another hydraulic challenge was storm run-off from the hills of Athens: drainage channels had to be built to protect the great structures of the Acropolis, and in their turn the Agora and its buildings (Crouch, 1993).

Under the Macedonian hegemony from the fourth century, Athens became a cultural centre, respected for its past glories, 'a petrified monument of its own past' (Hölscher, 1991, p.379). Gone was the political and economic supremacy which, through Greek labour and engineering, had been translated into a distinctive urban fabric.

2.6 Conclusion

The example of Athens exhibits a perennial feature of the relations between cities and technologies: the uniqueness of a city's location, topography and politics, set against the more generalized effects of diffusing technologies. Overall, the technologies of Greek cities had much in common with those of their Near Eastern predecessors: there was the same dominance of water transport in the system of cities, and of pedestrian and animal-powered movement within the cities. There were similarities in the energy sources, production processes and scale of urban manufacturing, and in its dissemination in the urban built environment. Building materials were of local origin, which lends character to Greek building methods, and emphasizes the localized nature of ancient technologies, and the consequent influence of the immediate natural environment – in particular the karst geology of Greece and its implications for water supply (Crouch, 1993). But in an important respect, Greek cities represented a departure from the Near Eastern types considered in the first chapter. Instead of channelling most of the surplus resources into the urban power structures of divine monarchs, the Greek city states deployed technologies in ways that benefited more of the urban residents. Thus the notion of urban public works and amenities occurs for the first time in Western history; it would remain part of the politics of the urban built environment, even after the return of monarchical rule. In this respect, it seems that social and political developments led the technologies of city-building.

Extract

2.1 Aristotle (1962) *Politics* (trans. T.A. Sinclair), Book VII, Chapter 11, Harmondsworth, Penguin Books, pp.278–80

We have already noted that a city should have easy access both to the sea and to the interior, and, so far as conditions allow, be equally accessible to the whole of its territory. The land upon which the city itself is to be sited should be sloping. That is something that we must just hope to find, but we should keep four considerations in mind. First and most essential the situation must be a healthy one. A slope facing east, with winds blowing from the direction of sunrise, gives a healthy site, rather better than one on the lee side of north though this gives good weather. Next, it should be well situated for carrying out all its civil and military activities. For the purposes of defence the site should be one from which defenders can easily make a sally but which attackers will find difficult to approach and difficult to surround. Water, and especially spring water, should be abundant and if possible under immediate control in time of war; alternatively a way has been discovered of catching rain water in large quantities in vessels numerous enough to ensure a supply when fighting prevents the defenders from going far afield.

Since consideration must be given to the health of the inhabitants, which is partly a matter of siting in the best place and facing the right way, partly also dependent on a supply of pure water, this too must receive careful attention. I mention situation and water supply in particular because air and water, being just those things that we make most frequent and constant use of, have the greatest effect on our bodily condition. Hence, in a state which has welfare at heart, water for human consumption should be separated from water for all other purposes, unless of course all the water is alike and there are plenty of springs that are drinkable.

In the matter of defensive positions it should be remembered that what is best for one type of government is not so good for another. A lofty central citadel suits both oligarchy and monarchy, a level plain democracy; neither suits an aristocracy, which prefers a series of strongly held points. In laying out areas for private dwelling houses, the modern or Hippodamean method has the advantage of regularity; it is also more attractive and for all purposes save one, more practical. For ease of defence, the old-fashioned irregular siting of houses was better, hard for foreign mercenaries to get out of and for attackers to penetrate. It follows that both methods should be used and this is quite possible: arrange the buildings in the same pattern as is used for planting vines, not in rows but in *quincunx*, and do not lay out the whole city with geometric regularity but only certain parts. This will meet the needs both of safety and good appearance.

As for walls, it is quite out of date to say, as some do, that cities that lay claim to valour have no need of walls; we have only to look at what in fact has happened to cities that made that boast. Doubtless there is something not quite honourable in seeking safety behind solid walls, at any rate against an enemy equal in numbers or only very slightly superior. But it may happen, and does happen, that the numerical superiority of the attackers is too much for the courage of the defenders, both of the average man and of a chosen few. If then we are to save our city and avoid the miseries of cruelty and oppression, we must concede that the greatest degree of protection that walls can afford is also the best military measure. The truth of this is emphasized by all the modern improvements in missiles and artillery for attacking a besieged town. Deliberately to give cities no walls at all is like choosing an easily attacked position and clearing away the surrounding high ground. It is as if we were to refrain from putting walls round private property for fear of rendering the inhabitants unmanly. Another thing that should not be lost sight of is that those who have provided their city with a wall are in a position to regard that city in both ways, to treat it either as a fortifed or an unfortified city. Those who have no walls have no such choice. And if this is so, then it is a duty not only to build walls but also to maintain them in a manner suitable both for the city's appearance and for its defensive needs, which in these days are very numerous. Just as the attacking side is always on the lookout for methods which will give them an advantage, so too the defenders must seek additional means of defence by the aid of scientific inquiry. An enemy will not even attempt an attack on those who are really well prepared to meet it.

References

ARISTOTLE (1962) *Politics* (trans. T.A. Sinclair), Harmondsworth, Penguin Books.

BURNS, A. (1974) 'Ancient Greek water supply and city planning: a study of Syracuse and Acragas', *Technology and Culture*, vol.15, pp.389–412.

CAMP, J.M. (1986) *The Athenian Agora: excavations in the heart of Classical Athens*, London, Thames and Hudson.

CARTER, H. (1983) *An Introduction to Urban Historical Geography*, London, Edward Arnold.

CARTER, J.C. (1980) 'A Classical landscape: rural archaeology at Metaponto', *Archaeology*, vol.33, no.1, pp.23–32.

CASSON, L. (1974) *Travel in the Ancient World*, London, George Allen and Unwin.

CHANT, C. (ed.) (1999) *The Pre-industrial Cities and Technology Reader*, London, Routledge.

CHILDE, V.G. (1966) *Man Makes Himself*, London, Fontana (first published 1936).

COULTON, J.J. (1974) 'Lifting in early Greek architecture', *Journal of Hellenic Studies*, vol.94, pp.1–19.

CRAWFORD, N. (1984) *The Karst Hydrogeology of the Cumberland Plateau Escarpment of Tennessee*, Nashville, State of Tennessee Department of Environment and Conservation.

CROUCH, D.P. (1993) *Water Management in Ancient Greek Cities*, Oxford, Oxford University Press.

DICKINSON, O. (1994) *The Aegean Bronze Age*, Cambridge, Cambridge University Press.

GARLAND, R. (1971) *The Piraeus from the Fifth to the First Century BC*, London, Duckworth.

HAMMOND, M. (1972) *The City in the Ancient World*, Cambridge, Mass., Harvard University Press.

HODGES, H. (1971) *Technology in the Ancient World*, Harmondsworth, Penguin Books.

HÖLSCHER, T. (1991) 'The city of Athens: space, symbol, structure' in A.K. Raaflaub and J. Emlen (eds) *City States in Classical Antiquity and Medieval Italy*, Ann Arbor, University of Michigan Press, pp.355–80.

JAMESON, M. (1990) 'Private space and the Greek city' in O. Murray and S. Price (eds) *The Greek City: from Homer to Alexander*, Oxford, Oxford University Press, pp.171–95.

LANDELS, J.G. (1978) *Engineering in the Ancient World: ancient culture and society*, London, Chatto and Windus.

LAWRENCE, A.W. (1973, 3rd edn) *Greek Architecture*, Harmondsworth, Penguin Books.

LITTAUER, M.A. and CROUWEL, J.H. (1996) 'The origin of the true chariot', *Antiquity*, vol.70, no.270, pp.934–9.

MORGAN, L.H. (1877) *Ancient Society, or Researches in the Lines of Human Progress from Savagery, through Barbarism to Civilization*, London, Macmillan.

MORKOT, R. (1996) *The Penguin Historical Atlas of Ancient Greece*, Harmondsworth, Penguin Books.

MORRIS, A.E.J. (1979, 2nd edn) *A History of Urban Form: before the Industrial Revolutions*, London, George Godwin (first published 1974).

MORRIS, I. (1987) *Burial and Ancient Society: the rise of the Greek city-state*, Cambridge, Cambridge University Press.

MUMFORD, L. (1966) *The City in History: its origins, its transformations and its prospects*, Harmondsworth, Penguin Books (first published 1961).

OWENS, E.J. (1991) *The City in the Greek and Roman World*, London, Routledge.

PIGGOTT, S. (1983) *The Earliest Wheeled Transport: from the Atlantic coast to the Caspian Sea*, London, Thames and Hudson.

POHL, E.R. (1955) *Vertical Shafts in Limestone Caves*, Ticuta, New Jersey, National Speleological Society.

POLIGNAC, F. DE (1995) *Cults, Territory and the Origins of the Greek City-State* (trans. J. Lloyd), Chicago, University of Chicago Press.

RENFREW, C. (1972) *The Emergence of Civilisation: the Cyclades and the Aegean in the third millennium BC*, London, Methuen.

RENFREW, C. (1987) *Archaeology and Language: the puzzle of Indo-European origins*, London, Jonathan Cape.

RIHLL, T.E. and TUCKER, J.V. (1995) 'Greek engineering: the case of Eupalinos' tunnel' in A. Powell (ed.) *The Greek World*, London, Routledge, pp.403–31.

SINGER, C., HOLMYARD, E.J. and HALL, A.R. (eds); assisted by JAFFE, E., THOMSON, R.H.G. and DONALDSON, J.M. (1954) *A History of Technology*, Oxford, Clarendon Press, vol.1.

SINGER, C., HOLMYARD, E.J., HALL, A.R. and WILLIAMS, T.I. (eds); assisted by JAFFE, E., CLOW, N. and THOMSON, R.H.G. (1956) *A History of Technology*, Oxford, Clarendon Press, vol.2.

SNODGRASS, A. (1980) *Archaic Greece: the age of experiment*, London, Dent.

SNODGRASS, A.M. (1989) 'The coming of the Iron Age in Greece' in M.L.S. Sørensen and R. Thomas (eds) *The Bronze Age–Iron Age Transition in Europe*, Oxford, BAR, vol.1, pp.22–35.

TRAVLOS, J. (1971) *Pictorial Dictionary of Ancient Athens*, London, Thames and Hudson.

WARD-PERKINS, J.B. (1974) *Cities of Ancient Greece and Italy: planning in Classical antiquity*, London, Sidgwick and Jackson.

WHITE, K.D. (1984) *Greek and Roman Technology*, London, Thames and Hudson.

WYCHERLEY, R.E. (1976, 2nd edn) *How the Greeks Built Cities*, New York, W.W. Norton (first published 1962).

YEGÜL, F.K. (1995) *Baths and Bathing in Classical Antiquity*, Cambridge, Mass., MIT Press (first published 1992).

Chapter 3: ROME

by Colin Chant

3.1 The pattern of Roman urbanization

After the beginnings of European urbanization in the Aegean, it took several more centuries of the diffusion of technologies, the movement of peoples and the extension of trade before cities grew up in the central Mediterranean region. Metallurgy, one of the main technological conditions for V. Gordon Childe's Urban Revolution, was probably introduced into the Italian peninsula in the second millennium, by migrants from the north. These included the Latins, on whose territory (Latium) south of the River Tiber Rome grew up. Some archaeologists put the beginnings of Italian urbanization as early as the end of the second millennium, during the Bronze Age; they believe that the features known as the Terremare – mounds of earth in the Po Valley – are the remains of planned cities (Holloway, 1994, p.14). A more lasting process of urbanization set in during Italy's Iron Age, partly stimulated by contact with the East. Among the first trading colonies in Italy were those founded during the eighth century BCE by Phoenician traders in western Sicily, and by Greek settlers in eastern Sicily and southern Italy. The first Greek trading post in the region was probably at Pithekoussai, on the island of Ischia in the Bay of Naples, giving these western Greeks access to the metal ores of Etruria, to the north of the Tiber. But urbanization in the region cannot be reduced to a process of diffusion from the East. The Etruscans, a people whose language was unrelated to the Indo-European family, had their own traditions of city-building, and exerted a strong cultural influence on the Latins. In the ninth century, before any contact with Eastern traders, the populations of both Latin and Etruscan settlements were becoming more concentrated. By the seventh and sixth centuries, cities were emerging throughout central and southern Italy. These included Rome (see Figure 3.1 overleaf).

According to one tradition, Rome was founded by Romulus in 753 BCE. Archaeological evidence indicates that the city resulted from the federation of a group of villages (the process known as synoecism), located on four hills (the Capitoline, Palatine, Esquiline and Quirinal) overlooking the Tiber flood-plain, with the Forum (once a cemetery) as communal ground. It probably reached an urban size and degree of complexity by the early sixth century, when it was ruled by kings. In 510 BCE, the Romans rejected kingship and, for the next 400 years and more of the Roman Republic, the city was governed by elected magistrates and popular assembly. Monarchy returned in 27 BCE in the form of the principate (from *princeps*, first citizen – a republican term) of Augustus, who would assume the title *imperator* (whence 'emperor', though in its original meaning, a military general); his successors continued to rule Rome until the collapse of its empire in the West, in the fifth century CE. Roman cities were evidently built within a shifting ideological context; the sequence of tyrannical, republican and imperial forms of government and administration is broadly comparable to developments in ancient Greece (see Table 3.1, on p.83, for a framework of ancient Roman history).

The author of this chapter is grateful for detailed advice and corrections from Janet DeLaine, and for the comments of colleagues at The Open University, especially Phil Perkins. The overall approach adopted, however, is the author's own.

Figure 3.1 Roman Italy

The Romans soon fought their way to dominance in Latium, as might be expected from the city's commanding position on the Tiber (Holloway, 1994, p.173). During the period of republican rule, Rome embarked on the conquest and colonization of the rest of the Italian peninsula; it extended its territory into Sicily and Spain during the third century, and then into southern Gaul (modern France). Following the destruction of Carthage and the sacking of Corinth in 146 BCE, North Africa and Greece became Roman provinces, and during the late republic Syria and Judaea were annexed. Clearly, there was a substantial Roman territorial empire before the city itself was subject to autocratic rule. Under the emperors, Rome's frontiers were stretched further, probably, than could be defended in the long term; at its greatest, Rome's dominion extended from Britain in the west to Mesopotamia in the east (see Figure 3.2).

Rome's enduring pre-eminence in the imperial system of towns and cities was unprecedented in the ancient world. The primacy of Rome is nowhere better seen than in its domination of the empire's food-production system (Garnsey, 1988). The city's nutritional requirements were met by the Mediterranean region's established methods of agriculture, the main products of which were grain, wine and olives. Agricultural technology was not entirely static in the empire as a whole: there were developments in iron tools,

***Table 3.1 Selected periods and (in italics) emperors
in ancient Roman history (adapted from Scarre, 1995, pp.136–7)***

Dates	Period
753–510 BCE	Rome ruled by kings
510–27 BCE	Roman Republic
27 BCE–476 CE	Rome ruled by emperors, including:
27 BCE–14 CE	*Augustus*
41–54	*Claudius*
54–68	*Nero*
69–79	*Vespasian*
79–81	*Titus*
81–96	*Domitian*
96–98	*Nerva*
98–117	*Trajan*
117–38	*Hadrian*
138–61	*Antoninus Pius*
193–211	*Septimius Severus*
211–17	*Caracalla*
270–75	*Aurelian*
284–305	*Diocletian*
306–12	*Maxentius*
307–32	*Constantine I*

Figure 3.2
The Roman Empire, c.120 CE

Figure 3.3 Reconstruction of a grain-harvester, based on descriptions by Pliny the Elder in the first century CE and Palladius in the fifth. Such machines were wheeled frames or, as in the illustration, carts – pushed by oxen, mules or donkeys, and fitted at the front with a row of knife-blades (Singer *et al.*, 1956, p.97; by permission of Oxford University Press)

including double-ended implements such as pick-axes and mattocks; there were also some regional innovations, notably the use of animal-powered grain-harvesters (see Figure 3.3) in Gaul, where level plains and reliable rainfall encouraged extensive monocultures. It is unlikely, however, that grain harvested by such machines reached Rome itself. The uneven topography and unpredictable rainfall of the Mediterranean region ruled out mechanization. Under these conditions, the practice of labour-intensive mixed farming regimes provided some insurance against crop failure. Another regional variant, though the technology is probably pre-Roman in origin, was the run-off agriculture practised in the marginal, 'pre-desert' conditions of North Africa, where the little rain that fell was captured by control walls, dams, terraces and cisterns (Mattingly, 1995, p.144). But the capital's gargantuan appetite failed to stimulate any widespread change in the agricultural practice of the regions that supplied it; this continued to be founded on the abundant muscle power of draught-animals and agricultural labourers, some of whom were probably slaves.

Instead of innovations in agricultural production, Rome looked to the transhipment of agricultural produce, much of it a tax in kind, from a growing collection of overseas provinces – Sicily, Sardinia, Spain, North Africa and Egypt. By the first century CE, an estimated 300,000 tonnes of grain – some 250 standard shiploads, or more than ten million sackloads – were unloaded by sack-men (*saccarii*) each year at Portus, the new port of Rome at the mouth of the Tiber. Even for 'what was by far the most massive handling operation in the classical world', the versatility and mobility of the human porter was preferred, though his carrying capacity was one-quarter that of a panniered mule (White, 1984, pp.127, 153). A typical grain-ship weighed between 340 and 400 tonnes, more than twice the usual weight of Hellenistic cargo-ships. The grain was either stored at Portus or transferred directly to lighters or river-boats. Some 8,000 of these boats each year were towed upstream, by manual labour or by oxen, to Rome's purpose-built riverside warehouses, or *horrea*, in the area known as the Emporium (see Figure 3.4). These warehouses were carefully designed to protect the product: raised floors kept the grain cool and dry, and also protected it from vermin. More generally, the location and design of the warehouses reflected the importance of water transport and manual labour in the Roman grain trade; warehouses were not built to allow wheeled carts to enter their courtyards (Rickman, 1971, p.8; see Figure 3.5).

Figure 3.4 Men hauling a river-boat (*navis codicaria*) upstream, probably on the River Durance in Provence. Typically, the towline is led to a mast set forward of the centre of gravity, an arrangement which kept the line clear of the water and the bank. From a relief in the Musée Calvet (Casson, 1965, plate 3(2); photograph: Giraudon)

Figure 3.5 Grandi Horrea, Ostia, south-east staircase and entrance to rooms with raised floors. There were precedents for granaries with raised floors at Hellenistic Pergamon and at Harappa in the Indus Valley. In Roman stone or concrete granaries, floors were raised either on brick piers, as in this photograph, or on tiles or stone slabs. Ventilators set into the walls encouraged a flow of air beneath the floors. Not all stone granaries had raised floors, though these may have been a standard feature of earlier wooden structures (Rickman, 1971, plate 22; by permission of the author)

Sea transport continued to be markedly more efficient than land transport for carrying bulk goods such as agricultural produce, metals and building stones: according to one estimate of prices in Diocletian's reign, it was cheaper to carry grain the entire length of the Mediterranean than to transport it 120 kilometres by road (White, 1984, p.131). The marine trading system and its prevailing modes of transport therefore helped shape the pattern of urbanization in the Roman Empire; they were also the means by which Rome could grow as large as it did.

Although the supremacy of marine trade stimulated Roman activity in harbour construction and shipbuilding, historians have often charged the Romans with a lack of technological creativity, and their contribution to innovations in shipping is a case in point. Henry Hodges, one of the Romans' harshest critics, was more forgiving in this case, conceding that they had no competitors on their seas, other than pirates, to stimulate innovation (Hodges, 1971, p.203). In both war and commerce, the Romans continued to rely on galleys and merchant ships of Greek and Carthaginian type, though built on a greater scale. Some substantial developments took place in ship construction and design, notably the introduction of ships with three masts. There is some pictorial evidence, including a tombstone relief of the second century CE, that, as well as the simple square-rig sails of their predecessors, the Romans used fore-and-aft sails, set parallel with the line of the keel, such as the usually triangular lateen rig; these would have enabled the ship to make steadier progress against a headwind (Casson, 1971, p.244). Without such equipment, it could take one or two months for a grain-laden ship to journey from Alexandria to Rome against the summer winds, and only two or three weeks to make the downwind return journey (White, 1984, p.154).

Following the precedent of the Assyrians and Persians, the Romans articulated their imperial system of cities with well-made roads. The network of roads, with its embankments, cuttings, tunnels and arched bridges, was one of their most enduring engineering achievements. Roman roads were constructed from a wide variety of materials, such as gravel, clay, timber and different kinds of stone paving, depending on local conditions (Chevallier, 1976, pp.86–93; see Figure 3.6). They were intended not for goods vehicles, which probably ran on verges alongside the road, but as all-weather surfaces for infantry. Paved main thoroughfares were also a definitive internal feature of Roman cities. But, apart from the private carriages of the wealthy, and litters or chairs for women, old people and dignitaries, there was no urban passenger transport, and, by a law of Julius Caesar, the use of wheeled vehicles for commercial and industrial haulage was severely restricted in daylight hours (Robinson, 1992, pp.73–6). The Romans' debt to others for developments in land transport is reflected in the names of their wheeled vehicles, which are almost all of Celtic origin. Hodges' hindsight was particularly keen in this area; he criticized the Romans for failing to take the first steps towards a more efficient mode of land transport by replacing yoked oxen with specially reared, heavy draught-horses, and developing shafts and collars (Hodges, 1971, p.202). In fact, there is evidence from Roman Gaul of the existence of shafts and collars, although none that they spread into the southern bulk of the empire. In any case, criticisms of this kind ignore ancient requirements in heavy transport; for work such as hauling building materials in particular, it was the slow strength of the yoked ox, the most economical working animal in the ancient Mediterranean context, that was valued, rather than the speed of the aristocratic horse (Burford, 1960).[1]

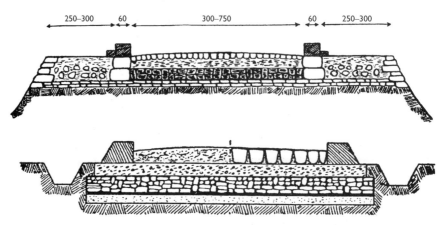

Figure 3.6 Cross-sections of two Roman roads. *Top* a principal highway. The layers, from the top down, are: cobbles or slabs set in mortar; concrete; stone blocks in mortar; flat stones. *Bottom* a subsidiary road. The layers, from the top down, are: stone setts or gravel concrete; concrete; slabs and blocks in cement mortar; mortar layer on top of sand course (Singer *et al.*, 1956, p.503; by permission of Oxford University Press)

[1]   Extracts from Burford (1960) can be found in the Reader associated with this volume (Chant, 1999).

3.2 Roman urban planning and morphology

The respective influences of Etruscan and Greek exemplars on Roman city-planning have long been debated (Anderson, 1997, pp.188–9). Some of this debate has centred on the urban form and structure of Pompeii, a city which, along with the nearby settlements of Herculaneum and Stabiae, was uniquely preserved by the eruption of Vesuvius in 79 CE – the ultimate demonstration of the importance of calamity in the history of the urban built environment. The original settlement of the sixth century BCE dates from a time when Etruscans and Greeks were neighbours in Campania; scholars remain undecided about which tradition was more influential.

The Romans may owe to the Etruscans the religious rituals associated with the founding of a planned provincial city. The chosen site was bounded by a furrow (the *sulcus primigenius*) created with a bronze plough drawn by a male and a female ox, along the line of which was later built an encircling wall. Burials were prohibited inside the wall, and on either side of it was established a sanctified area (the *pomerium*), on which no building was permitted. The gridiron street plan enclosed by the walls may have been inspired by the Greek colonies of the seventh and sixth centuries BCE, which had orthogonal layouts, with walls of regular cut stone. The plans of these western Greek colonies were distinctive, in that there were a few longitudinal avenues crossed by numerous narrower streets; as a result, building blocks were very elongated (see Figure 3.7).

As with Athens and the Greek colonies, there was a sharp contrast between the planned cities of the provinces and the organic growth of Rome, at least up to the first century BCE. The most that can be said for Rome is that from the late republican period, and especially under the empire, certain city precincts were internally planned, though only as blocks within the existing street framework. Among these were the Campus Martius (the Field of Mars), laid out by Augustus' right-hand man, Agrippa, and the succession of imperial fora added to the original Forum Romanum. However, fragments of a plan of the city carved in marble, which was made at the time of the emperor Septimius Severus in about 200 CE for the walls of the Temple of Peace (there is evidence of a version from the late first century too), suggest that there was an administrative need to conceive the city as a whole during the imperial period (Anderson, 1997, pp.225–6; see Figure 3.8 overleaf).

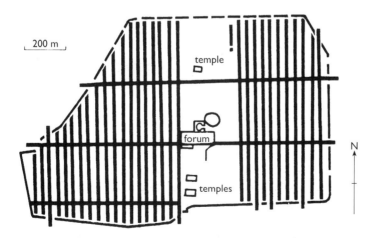

Figure 3.7 Plan of Poseidonia, on the coast of Campania, a settlement that later became the Roman colony of Paestum; the layout probably dates from the late sixth century BCE. The walled settlement had two residential areas separated by a large strip of public land on which stood all the main civic and religious buildings. The layout is typical of a western Greek colony, having three avenues running east–west and some thirty streets crossing them north–south, to create elongated rectangular housing blocks measuring thirty metres by 300 metres (Owens, 1991, p.40)

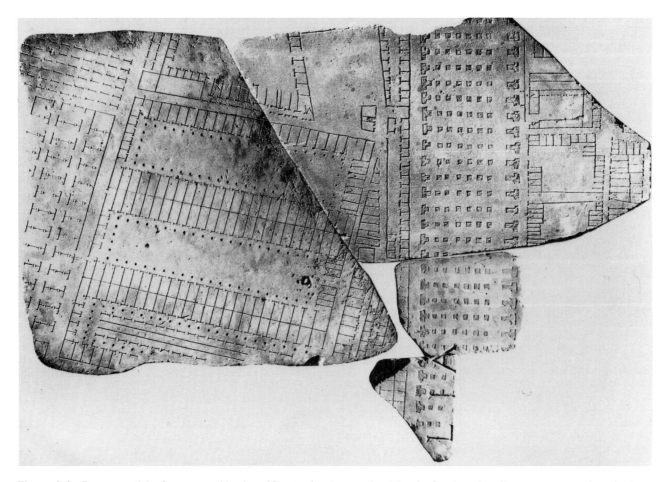

Figure 3.8 Fragment of the Severan marble plan of Rome, showing, on the right, the Porticus Aemilia, a great covered, vaulted bazaar in the city's Emporium district, dating from the early second century BCE, and one of the earliest monuments of Roman concrete architecture. On the left are the Horrea Galbae, with their three rectangular courtyards. These were extensive warehouses, providing storage for a variety of goods, including the public grain dole (*publica annona*) (MacDonald, 1982, plate 2a; photograph: Fototeca Unione, American Academy in Rome)

Beyond the capital, the Romans began as early as the fourth century BCE to build colonies with orthogonal street plans. The plan was usually based on two main intersecting roads running east–west and north–south. Archaeologists customarily refer to these streets as the *decumanus maximus* and the *cardo maximus*, though it should be noted that *decumanus* and *cardo* are not found in ancient lay descriptions of cities, but were technical terms used by Roman land-surveyors. This basic street pattern has been attributed to the Greeks and the Etruscans, and also to the Roman military camp, or *castrum*, which was a fortified rectangular area with gates in the middle of each wall. The streets ran between each pair of opposite gates, and at their intersection was an assembly area, with the main buildings of the camp headquarters around it. This area is analogous to the forum of a provincial city, with its temple and basilica; the forum was essentially an enclosed and more self-contained version of the open Greek agora. The influence of the military camp on the development of early colonies is disputed, though camps and colonies were undeniably intertwined as the Roman Empire grew. Many colonies, indeed, were founded specifically for army veterans, and were probably laid out by army surveyors: a well-preserved example is that of Timgad, in North Africa (see Figure 3.9).

Unplanned Rome itself was heavily built-up; this density of building resulted from population growth within the twin constraints of ancient technology and Rome's socio-economic structure. The limited horizontal spread of the city

Figure 3.9 Plan of Timgad, a Roman colony in Numidia with an estimated population of 10–15,000 at its peak. It was founded by Trajan in 100 CE, specifically for retired Roman legionaries. The rigid geometrical layout of the original walled settlement – a square with sides of 355 metres – testifies to its military origin. The forum, with its basilica and temple, was, as usual, incorporated within the original grid near the intersection of the two main roads. Later public buildings were located in the unplanned suburbs; these included many of the settlement's fourteen public baths (Scarre, 1995, p.107)

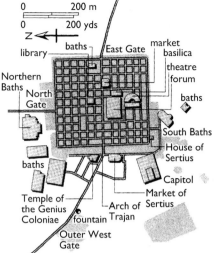

reflected above all the fact that most people needed to live within reasonable walking distance of the political and economic centre. The same consideration applied to the movement of goods within the city, much of which was undertaken by human porters, as well as by pack-animals. In Rome, exceptions to the ban on wagons during the day included the carriage of building materials for temples and public works, demonstrating the high priority given to such projects by the political elite. On the hills of Rome, there was also a green belt of villas, estates and gardens (*horti*) belonging to the senatorial class, sometimes used for entertaining only. This loop has been seen as a barrier of privilege to the expansion of the city, but a more recent view is that it was a fluid 'zone of transition' (Purcell, 1987, p.187). Less open to doubt, however, is the effect of the great tracts of land within the city given over to public buildings and spaces, or to enclaves of the wealthy; these served to squeeze the majority of the population into high-rise tenement blocks on the remaining land, such as on the eastern side of the Campus Martius, and in the valleys between the principal hills (see Figure 3.10 overleaf). This upward thrust was checked in the early imperial period by emperors fearful of fires and of the buildings' collapse: the height of the tenements was limited to seventy Roman feet (20.72 metres) by Augustus, and to sixty (17.76 metres) by Trajan.[2]

Despite the density of settlement in the centre, some suburban development took place in Rome during much of the empire, and in some of the larger provincial cities, such as Bononia (Bologna) (see Scagliarini, 1991). Roadside tombs and cemeteries had always been outside the *pomerium*, but, in addition to these, the suburbs became the location of industrial premises with unpleasant by-products, such as fulleries and pottery workshops, and of large recreational buildings, such as stadia, amphitheatres and circuses (Owens, 1991, p.152). This encroachment on the countryside reinforces the point, already evident from the discussion in Chapter 1 of the hydraulic engineering schemes of the Mesopotamians, that the ecological impact of urbanization has ancient roots (Hughes, 1994). Certainly, the growth of Rome had notable implications for its rural hinterland of Latium. At first, Rome drew on this area for its grain supplies, but, as its population increased, the city's needs had to be met by the collection of tithes from its overseas provinces, and the surrounding countryside (the *suburbium*) was increasingly turned over to market gardens and villas (Morley, 1996).[3]

The suburban or rural villas of the wealthy were the most noticeable urban incursion into the countryside of Rome and some of the provincial cities. A grandiose manifestation of this desire on the part of the wealthy to escape from the city is the emperor Hadrian's villa at Tibur (Tivoli), near Rome; a more typical example is the Villa of Mysteries at Pompeii. Few villa sites in Rome's immediate hinterland have been excavated; one is a site on the ancient Via Gabina, occupied from the third century BCE to the third century CE. Its history encapsulates changing relations between the metropolis and its hinterland: at first a rural villa and agricultural smallholding, it was rebuilt in the imperial period as an 'up-to-date suburban residence', with, typically, an

---

[2] A Roman foot (29.6 centimetres) was a little shorter than an imperial foot (30.1 centimetres).

[3] There are two extracts from Morley (1996) in the Reader associated with this volume (Chant, 1999).

internal courtyard, or atrium, with a shallow basin in the centre (*impluvium*) to collect rain from the roof, a heated suite of bathing-rooms and an outdoor swimming-pool; later, a facility for olive-oil production was added (Widrig, 1980). Despite the pre-eminence of Rome, a great deal of the empire's resources were invested in country estates, which were often built to very high standards: some were virtually self-sufficient, with their own mills, ironworks and perhaps their own pottery.

Other distinctive developments in urban morphology and the urban fabric took place in late antiquity. Cities contracted, partly as a result of economic and demographic decline, and partly because of the increasing need to live within defensive walls, as the threat of nomadic raids grew, and the countryside became more hazardous. In many provincial cities, large houses with peristyle gardens were converted into multiple dwellings and workshops. Public buildings and aqueducts decayed and, following the adoption of Christianity by Constantine I, churches became the most prominent buildings in the urban landscape (Rich, 1996).

Figure 3.10 Plan of Rome in the reign of Constantine I

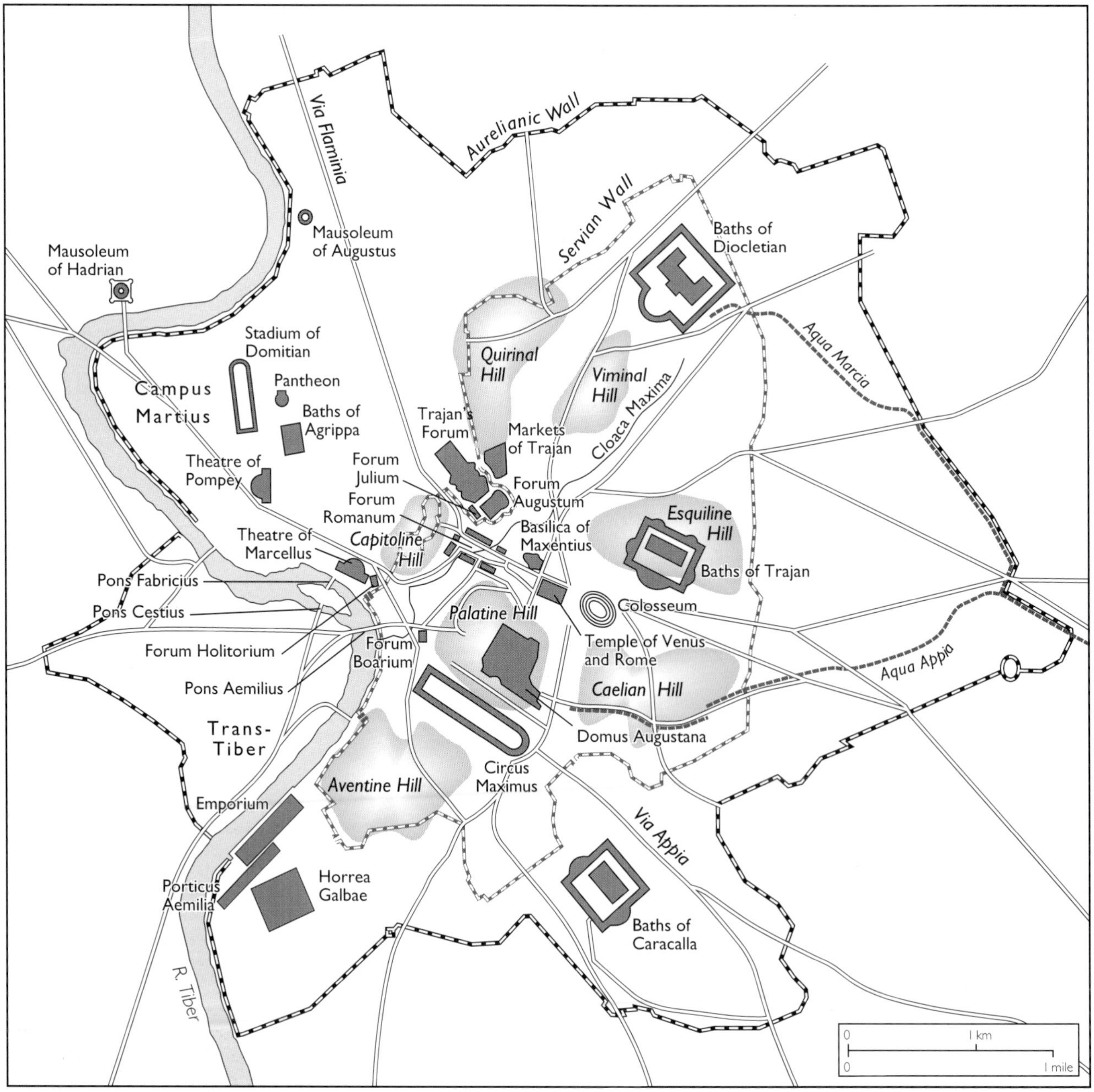

3.3 Technology and Roman city-building

What did Roman urban morphology owe to the technologies of city-building?
An issue that cuts across this question is the common view that the Romans
were technologically uninventive, and that the vast Roman achievement in
structural and hydraulic engineering was heavily dependent on the
innovations of their captive people (Hodges, 1971, pp.187–9; Smith, 1976,
p.48). According to Hodges, one of the few Roman innovations of industrial
production was that of blowing glass. In fact, glass-blowing was probably
invented in Syria at the start of the first century CE; within a matter of decades it
had spread throughout the Roman world. It was an innovation that depended
on iron technology, for only an iron blowpipe would allow glass to be blown.
It had a direct bearing on the urban built environment: apart from its
decorative applications (glassware made by other methods dates back to third-
millennium Egypt and Mesopotamia), glass-blowing was directly related to the
manufacture of window-glass. Iron was applied in innovative ways in Roman
city-building; it was used to reinforce the massive concrete vaults of some of
the major public buildings (see Figure 3.11). But the Romans undeniably failed
to anticipate that, with improved equipment, iron itself could be produced in
quantities sufficient for its use as a bulk structural material. Hodges noted that
the Romans used water-wheels to power corn mills, but that it did not occur to
them to apply this energy source to the production of iron, whether to drive
bellows or to lift heavy hammers (Hodges, 1971, pp.177, 196).

Hodges traced the Romans' alleged technological uninventiveness, offset by
a particular predilection for engineering, to their social institutions, including
the widespread use of slave-labour, an abundance of labour power in general,

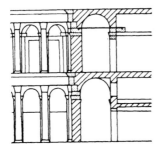
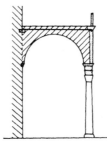

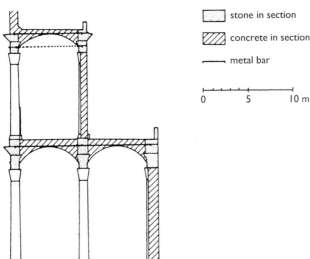

stone in section

concrete in section

metal bar

0   5   10 m

Figure 3.11 Iron-tie
construction. *Top left* in the
Horrea Agrippiana – a large,
Augustan tufa structure where
clothing and luxury fabrics were
sold – cuttings in the surface of
travertine blocks suggest the
use of exposed iron tie-bars at
the level of the springing of the
vaults, to relieve outward
thrust from the vault.
Bottom left in the Basilica Ulpia,
in Trajan's Forum, tie-bars may
have been concealed from view
in the barrel-vaulting of the
lower colonnaded aisles; there
is some evidence of exposed
tie-bars in the upper level.
Top right there is evidence of a
related concealed system in the
colonnades of the Baths of
Caracalla (DeLaine, 1990, p.419,
figure 8; by permission of the
author and publisher)

a bureaucracy that had no incentive to replace humans with machinery, and a political and intellectual elite that disdained commerce and industry (Hodges, 1971, pp.177–80). This picture of the Roman political elite evincing the Greeks' disdain for commerce and industry, and preferring *otium* (leisure) to *negotium* (business), was most influentially argued by Moses Finley (1973) in *The Ancient Economy*. Finley revived a portrayal of the ancient city as a 'consumer city', a concept introduced by the German sociologist Max Weber to distinguish it from the medieval 'producer city' out of which capitalism developed. But Finley's generalization of a non-market model for the entire ancient economy neglected the extensive and increasing involvement of Roman patricians in business through the medium of private contracting and business agents. The economy of the Classical Greek city state, and that of its Mediterranean successors, should be distinguished from the Hellenistic monarchies and the Roman Empire. In Athens, commerce and industry was mainly conducted by resident aliens and slaves, who had no access to political decision-making. The state's activities were restricted to controlling the routes by which the corn reached Athens, and the operations of the corn-dealers who handled the cargoes on arrival:

> A society of this type, the economic interest of whose members was confined to consumption, and in which productive industry and commerce were in the hands of outsiders, will scarcely have provided a favourable climate for technical invention or innovation. The Roman economy, on the other hand, contained elements which do not fit this framework, and belong to a different economic milieu.
>
> (White, 1984, pp.20–21)

Negative assessments of the Romans' technological creativity therefore rest on questionable generalizations about the economy and society of the ancient Mediterranean; such derogatory views should also be measured against Roman contributions to some urban technologies.

Surveying and fortifications

The planning of colonial towns included the division of the surrounding land among the colonists; this procedure was called 'centuriation', because land was initially divided into squares equivalent to 100 smallholdings. Both the laying out of the town and centuriation presupposed the technical skills of land-surveying and military engineering. In this instance, negative judgements about the Romans' technological creativity seem appropriate: the procedures have a lot in common with earlier Greek practice. The planning process was overseen by three land commissioners of high rank, the *tresviri coloniae deducendae*. Among their staff were surveyors known as *mensores* or *agrimensores*. They were trained in Greek geometry and, in order to establish the *limites*, or dividing lines of the grid, they used instruments of Greek origin, notably the *groma* (see Figure 3.12).

Fortifications, gates and other aspects of military engineering were among the first considerations in the planning of Roman cities in the republican period, a time of near continuous warfare (Cornell, 1995). As the empire grew in scope and confidence, walls and gateways became a mark of civic status rather than defensive need; this was testimony to the security of Roman imperial rule – the *pax Romana*. In this respect the Roman Empire had more in common with ancient Egypt than with Mesopotamia or Greece, though the renewed threat of nomadic invasion led to the hasty construction of fortifications in many Roman cities from the third century CE.

The first defence of Rome itself was a ditch and an earthwork (*agger*) across the hills to the north-west, constructed during the period of kingship. The first stone defences are known as the Servian Wall. Although they were named after

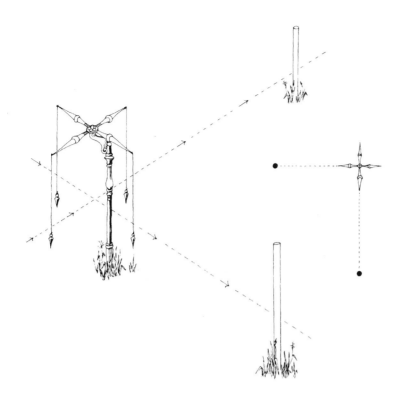

Figure 3.12 A *groma*, consisting of a horizontal cross suspended on a vertical staff with a point for fixing in soft ground. The instrument allowed the accurate setting out of straight lines and right angles, fundamental measurements in the laying out of orthogonal town plans and land centuriations. The four plumb-lines (*perpendicula*) attached to the arms of the directional square gave the *agrimensor* his two sighting-planes for the right angle (Hodges, 1971, p.158)

one of the kings, they were almost certainly put up under the republic, shortly after the city had been sacked by the Celtic Gauls in 390 BCE. The main building material was yellow tufa, a local volcanic stone, in the form of large, rectangular hewn blocks (ashlar masonry). These defences were a typical example of the dry-stone masonry construction prevalent in the ancient world at this time and, to underline Hodges' criticism, were built in the traditions of Greek masonry. They were usually set against natural slopes to form terraces, or built as a pair enclosing a sloping earthwork.

There were no further fortifications in Rome until the Aurelianic Wall of the third century CE, by which time the city had expanded considerably beyond its republican defences. At some nineteen kilometres, the second main fortification ring was nearly twice the length of the Servian Wall, and much more substantial, reflecting developments in the technologies of siege warfare. The wall was more than six metres high and some four metres thick; projecting from it at regular intervals were square towers, with windows from which stone-throwing weapons (*ballistae*) could operate. By the time of the new wall, concrete building technology was well established, and concrete faced with brick was used throughout. The wall was punctuated by well-fortified gates, allowing access to and from the city by the roads that linked it with the rest of the empire. But there is evidence too of haste, as existing arches, aqueducts and monuments were incorporated wholesale into the ring.

Building construction: a concrete revolution?

The charge against the Romans of a lack of invention has been levelled at their buildings in particular: 'It has been said that the Romans had no architecture; what their builders produced, according to this view, were imitation Greek façades, backed by brick and rubble' (White, 1984, p.83). And yet there were important Roman contributions to building technology: these included the invention of a new kind of concrete (a building material consisting of rubble set in mortar) and notable developments in architectural technology, particularly the extensive use of arches, vaults and domes, and of the tie-beam roof-truss. Although the tie-beam truss may date from Hellenistic Greece, and arches, vaults and domes were known in the Near East, the Romans used them on a far greater scale. Taken together, these developments enabled the Romans to transcend the limitations of Greek post-and-lintel construction.

At first, Roman buildings were constructed from materials immediately to hand – wood, thatch and mud. But the wider geological setting of the city (see Figure 3.1) afforded a wealth of construction materials – building stones, clay for bricks and lime for mortar. The first permanent bulk materials were local volcanic stones, in particular the various kinds of tufa, an easily worked sedimentary volcanic ash. Some tufas are very soft when quarried, and have to be left to harden; in his treatise on architecture, the Roman writer Vitruvius recommended a hardening period of two years (Vitruvius, 1960, p.50). Tufa was used at first as a cut stone, and for foundations, and later as an aggregate for concrete. Basalt, a much harder stone, provided the paving for roads in and around Rome. These local materials were transported to Rome by ox-cart or pack-animals. Bulk materials from further away, such as travertine, an off-white, hard form of limestone from quarries near Tivoli, were carried by water. Bricks were usually sun-dried, until the imperial period. From this time, most of Rome's bricks were transported downriver from the valleys of the Tiber and the Sabine Hills, where wood-fuel was available for baking them (Adam, 1994; DeLaine, 1995).

Under the empire, prestigious public buildings were increasingly decorated with valuable materials, which often had to be carried over long distances. In the middle of the first century BCE, the Romans began to quarry white marble at Carrara, in the north of the Italian peninsula. This renowned stone became the standard facing material for imperial projects; Augustus, the first emperor, was said to have boasted: 'I found Rome built of sun-dried bricks; I leave her clothed in marble' (Suetonius, 1957, p.66). Augustus' successors were even more ambitious: marbles, porphyries and granites, often in the form of great monoliths (single blocks of stone), were transported from quarries at the extremities of the empire, right across the Mediterranean to Ostia, Rome's main port. Most of the stones used in imperial building projects came from imperial quarries in the Eastern Roman Empire. At the top of the hierarchy of stones were the imperial porphyries: red porphyry from the Egyptian desert, and green porphyry from Sparta. Also highly prestigious were yellow marbles (*giallo antico*) from Chemtou in Numidia, purple-veined white marbles (*pavonazetto*) from Docimium in Turkey, and dark red-and-black marbles, now known as *africano*, though they came from Teos, on the coast of Anatolia. The increasingly widespread transport of marble throughout the Mediterranean basin brought with it greater uniformity of architectural design.

The most prized granites were red granite from Syene (modern Aswan), in southern Egypt, and, above all, grey granite from Mons Claudianus, in the eastern desert of Egypt. Because of the costs and hazards of transporting large blocks of stone across the Mediterranean, the use of monolithic columns of granite and marble was restricted to imperial buildings, baths and the private houses and villas of the wealthy. A granite column from the imperial quarries of Mons Claudianus might have weighed over 200 tonnes, and would have had to be dragged by ox-drawn sledge some 150 kilometres to the Nile (Ward-Perkins, 1971, p.142). Once the block of stone had arrived at Rome, it would have been carried through the streets to the building site, where gigantic cranes would have tilted it into place. These cranes, which also lifted smaller stone blocks to the required level, were similar to the lifting apparatus used by the Greeks (see Figure 3.13).

Why did the emperors go to such trouble? In this respect they followed the tradition of the Pharaohs, who monopolized Egyptian building stone for their temples and tombs. Most sources of precious building materials in the empire

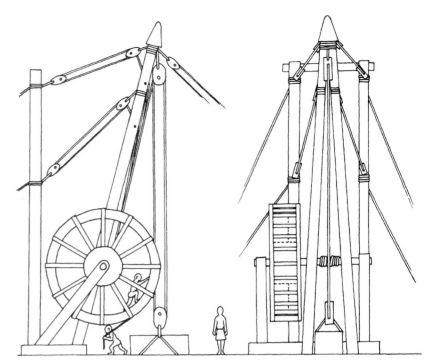

Figure 3.13 *Left* Roman bas-relief showing a particularly powerful wooden crane being used in the construction of a large building. From the family tomb of the Haterii, second century CE. *Right* diagram showing the probable elements of the crane. Compound pulleys were used both as supporting guys for the framework of the crane and to raise large stone blocks. The system of pulleys was in this instance powered not by a windlass, but by a human-operated treadmill (Hodges, 1971, pp.188–9; photograph: Alinari)

were either owned or controlled by the Roman emperors. By putting the imperial technological systems to the purpose of decorating the capital's monumental buildings, they demonstrated to the citizens of Rome their dominion over the rest of the world. Extravagance on this scale reinforces the widespread estimate of the ancient Roman elite as parasitic consumers, unlikely to foster technological creativity. There is much in this view; nevertheless, these projects in stone architecture required a great input of skilled labour, both at the quarry and on site, whether in lifting the blocks with levers, pulleys and cranes, or in the skilled positioning and jointing of them. The sheer extent of the market in building materials meant that the Roman quarrying industry was highly organized compared with its Greek and Egyptian forerunners; Ward-Perkins (1971, pp.145–6) even saw evidence of sets of columns cut to identical lengths as indicative of standardization and prefabrication in the industry, an interpretation now widely doubted (Anderson, 1997, p.177).

There were practical limits to the display of imperial power through decorative stones. For the sheer bulk of their building ambitions and needs, Augustus' successors resorted to materials closer to home. The structure of imperial Rome was increasingly a combination of baked brick and concrete. Roman concrete was made from various kinds of rubble (*caementa*) bonded with mortar. The first mortars were a mixture of lime from the Apennine Mountains, and sand from quarries, rivers or the sea. The discovery of a local volcanic sand, now known as pozzolana (after Puteoli, in the Naples area), was crucial. Pozzolana was the vital ingredient that imparted much greater strength to the mortar, matching that of building stone and brick; the mortar also possessed the invaluable property of setting under water. Unlike the crushed aggregates of modern concrete, the materials the Romans mixed with the mortar were fist-sized lumps of stone, usually tufa, but also pieces of brick and pumice; part of the skill of the engineer was to select materials appropriate to the role of the concrete within a structure. Pozzolana concrete proved to be a material of unprecedented versatility; its widespread applications in the early imperial period have been described as an 'architectural revolution' (MacDonald, 1982), or a 'concrete revolution' (DeLaine, 1990).

Concrete construction involved skills quite different from those needed for masonry work. Building a brick-faced concrete wall was a piecemeal operation, not unlike the building of a modern brick wall. A few courses of the two brick-and-mortar facings were set in place, and the space between was then filled with concrete; by this method, the wall was built up in stages (see Figure 3.14). The brick normally visible in Roman remains was the most common facing material for concrete in the imperial period: the bricks were actually triangular, with the apex set into the concrete. The technique of brick facing, known as *opus testaceum*, was often combined with the earlier method of *opus reticulatum*, which was concrete faced with small pieces of tufa, cut into squares and set diagonally to form a network pattern (see Figure 3.15). A distinctively Roman decorative technique associated with concrete was mosaic, which was used not only for floors, but also for the walls and vaults of imperial buildings. True mosaic, consisting of cut cubes of stone, terracotta or glass set in concrete, was 'a Roman development with a Hellenistic ancestry' (White, 1984, p.43).

Figure 3.14 Wall-painting of men bricklaying, showing the wooden scaffolding used in the construction of brick-faced buildings. The man on the right is using a hoe to mix mortar, which will be put in the trough on the stand beside him. The man on the left carries a basket of bricks, the man on the ladder a trough of mortar. The men on the scaffolding are laying bricks, and covering them with mortar using trowels. From the tomb of Trebius Justus, on the Via Latina in Rome (MacDonald, 1982, plate 130b; photograph: Fototeca Unione, American Academy in Rome)

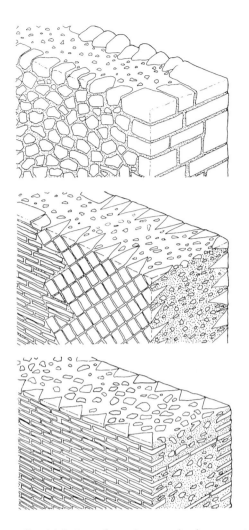

Figure 3.15 Concrete walls with facings of *opus incertum* (top), *opus reticulatum* (middle) and *opus testaceum* (bottom). The random setting of stones that characterized *opus incertum* was a feature of concrete construction of the later republican period. Its replacement by reticulate patterns, and then by brickwork, reflects the growing standardization and prefabrication of stones, and the mass production of bricks in the imperial period (Adam, 1994, pp.128–46; illustration from Grew and Hobley, 1985, p.24; drawing: R. Lea after Leacroft and Leacroft, 1969, figure 10; by permission of Hodder Educational)

Concrete was at first used only in foundations and defensive walls. As its quality improved, however, the Romans applied the new building technology to an increasingly diverse range of structures. Its waterproof property was suited to the construction of bridges and harbours: the harbour of Claudius, at Portus near Ostia, and the later inner hexagonal harbour of Trajan are the outstanding examples from the imperial period. The plasticity of concrete enabled imperial Roman architects and engineers to make unprecedented use of the arch, vault and dome, even though none of these forms was a Roman invention. An arch formed with brick facing and concrete infill proved as strong as a masonry arch, and easier to build. Compared with arches in which separate stone wedges (voussoirs) are held together in compression, the continuous mass of the concrete arch exerted very little lateral thrust; this property was exploited in the great arcaded structures that carried Roman aqueducts overground (see Figure 3.16 overleaf). A concrete arch could be supported by a relatively light and simple wooden centring while it set, though more elaborate carpentry was needed for vaults and domes.

Figure 3.16 The stresses involved in a Greek post-and-lintel structure, a masonry arch and a Roman brick-faced concrete arch. Once the concrete arch had set, it behaved like a massive lintel, exerting very little lateral thrust (though there was some outward thrust in Roman vaulting; see Figure 3.11). There was, therefore, no need to buttress the pillars supporting the arch (Hodges, 1971, p.198)

The barrel vault, effectively an extended series of arches, permitted a considerable increase in the area of uninterrupted floor space, and was much used in warehouses. During the first century CE, Roman engineers discovered that two barrel vaults could be made to intersect without loss of strength (see Figure 3.17); these groined vaults, or cross-vaults, were used to magnificent effect in the great public buildings of imperial Rome, notably the baths. The largest known vaulted hall in antiquity was the space covered by the three cross-vaults of the Basilica of Maxentius, an imposing brick-faced concrete structure on Rome's Sacred Way, which was completed by Constantine in the early fourth century CE.

The use of concrete rather than stone enabled the Romans to construct monumental public buildings, notably theatres, amphitheatres and bathhouses, on a greater scale or in a shorter period, or both. It also enabled them to build more utilitarian structures to keep pace with the capital's demographic and economic expansion. There was, however, no corresponding saving of labour in construction methods; indeed, there was resistance to the idea of labour-saving machinery, if a story told by Suetonius about the emperor Vespasian can be regarded as typical:

> An engineer offered to haul some huge columns up to the Capitol at moderate expense by a simple mechanical contrivance, but Vespasian declined his services: 'I must always ensure,' he said, 'that the working classes earn enough money to buy themselves food.'
>
> (Suetonius, 1957, p.283)

Figure 3.17 Roman vaults. *Left* single barrel vault. *Right* groined or cross-vault, comprising two intersecting vaults springing out at the same level. The relative ease with which the Romans could construct cross-vaults in concrete, using wooden centring, meant that they were able to avoid the problems involved in building such structures with stone blocks (Adam, 1994, p.191; White, 1984, p.83)

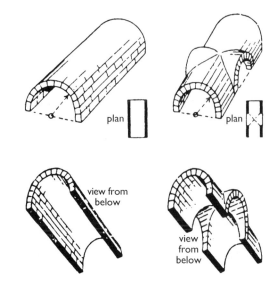

Although evidence like this is scanty, it favours the view that public construction in ancient Rome was undertaken by free-labourers rather than by slaves (Brunt, 1980).

The most common use of brick-faced concrete was in the construction of residential buildings. The upper-class house (*domus*), well known from the remains of Pompeii and Herculaneum, with its atrium and perhaps a peristyle garden, certainly existed in the capital: three such houses have been identified on the surviving fragments of the marble plan of Rome. But such ample dwellings, with no more than two storeys, were prodigal of space in rapidly growing cities such as Rome and its port of Ostia, where most people were accommodated in concrete tenement blocks, or *insulae* (literally 'islands'), up to five or six storeys high. In social structure they were the reverse of the modern apartment block, where the richer inhabitants occupy the higher levels. In the *insulae*, the higher a dwelling was, the poorer were its inhabitants; there were, of course, no lifts to the upper floors, and water was almost certainly available at ground-floor level only, if at all.

Many of the early *insulae* were poorly constructed, if references to their collapse in Roman satire are to be taken seriously. Their structural inadequacy may have been one reason for limits being placed on the heights of private buildings. Another may have been the ever-present risk of fires, one cause of which was the use of charcoal braziers for cooking and heating. The great fire of Rome in 64 CE prompted Nero to issue a set of building regulations: streets were to be widened and, wherever possible, new buildings were to be detached, rather than joined on to other buildings; the use of wood in construction was banned, and balconies were to be built on the first floor, perhaps for fire-fighting purposes.

The first *insulae* were made of mud-brick and timber, but from the first century CE, brick-faced concrete was the principal building material. Concrete vaulting was an integral construction feature, serving both to strengthen the building as a whole and to provide some of its floors (see Figure 3.18). The floors of the upper storeys were more likely to be of wooden construction; the roofs were normally timber-framed and tiled in the typical Roman style, with *tegulae* (flat roofing-tiles) joined together, and the joins covered by semicircular *imbrices* (overlapping tiles). In some *insulae*, however, vaulted construction may have been used for the whole block, which may have had a flat roof covered with a waterproof material.

The wooden centring necessary for the formation of concrete arches and vaults presupposes an existing tradition of building in wood. Indeed, the Romans have been credited with the development of five new carpenter's tools – the auger, the brace, the countersinker, the frame-saw and the gimlet (White, 1984, p.90). The early Roman built environment featured some quite complex structures in timber, including temporary wooden amphitheatres. Giant timber roof-trusses were used in large structures, such as the Basilica Ulpia in Trajan's Forum, built at the beginning of the second century; the span of this building was over thirty metres. As with so many aspects of building technology, although the Romans probably did not invent the timber tie-beam roof-truss, they greatly extended its use (see Figure 3.19 overleaf).

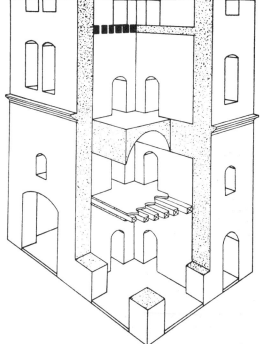

Figure 3.18 *Insula* in Ostia, with plank floors (*contignationes*) and vaults as floor supports on alternate levels (Packer, 1968, p.363; by permission of the author and University of Chicago Press)

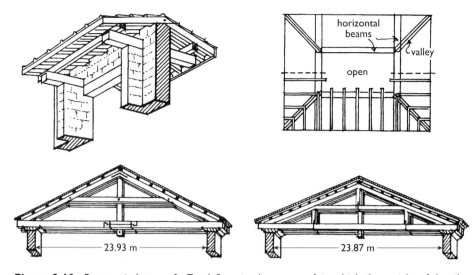

Figure 3.19 Roman timber roofs. *Top left* a simple span roof, in which the weight of the tiles and rafters is borne by walls and horizontal wooden beams. *Top right* a downward-sloping roof around the open court of a house, based on a description by Vitruvius. Again, the weight of the roof is borne by horizontal beams. *Bottom left* a roof-truss used in the original roof of the Basilica of St Peter at Rome, fourth century CE. *Bottom right* a roof-truss used in the original roof of the Basilica of St Paul-without-the-Walls, fifth century CE. In the triangular truss form, the forces transmitted by the two principal rafters serve to hold the main horizontal beam (tie-beam) in tension, giving a structure of great rigidity that is capable of spanning wide gaps (Singer *et al.*, 1956, p.414; by permission of Oxford University Press)

Is talk of a concrete revolution in architecture justified? The idea at least offers a counter-weight to the view that the Romans were uninventive, but there are some difficulties with it. Pozzolana concrete is rarely found outside Italy, one exception being its use in the construction of the harbour at Caesarea, in Judea. This may have been because of transport problems, or because pozzolana lost its strength unless used within a limited time (White, 1984, p.108). Consequently, the built forms and building types (palaces and apartment buildings, amphitheatres and baths) pioneered in concrete by Roman architects were imitated outside Italy using other materials; often these were local stones, though ground terracotta was used as a substitute for pozzolana in many areas. The fact that such forms could be imitated in other materials is beside the main point, which is that the use of concrete allowed Roman builders and architects to experiment, and come up with new architectural forms. It undoubtedly encouraged innovation, and allowed the rapid and economical construction of large-scale structures in Rome and its environs, but it was a relatively localized phenomenon. Although localization is a characteristic of ancient technology, the limited range of Roman concrete in both space and time somewhat diminishes its candidacy as one of the key innovations of ancient building technology. Little or no diffusion of concrete technology took place even within the empire; and although pozzolana mortar continued to be used for walls in Rome after the fall of the Roman Empire in the West, it was no longer the basis of large vaulted structures. There was to be no more concrete architecture as such in Europe until the late nineteenth century.

Evidence is already accumulating that the technologies of Roman city-building were strongly influenced by social and political conditions. The deployment of building materials and technology, from the transporting of monoliths across the Mediterranean to the construction of brick-faced concrete tenements, clearly reflected inequalities of status and wealth in Roman society. A general issue in the development of the built environment is the development of property rights and legal relationships between the various agents. Throughout the republic, public and private building was governed by contracts between, on the one side, self-employed architects and building contractors and, on the other, either magistrates (censors, aediles) or landowners; but, in the imperial period, public building was increasingly controlled by the imperial bureaucracy, and to some extent carried out by its permanent staff of slave and freedmen specialists (Anderson, 1997, pp.88–95).[4] Private property was an important determinant of Rome's urban morphology, and especially of its residential patterns; but, from the time of the late republic, the government assumed powers to ensure that its view of the public interest prevailed, as can be seen in the measures it brought in to control traffic and the height of buildings, and above all in the great public building programmes that punctuated the capital's history.

Water supply and sanitation

The Romans' contributions to the technology of water supply are well known, but how innovative were they? Of their early neighbours, the Etruscans seem to have been ahead of the Greek settlers in water-supply technology and, especially, in land-drainage: their underground channels (*cuniculi*), cut into tufa rock and up to four and a half kilometres in length, were intended mainly to make waterlogged land fit for cultivation (Hodge, 1992, pp.46–7). The early Romans built on this tradition: draining the site of the Forum Romanum, an area of marshy land between the main hills, was the first notable technological achievement in the city's history. As early as the sixth century BCE, an existing stream was channelled for this purpose; later, during the republic, the open ditch was lined with stone and vaulted over. The result was Rome's main drain and sewer, the Cloaca Maxima.

Rome's first sources of drinking-water were springs, wells and rainwater cisterns. Responding to the growth of the city, its republican rulers initiated arrangements to bring in water from distant sources, predominantly from the Anio river system and the Sabine Hills. The first aqueduct, the Aqua Appia, was commissioned in 312 BCE. The longest, the Aqua Marcia, was more than ninety kilometres in length, and eleven aqueducts supplied imperial Rome at its height, bringing in each day an amount estimated at nearly one million cubic metres (Hodge, 1992, p.273). The later the date of the aqueduct, the higher was the point of delivery in Rome; this reflected a growing demand from elite residences on the hillsides of Rome, and the increasing confidence of Roman hydraulic engineers in constructing longer aqueducts from higher sources.

Some provincial cities could also afford to invest in substantial aqueducts, such as the one carried by the majestic Pont du Gard, near Nîmes in Provence (Smith, 1990–91; see Figure 3.20 overleaf). Cologne's aqueduct was ninety-five kilometres long, and one of the four supplying Lyons was seventy-five kilometres (Hodge, 1992, pp.347–8). Reservoirs formed by dams were essential for the water supply of some provincial cities in the semi-arid areas of the empire, notably Spain, North Africa and the Near East. The Lake of Homs, which supplied the city of Homs, in Syria, with both domestic and irrigation water, was the largest artificial lake of its time. It was created by a dam on the River Orontes that was nearly two kilometres long; this great structure was made of pyramidal blocks of local basalt, with a basalt rubble core – a characteristic example of Roman engineers using local resources to solve local problems, but on an unprecedented scale (White, 1984, p.102). Pompeii provides one of the best-preserved examples of a provincial water-supply system, including an aqueduct, water-towers, public fountains, lead pipes and public baths. There was also a system of covered conduits to drain the forum, and storm sewers near the walls; but, otherwise, the continuous flow from the baths and the fifty to sixty public fountains washed through the streets, and was the main reason for the stepping-stones that crossed them (Hodge, 1992, pp.335–6).

---

[4] There is an extract from this passage of Anderson's book in the Reader associated with this volume (Chant, 1999).

Figure 3.20 Pont du Gard, near Nîmes, built in 19 BCE. The bridge is 275 metres long and nearly fifty metres high. The stone blocks for the facings were extracted from the banks of the River Gard, a few hundred metres from the site. Ledges used to support the wooden centring are visible in the middle series of arches (photograph: Giraudon)

The aqueducts were among the most notable Roman engineering achievements. They not only supplied water to cities, but also carried it to mining areas, such as north-western Spain, where it was used in technologically complex operations for extracting and washing valuable metal ores (White, 1984, pp.116–20). Traditional Greek surveying instruments were sufficient to ensure that each aqueduct followed a gentle incline across irregular territory. Vitruvius, the principal Roman source on architecture, recommended the *chorobates* for determining levels, and thereby measuring uneven ground (see Figure 3.21). The Romans strongly preferred to bring the water by gravity flow from a source above a city, and consequently built numerous arcades, or arched structures, sometimes with several tiers, to carry the aqueducts over significant and unavoidable dips in the ground, such as deep river valleys. Closed-pipe inverted siphons were occasionally used to cross such gaps, and at Lindum (Lincoln), in England, water was piped under pressure to a level twenty metres above a perennial spring. For the supply of Lugdunum (Lyons), closed pipes and siphons were combined with open conduits and arcades: the valleys to be crossed were so deep that the pressures in pipes alone would have been too great, and arcades on their own impossibly high. According to de Montauzan, an expert on the system: 'Of all Roman aqueducts this is perhaps the one that brings the greatest credit to the scientific knowledge and engineering skill of the Romans' (quoted in White, 1984, p.164).

Figure 3.21 *Chorobates.* This instrument was essentially a table with sights (A), plumb-lines and plummets (C) and (D), and a water channel (B), which functioned like a spirit level. According to Vitruvius, the water channel was used to determine the level only if wind was interfering with the plumb-lines (White, 1984, p.170)

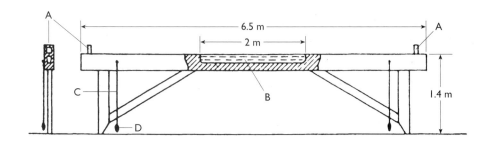

The arcades should not be confused with the aqueducts themselves, which, for security reasons, went underground most of the way: only twelve per cent of the total length of the seven main aqueducts of Rome was above ground (White, 1984, pp.162–3). Within the city itself, arcades were a prominent aspect of the built environment, and, indeed, could be an obstacle to subsequent urban development. The first were built of stone, though from the first century CE concrete was used, faced with masonry or brick. Upper tiers were often narrower than those below, allowing the wooden centring for the concrete to be supported, without the expense and inconvenience of scaffolding.

A good deal is known about the water supply of Rome and its administration because of the survival of a report prepared by Sextus Julius Frontinus, who became water commissioner in 97 CE, and had a staff of slave workmen and technicians (Evans, 1994).[5] Once the water entered the city, an aqueduct typically fed it into settling tanks to remove as much sediment as possible; from these, it was directed to a water-tower known as a *castellum divisorium*. Sluices and taps in the *castellum* distributed the water to secondary tanks, and from them it passed through pipes to public fountains, public baths, industrial premises and the villas of the wealthy. Villa owners paid for the water according to the diameter of their supply pipe, a rather crude indication of the water flow. The pipes were made of earthenware or lead, which was more durable. Lead was known by Vitruvius and other Roman writers to endanger health, but because the pipes became lined with calcium carbonate, and there were few taps to reduce the flow, there was little risk from drinking the water that passed through them. Taps were used, but most of the water ran continuously through open outlets, both public and private, straight into the drains – partly because of the practical difficulties of shutting off an aqueduct:

> This also had the advantage that the sewers were kept constantly flushed with a process of continuous dilution of solid sewage that no doubt played its part in helping the Romans to live in large cities wholly innocent of any sewage treatment.
>
> (Hodge, 1992, p.3)

The pipes for private users, including industrial establishments, were higher on the tank than those for public use, and would be the first to be cut off if the water supply were reduced. Frontinus was frequently indignant about abuses of the supply by private individuals, who might tamper with their pipes to get more water than they were paying for, or divert it for unauthorized use, such as in shops, brothels and suburban horticulture (Evans, 1994).

Very few traces have been found of water being supplied to individual apartments in tenement buildings, though this may be because the precious lead used for the pipes has been stripped away. Some blocks had a communal supply in their internal courtyards, but it is unlikely that water was piped to the upper floors. Most residents went to a public fountain for their water, though the better-off might pay a water-carrier to bring it to them. Evidently, the Roman water-supply system was socially shaped, access to it being governed by political power, wealth and social status.

Similar considerations apply to sanitation. In *insulae*, there were latrines on the ground floor under the main stair; some had an upstairs latrine. There were also public latrines near public baths and fountains. Urinals in the form of large pots were sunk into the ground at the side of a street, or stood in a passageway or by the entrance to a shop, and the urine collected from them was passed on to the fullery, where it was valued as a cleaning agent. As in Greek cities,

[5] Translated extracts from Frontinus' report are included in the Reader associated with this volume (Chant, 1999).

Figure 3.22 Schematic section of hypocaust system of Central Baths, Pompeii, showing the hot pool heated by furnace gases. The illustration also shows the furnace opening (*praefurnium*), and the standard arrangement of a suspended floor (*suspensura*) consisting of concrete laid on tiles, brick piers (*pilae*) and terracotta pipes (*tubuli*) (Yegül, 1995, p.359; courtesy of J.-P. Adam)

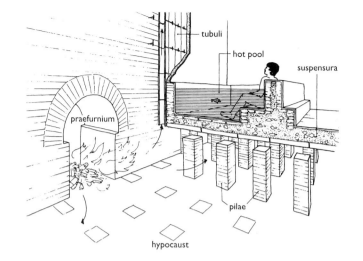

fulleries played an important part in Roman urban life. They were places where laundry was done, as well as where wool was cleaned, woollen cloth felted, and new cloth or old garments dyed. The fulleries in Rome gave as well as received: along with the public baths, they provided water to flush out the public latrines. As in many pre-industrial cities, night-soil was collected from private latrines, and served to raise the fertility of nearby market gardens.

Heating and lighting

Another innovation that needs to be considered, before the Romans are dismissed as uninventive, is their system of central heating – the hypocaust. The invention of the hypocaust was attributed by ancient authors such as Pliny the Elder to a Roman, Caius Sergius Orata, in the first century BCE; according to Pliny, its original purpose was to ensure a year-round supply of oysters. In fact, although Orata made money by adding hypocausts to renovated rural villas, he was not their inventor; true hypocausts dating from the second century BCE have been identified at Pompeii's Stabian Baths (Anderson, 1997, p.77; Yegül, 1995, p.379). The principle of the system was simple enough, though more complex in construction than the underground heating channels found at some Greek baths of the fourth or third centuries BCE (see Chapter 2, Section 2.4). In a hypocaust, hot gases from a furnace were circulated under suspended floors supported by brick piers, and sometimes also through terracotta pipes set in the walls (see Figure 3.22). Apart from its use in public baths, the hypocaust was mainly a luxury feature of wealthy villas.

In poorer houses, and for most of the inhabitants of *insulae*, amenities were basic, at least by modern standards. For lighting, the Romans used oil lamps, which were small covered bowls with a circular hole for the wick. The continued use of these simple lamps needs to be set in context. The Romans, like the Mesopotamians, Egyptians and Greeks, were content to divide up their day so as to make the best use of natural light; for them, the daylight hour was one-twelfth of the time from sunrise to sunset, and so varied with the season: hours in the summer were therefore longer than they were in winter. They concentrated on maximizing natural light: tenement blocks had large glazed windows, and often had an internal courtyard that acted as a light-well. For such heating of domestic space as was needed in the Mediterranean climate, the residents of tenement buildings made do with charcoal braziers. Heat may also have been conserved by shutters, as it certainly was in bath buildings, the remains of which also suggest that a form of double-glazing served this purpose.

Urban industry

It would be difficult to argue that any differences between Roman and Greek cities were due to manufacturing processes. Opportunities to substitute animal power for human muscle were necessarily limited to simple repetitive operations, a good example being the harnessing of donkeys, mules and worn-out horses to rotary grain mills in the bakeries that were one of the main urban industries of imperial Rome (White, 1984, pp.64–5). Despite the fact that Vitruvius described more than one type of water-wheel in the first century BCE, providing a suitable flow of water was problematic in the Mediterranean region, and both hand-mills and animal-powered mills remained competitive. Unusually, a water-powered corn mill was installed beneath one of Rome's great imperial baths, the Baths of Caracalla, and was turned by the flow from its aqueduct. Only in the later empire did water-mills for grinding corn became widespread; a shortage of labour has been suggested as the reason. This lag in the spread of an innovation was described by White as 'a good example of failure to exploit an invention for lack of the necessary motivation':

> Where the demand existed, technical resources were more than adequate to supply what was required, as in the military sphere, where such missile weapons as the torsion catapult and the stone-throwing ballista were developed to a high technical standard, and produced in large quantities … Throughout Greek and Roman antiquity, however, the demand for increased mechanization was negligible, except in the areas of state service and public works; the demand for clothing, footwear and domestic utensils which, in modern times has led to mechanization and mass production, was, with some exceptions met by local craftsmen, working in small workshops to satisfy local needs.
>
> (White, 1984, p.56)

3.4   *Rome: building the metropolis*

Part of the Romans' achievement in building, defending, maintaining and supplying a great metropolis has already been dealt with in Section 3.3. In summary, it involved the implementation on an unprecedented scale of the technologies of water transport, hydraulic and civil engineering, and building construction. The purpose of this section is to give an idea of the development of Rome's built environment, and the influence on that development of its various forms of administration (Patterson, 1992; Stambaugh, 1988).

Situated between the hills on which Rome's first inhabitants lived, the Forum Romanum was the city's first distinctively urban space. Apart from its religious and royal significance, it was the place where political, legal, financial and commercial transactions were conducted. As the city grew in size and in economic and political importance, many of these functions transferred to other public areas of the city, including the Forum Holitorium (vegetable market) and Forum Piscarium (fish market). Very early in the history of the city, the Forum Boarium (cattle market), because of its location on the banks of the Tiber, became the centre of commerce for external markets. Buildings were put up in these areas, both to perform specific functions and to give protection from the typically intense heat or heavy downpours of the region. An early innovation in building types was the basilica, a large hall in which legal, financial or commercial business was transacted. The earliest known basilicas in Rome, first built in the second century BCE, were at the Forum Romanum; they were similar in their post-and-lintel construction to the stoas found in a Greek agora. The Basilica Julia, which replaced one of these early structures in the late first century BCE, represented a departure from Greek architectural forms. The traditional trussed roof enclosing the main hall was now supported by two storeys of arcading; external columns were retained for decorative rather than structural reasons.

The Forum Romanum was the focal point of an irregular street system, adapted to the organic growth of the original villages. Bridges were essential to the internal communications and built environment of Rome, facilitating the expansion of the city into the Trans-Tiber region. The first bridges were wooden, but for stronger and more durable structures, stone and then stone-faced concrete were used. Both stone and concrete are strong in compression but not in tension, and can span a sizeable gap only when formed into arches. The first stone bridge, the Pons Aemilius, also served to carry aqueducts to the Trans-Tiber region.

By the imperial period, the transfer of many of its functions to other parts of the expanding city had left the Forum Romanum as something of a museum. Following the lead of Julius Caesar, emperors from Augustus to Trajan added their own enclosed monumental fora (see Figure 3.23). Out of respect for their predecessors, the architecture of these precincts was overwhelmingly traditional. Even so, their walls, temples and colonnades show the varying structural utility of local building materials, such as tufa and travertine, as well as of prestigious materials, such as marble and granite. The largest of the imperial fora, the Forum of Trajan, was designed by Trajan's chief architect, Apollodorus of Damascus, and was in antiquity one of the most admired precincts of Rome. Though traditional in conception and construction, it boasted the Basilica Ulpia, at that time the largest example of this type of functional building.

Figure 3.23 Imperial fora, Rome

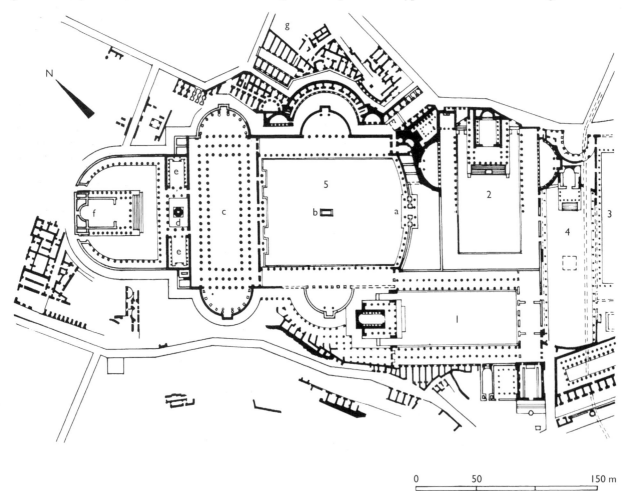

Key 1: Forum Julium and Temple of Venus Genetrix; 2: Forum Augustum and Temple of Mars Ultor; 3: north-west end of the Temple of Peace (built by Vespasian in the first century CE); 4: Forum Transitorium and Temple of Minerva (built by Domitian, but dedicated during the short reign of his successor, Nerva, near the end of the first century); 5: Forum of Trajan: (a) monumental entrance; (b) equestrian statue of Trajan; (c) Basilica Ulpia; (d) Trajan's column; (e) libraries; (f) Temple of Divine Trajan (built by Hadrian after Trajan's death); (g) Markets of Trajan (Ward-Perkins, 1974, p.124; by permission of the Electa Archive)

By the time of Trajan, more experimentation was acceptable in other kinds of public building. Behind the semicircular recess (hemicycle) at the rear of the Forum was a complex structure, which archaeologists in the 1920s dubbed the 'Markets of Trajan'. This building is now in full view as a result of the decay of the Forum proper. With its vaulted and domed halls, it demonstrated the power of the new concrete building techniques, and also their versatility, which enabled its builders to solve the problems of building a six-storey structure into a large cut in the Quirinal Hill, and around the Forum's rear hemicycle (see Figure 3.24). An inscription on the famous column in Trajan's Forum declares that its height was the depth to which the hill was excavated, presumably to clear a flat area for the Forum. The use of concrete was acceptable, because the Markets complex was clearly a utilitarian structure of some kind, obscured by the rear wall of the Forum. However, despite its shop-like units, with their distinctive travertine door-frames, it may not have been a market as such; instead, or in addition, it may have performed some administrative function.

The use of brick-faced concrete and vaulting for monumental structures was pioneered in the great imperial residences, beginning with Nero's Domus Aurea (Golden House), a vast complex built after the great fire of 64 CE, and culminating with Domitian's Domus Augustana, the remains of which still dominate the Palatine Hill (the word 'palace' derives from the emperor's residence here). These palaces not only accommodated the royal household, but also the increasingly complex imperial bureaucracy.

Temples were ubiquitous in the Roman built environment. As early as the sixth century BCE, two temples were built on the citadel, the Capitoline Hill. They were probably constructed in the local Italic style, with a deep portico leading to a rectangular main room *(cella)* built of tufa stone blocks, with wooden columns, pediment and roof. From the second century BCE, Hellenistic designs were increasingly adopted. As in subsequent European history, traditional materials and architectural styles were preferred for such venerated structures. Vaulting was used in some later temples, as can be seen in the now exposed *cellae* of the massive Temple of Venus and Rome, which overlooks the Forum on one side and the Colosseum on the other. These vaults date from the rebuilding of the temple by Maxentius in the early fourth century CE; even then, the underlying structure was disguised with external columns to make it look like a classical temple, and its brick-faced concrete walls were covered over with white marble.

The techniques of concrete construction left an indelible mark on Rome even in the conservative area of temple-building, in the form of the Pantheon, situated in the Campus Martius. The first Pantheon was built by Agrippa in 27 BCE, as proclaimed by the inscription on the front of the extant building. The first temple burnt down in around 80 CE and, having been rebuilt by Domitian, burnt down again at the beginning of the second century. The present structure was completely rebuilt by Hadrian (from 126 CE), though this was proved only relatively recently by archaeological dating of the stamps that brick-makers frequently impressed on their products.[6] Concrete was chosen as the main material, perhaps because of its fireproof property, though it was also clearly necessary for the construction of the temple's great dome.

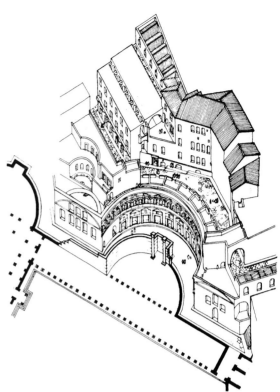

Figure 3.24 Markets of Trajan, restored axonometric view. The complex was at least six storeys high and built on terraces. It had several halls and more than 170 rooms, which were grouped at several levels, and connected by corridors and stairways. External access was from two streets, one behind the wall of the Forum of Trajan, and the second halfway up the slope known by its medieval name, the Via Biberatica (Ward-Perkins, 1974, p.126; by permission of the Electa Archive)

---

6   There is an illustration of brick-stamps, and discussion of them, in the extract from Anderson (1997) in the Reader associated with this volume (Chant, 1999).

Figure 3.25 Pantheon, section and plan. In the section, the coffering on the inside of the dome is clearly depicted. The plan reveals the numerous cavities in the drum, which enabled the concrete to dry out, and the pattern of precious marbles making up the floor (Ward-Perkins, 1974, p.134; by permission of the Electa Archive)

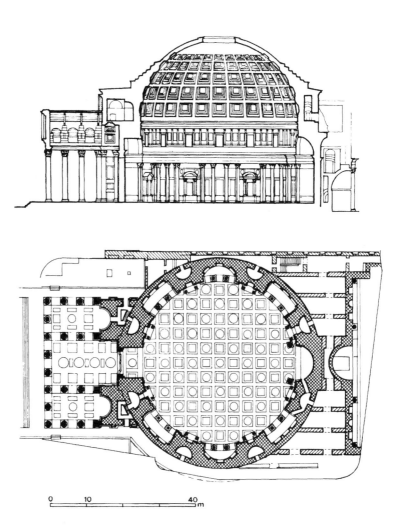

Taken as a whole, the Pantheon was a hybrid between the new concrete construction technology and traditional temple-building techniques. The front of the building is a classical portico, with huge monolithic granite columns, but the bulk of the building consists of a rotunda and dome, built in Roman concrete. The structure as a whole brilliantly demonstrates the versatility of the material. The engineers were careful in their choice of aggregates for different parts of the structure: basalt made it heavy at the bottom; the materials were then graded up, through travertine and other types of stone and brick, to light pumice, possibly from Mount Vesuvius, at the top of the dome. The central opening, or *oculus*, of the dome was surrounded by almost pure cement. The dome is one of the great Roman contributions to the engineering challenge of roofing large enclosures, and was the largest masonry dome built in antiquity. Its internal height of 43.2 metres exactly matched the diameter of the rotunda (see Figure 3.25).

The growth of Rome's population, which peaked at about one million in the late republican and early imperial period, made land increasingly scarce, and multi-storey *insulae* became the predominant form of housing. Economic specialization generated a proliferation of shops and workshops on the ground floor of many of these buildings. Expansion of the economy and trade under the republic had already led to the clustering of retail traders round the fora, and to the development of the Emporium, the city's wholesale market. The Emporium was located outside the city walls, and stretched narrowly alongside the Tiber, reflecting its dependence on river transport. The area became crowded with barrel-vaulted concrete warehouses (*horrea*) for the storage of goods such as grain, oil and wine. Its wharves were also of concrete, faced with *opus reticulatum* and bands of brick, and paved with travertine (Richardson, 1992, pp.143–4).

Apart from its monumental public precincts and buildings, and its utilitarian structures, Rome was notable for its places of leisure and entertainment: its stadia, theatres and public baths. The poverty of the domestic amenities of *insulae*, compared with the villas of the wealthy, was mitigated, in the words of the satirist Juvenal, by 'bread and circuses'. Bread was provided for the urban poor by the corn dole, and circuses through an emphasis on entertainments in everyday life that was probably unparalleled in world history until the twentieth century. Many of the most striking structures were places for games and spectacles, such as the Circus Maximus, which could seat a quarter of a million spectators, and where the chariot races took place. The place of entertainment in the life of the metropolis raises the wider debate about Rome's contribution to the economy of the empire. The building priorities of the ruling elite lend some support to the increasingly challenged, but tenacious view of Rome as a parasitic city, in which conspicuous consumption and decadent spectacle was the city's *raison d'être* (Parkins, 1996; Whittaker, 1995).

Other outstanding places of entertainment were the permanent, self-standing stone theatres, such as the Theatre of Marcellus, begun by Julius Caesar and completed by his adoptive son, Augustus. Its travertine exterior can now be seen incorporated into a later structure. The self-standing theatres were a notable architectural advance on their Greek predecessors. Instead of utilizing a natural slope, as the Greek theatres had done, the tiered seating of these places of entertainment rested on a hidden substructure of concrete vaulting. The first such theatre, the Theatre of Pompey, was dedicated in 55 BCE. A temple was built into the permanent seating, supposedly to placate those who thought such theatres a decadent luxury; but this moral posture was probably a cloak for the power politics usually involved in Roman building programmes.

No such compunctions affected the building of the Flavian Amphitheatre, now known as the Colosseum, and site of the famous gladiatorial contests and animal hunts, during which countless thousands of animals, convicts and prisoners of war met their end. The amphitheatre, a structure with continuous seating around a central open space, was an original Roman building type. The Colosseum, the most famous amphitheatre of all, was built in the basin beyond the Forum Romanum, between the Caelian and Esquiline Hills, on the site of an ornamental lake which was part of Nero's lavish Domus Aurea. Indeed, in the building's early years, it was reputedly flooded, so that naval battles could be enacted, though there is no proof of this. The building of the basic structure, initiated by Vespasian and dedicated by him in 79 CE, continued into the reigns of Titus and Domitian.

As a structural achievement, it was impressive, though very much in the tradition of the first stone theatres, the Theatres of Pompey and Marcellus. Concrete played an indispensable but hidden role in the structure; annular tiers of vaulted corridors supported the great four-storey seating structure (*cavea*), and provided access to it. Concrete was also used for the foundations, and for the system of subterranean passages beneath the arena. For the visible superstructure – the walls and the *cavea* – masonry commensurate with the monumentality of the building was chosen. The internal wall was fashioned from blocks of the highest-quality travertine; so too was the external wall with its impressive arcading, each storey featuring columns of a different classical order (see Figure 3.26 overleaf).

Figure 3.26 Colosseum, plan and sections, showing the main construction materials and the pattern of communication. There were five annular corridors on the ground floor, from which a system of stairs led as many as 50,000 spectators to entrances (*vomitoria*) at the level appropriate to their social standing – from the emperor and senators at the level of the arena itself, to women at the very top (Ward-Perkins, 1997, p.134)

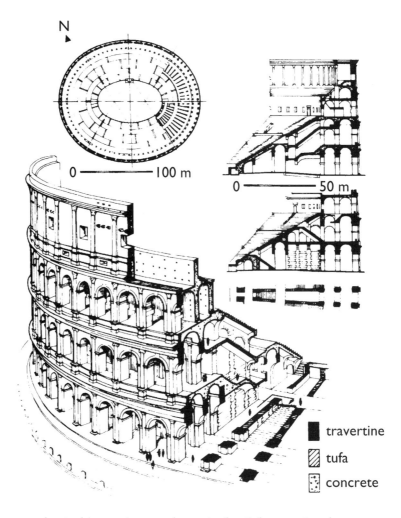

travertine

tufa

concrete

Some mechanical ingenuity was shown in the Colosseum's subterranean arrangements; exotic animals were hoisted in cages up to the arena, for the animal hunts (*venationes*) that were a popular part of Roman games. Naval technology was adapted to shelter the spectators from the elements. A sail-like awning was devised for this purpose; it was suspended from masts braced by travertine corbels in the solid wall of the uppermost storey, and sailors were permanently stationed in the capital to operate it.

A more everyday feature of life in Rome, and Ostia too, was communal bathing. Public and private baths, of varying sizes, were a common feature of the urban built environment. There were large public baths (*thermae*) and smaller privately owned baths (*balnea*), both types in fact being open to the public. By the fourth century CE, Rome is thought to have had well over 800 *balnea*. At the apex of the hierarchy of bathing establishments were the imperial baths of Rome. Among the most imposing of the structures of the built environment, they were far from being purely functional buildings; they were the gift of the emperor to the people of Rome, serving perhaps to reconcile them to the general inadequacy of their own accommodation, compared with the villas of the rich. The social shaping of ancient technologies is evident in these structures: rather than being used to provide water and bathing facilities for individual residences, the most innovative aspects of Roman engineering (water supply, concrete vaulting and central heating) were geared to the display of imperial public benefaction.

By the fourth century CE, there were eleven imperial *thermae* in Rome; the three largest were the Baths of Trajan, the Baths of Caracalla and the Baths of Diocletian (DeLaine, 1997). Their design and internal layout followed the general lines laid down by the Baths of Trajan, which were built by Trajan's architect, Apollodorus of

Damascus, on the crest of the Oppian Hill (the south-western lobe of the Esquiline), over the foundations of Nero's Domus Aurea (see Figure 3.27). The bath complex was set in extensive grounds and, as in most of the imperial baths, there was an open courtyard (*palaestra*) on either side of the main block, intended for walking, exercise and ball-games. The central axis of the baths building consisted of a *caldarium*, or heated area with hot baths, a *tepidarium*, or warm area, a central *frigidarium* with cold plunge-pools and, finally, a *natatio*, or swimming-pool. Water for the hot baths came from metal boilers, and the hot rooms were heated to the required temperature by hollow terracotta tubes in the walls that were linked to a hypocaust underneath.

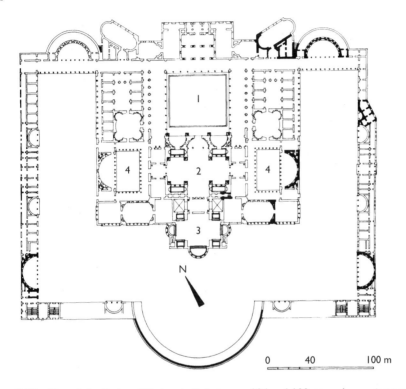

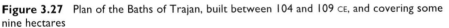

Figure 3.27 Plan of the Baths of Trajan, built between 104 and 109 CE, and covering some nine hectares

Key 1: swimming-pool (*natatio*); 2: central hall (*frigidarium*), with three vaulted bays; 3: hot room (*caldarium*), also with three vaulted bays; 4: exercise yards (*palaestrae*) (Ward-Perkins, 1974, p.124; by permission of the Electa Archive)

The baths carried little of the traditional Greek architectural baggage of the fora and temples, and (to some extent) of the theatres. Their architects were able to explore fully the potential of the new concrete construction technology; in particular, the great unimpeded, covered spaces of the *frigidaria* were made possible by massive cross-vaults. The strength of concrete also permitted large glazed openings in the walls, so that a light, airy effect could be created internally; in the *frigidarium*, this was accentuated by large semicircular windows let into the cross-vaults. The need to conserve heat influenced the mature design of the baths: the *caldarium* faced south-west to catch the afternoon sun; the *laconicum*, or sweat-room, might be positioned over the boiler; and flue gases from the furnace heating the boilers were directed into the hypocaust.

It is appropriate to end this section on ancient Rome with the great imperial baths. They were at once an embodiment of Roman imperial ideology, and the culmination of the Romans' achievements in urban technology, bringing together their substantial contributions to building construction, glass-making, central heating and water supply.

3.5 *Conclusion*

Rome grew far bigger than Babylon, Athens or Alexandria. Under ancient conditions, for such a giant city as this to be sustainable, it had to have a favourable location with respect to water transport, and a ready supply of fresh water and building materials. But although technology and resources were necessary for its growth, they cannot explain why Rome grew so big in relation to the other cities of the empire. In this case, the motor seems to have been the political environment, rather than any technologies of city-building.

In this chapter there are two assessments of the contribution of the Romans to the history of technology, which rest on divergent historiographical assumptions. First, there is the view that the Romans were technologically uninventive. Even if many of the technologies of ancient Rome were not actually invented by the Romans – and some important ones were – they were developed and put together in ways that far outstripped their original applications. Whether or not the Romans were uninventive, only a history of technology that is preoccupied with historical 'firsts' would seek to depreciate them in this way. Tracing innovations such as arches, vaults, aqueducts and hot baths to their first glimmerings is no way to place the Romans' contribution to technology in its full historical context. This is much better achieved by comparing the total built environment of ancient Rome with that of earlier ancient cities.

Another kind of historiographical pitfall awaits more charitable assessments of Roman technological creativity. Only a history of technology that is blind to its social, political and economic context would celebrate uncritically the Romans' most innovative contribution to building technology as a 'concrete revolution'. The uses of concrete were far from being dictated by the material's remarkable new properties. They also bore eloquent testimony to the social divisions of ancient Rome, and the ideological priorities of its rulers.

The concept of a 'concrete revolution' raises the question of technological determinism – one of the first issues we examined in this part of the book. The generalization that major changes in urban form and fabric are technology-driven has a surface plausibility, given the great transformations in the urban built environment over the nineteenth and twentieth centuries that are associated with such innovations as steam-powered factories and railways, structural steels and reinforced concrete, electric lighting and transport, and the internal combustion engine. The advantages of beginning a study of the relations between urban and technological change in antiquity is that there is no such surface plausibility. It seems clear that applications of technology had to serve the political and ideological ends of the ruling elites of ancient Mesopotamia, Egypt, Greece and Rome, who relegated technology and its development to a subservient social class. But it remains to be seen whether this perspective is fatal to technological determinism. It may be, as some scholars have argued, that the power of technologies has subsequently increased, to such an extent that technological determinism can be seen as a compelling theory for the industrial era.

References

ADAM, J.-P. (1994) *Roman Building: materials and techniques* (trans. A. Mathews), London, B.T. Batsford.

ANDERSON, J.C., Jr (1997) *Roman Architecture and Society*, Baltimore, Johns Hopkins University Press.

BRUNT, P.A. (1980) 'Free labour and public works', *Journal of Roman Studies*, vol.70, pp.81–100.

BURFORD, A. (1960) 'Heavy transport in classical antiquity', *Economic History Review*, 2nd series, vol.18, pp.1–18.

CASSON, L. (1965) 'Harbour and river boats of ancient Rome', *Journal of Roman Studies*, vol.55, pp.31–9.

CASSON, L. (1971) *Ships and Seamanship in the Ancient World*, Princeton, Princeton University Press.

CHANT, C. (ed.) (1999) *The Pre-industrial Cities and Technology Reader*, London, Routledge, in association with The Open University.

CHEVALLIER, R. (1976) *Roman Roads* (trans. N.H. Field), London, B.T. Batsford.

CORNELL, T.J. (1995) 'Warfare and urbanization in Italy' in T.J. Cornell and K. Lomas (eds) *Urban Society in Roman Italy*, London, UCL Press, pp.121–34.

DeLAINE, J. (1990) 'Structural experimentation: the lintel arch, corbel and tie in Western Roman architecture', *World Archaeology*, vol.21, pp.407–24.

DeLAINE, J. (1995) 'The supply of building materials to the city of Rome' in N. Christie (ed.) *Settlement and Economy in Italy 1500 BC to AD 1500*, Oxford, Oxbow Press, pp.555–62.

DeLAINE, J. (1997) 'The Baths of Caracalla: a study in the design, construction and economics of large-scale building projects in imperial Rome', *Journal of Roman Archaeology*, Supplement 25.

EVANS, H.B. (1994) *Water Distribution in Ancient Rome: the evidence of Frontinus*, Ann Arbor, University of Michigan Press.

FINLEY, M.I. (1973) *The Ancient Economy*, London, Chatto and Windus.

GARNSEY, P. (1988) *Famine and Food Supply in the Graeco-Roman World: responses to risks and crisis*, Cambridge, Cambridge University Press.

GREW, F. and HOBLEY, B. (eds) (1985) *Roman Urban Topography in Britain and the Western Empire*, London, Council for British Archaeology.

HODGE, A.T. (1992) *Roman Aqueducts and Water Supply*, London, Duckworth.

HODGES, H. (1971) *Technology in the Ancient World*, Harmondsworth, Penguin Books.

HOLLOWAY, R.R. (1994) *The Archaeology of Early Rome and Latium*, London, Routledge.

HUGHES, J.D. (1994) *Pan's Travail: environmental problems of the ancient Greeks and Romans*, Baltimore, Johns Hopkins University Press.

LEACROFT, H. and LEACROFT, R. (1969) *The Buildings of Ancient Rome*, Leicester, Brockhampton Press.

MacDONALD, W.L. (1982, rev. edn) *The Architecture of the Roman Empire*, New Haven, Yale University Press (first published 1965).

MATTINGLY, D.J. (1995) *Tripolitania*, London, B.T. Batsford.

MORLEY, N. (1996) *Metropolis and Hinterland: the city of Rome and the Italian economy 200 BC to AD 200*, Cambridge, Cambridge University Press.

OWENS, E.J. (1991) *The City in the Greek and Roman World*, London, Routledge.

PACKER, J.E. (1968) 'Structure and design in ancient Ostia: a contribution to the study of Roman imperial architecture', *Technology and Culture*, vol.9, pp.357–88.

PATTERSON, J.R. (1992) 'The city of Rome: from Republic to Empire', *Journal of Roman Studies*, vol.82, pp.186–215.

PARKINS, H. (ed.) (1996) *Roman Urbanism: beyond the consumer city*, London, Routledge.

PURCELL, N. (1987) 'Town in country and country in town' in E.B. MacDougall (ed.) *Ancient Roman Villa Gardens*, Washington, D.C., Dumbarton Oaks Research Library and Collection.

RICH, J. (ed.) (1996) *The City in Late Antiquity*, London, Routledge.

RICKMAN, G.E. (1971) *Roman Granaries and Store Buildings*, Cambridge, Cambridge University Press.

RICHARDSON, L., Jr (1992) *A New Topographical Dictionary of Ancient Rome*, Baltimore, Johns Hopkins University Press.

ROBINSON, O.F. (1992) *Ancient Rome: city planning and administration*, London, Routledge.

SCAGLIARINI, D. (1991) 'Bologna (Bononia) and its suburban territory' in G. Barker and J. Lloyd (eds) *Roman Landscapes*, London, British School at Rome.

SCARRE, C. (1995) *The Penguin Historical Atlas of Ancient Rome*, Harmondsworth, Penguin Books.

SINGER, C., HOLMYARD, E.J. and HALL, A.R. (eds); assisted by JAFFE, E., CLOW, N. and THOMSON, R.H.G. (1956) *A History of Technology*, Oxford, Clarendon Press, vol.2.

SMITH, N.A.F. (1976) 'Attitudes to Roman engineering and the question of the inverted siphon' in A.R. Hall and N. Smith (eds) *History of Technology*, London, Mansell, vol.1.

SMITH, N.A.F. (1990–91) 'The Pont du Gard and the Aqueduct of Nimes', *Transactions of the Newcomen Society*, vol.62, pp.53–80.

STAMBAUGH, J.E. (1988) *The Ancient Roman City*, Baltimore, Johns Hopkins University Press.

SUETONIUS (1957) *The Twelve Caesars* (trans. R. Graves), Harmondsworth, Penguin Books.

VITRUVIUS (1960) *The Ten Books on Architecture* (trans. M.H. Morgan), New York, Dover Publications.

WARD-PERKINS, J.B. (1971) 'Quarrying in antiquity: technology, tradition and social change', *Proceedings of the British Academy*, vol.17, pp.137–58.

WARD-PERKINS, J.B. (1974) *Cities of Ancient Greece and Italy: planning in Classical antiquity*, New York, George Braziller.

WARD-PERKINS, J.B. (1977) *Roman Architecture*, New York, Harry N. Abrams.

WARD-PERKINS, J.B. (1997) *Roman Imperial Architecture*, New Haven, Yale University Press.

WHITE, K.D. (1984) *Greek and Roman Technology*, London, Thames and Hudson.

WHITTAKER, C.R. (1995) 'Do theories of the ancient city matter?' in T.J. Cornell and K. Lomas (eds) *Urban Society in Roman Italy*, London, UCL Press, pp.9–26.

WIDRIG, W.M. (1980) 'Two sites on the ancient Via Gabina' in K. Painter (ed.) *Roman Villas in Italy*, London, British Museum Press, pp.119–40.

YEGÜL, F.K. (1995) *Baths and Bathing in Classical Antiquity*, Cambridge, Mass., MIT Press.

Part 2
MEDIEVAL AND EARLY MODERN CITIES

Chapter 4: MEDIEVAL CITIES

by David Goodman

4.1 The barbarian invasions and the fate of cities

The Middle Ages began with the Eurasian world in turmoil – the Huns were on the move. In the fourth century CE there erupted one of those mass movements of nomadic horsemen from their pastoral homeland in the steppes of central Asia (Mongolia), in search of fresh pastures and booty from settled populations. In a world-wide confrontation with settled agriculturists and city-dwellers, these nomads made lightning strikes over vast distances and in all directions. Westwards from the Mongolian steppes, thousands of miles of continuous plains stretched to Hungary, the gateway to urban Europe. This vast terrain was ideal for rapid movement by skilled horsemen, and the horsemanship of the tribal cavalry of the steppes was unmatched anywhere in the world. The fourth-century invasions of the nomads were nothing new. Similar invasions had occurred in the previous millennium; and they would recur a millennium later with the thirteenth-century incursions of the terrifying Mongol horde led by Genghis Khan. For a long time the civilized world had no answer to the mobility, horsemanship and archery skills of the nomads. Only a failure to find pasture for their horses or family tribal discord in their homeland could force their withdrawal to the steppes. In the fourth and fifth centuries, these confederations of Turkic and Mongol nomads struck against the settled populations of China, India, Persia and Europe (see Figure 4.1 overleaf). The hordes attacking Europe were called the Huns, and their leader was Attila. The monumental consequence was the collapse of the Roman Empire in the West.

  In the fifth century, Attila and his warriors invaded Gaul and Italy, but these were not the raids that brought down the Roman Empire. After Attila's death (in 453) the Huns disintegrated, but their movements had set off a chain reaction, a succession of displacements of the various Germanic tribes whom they encountered in their westward progress (see Figure 4.2, p.118). Those Germanic peoples, originating in Scandinavia and the southern shores of the Baltic, had been nomadic some centuries before, but were now mostly settled agriculturists living in regions north of Rome's Rhine (the Franks and Saxons) and Danube (the Ostrogoths) frontiers. It was the Germanic Ostrogoths, in the Ukraine, who were first pushed back by the Huns; they in turn drove the Visigoths across the Danubian frontier. Other Germanic tribes were subsequently driven west and south across the frontiers of the empire, causing havoc: the Alans, the Suebi and the tribe whose

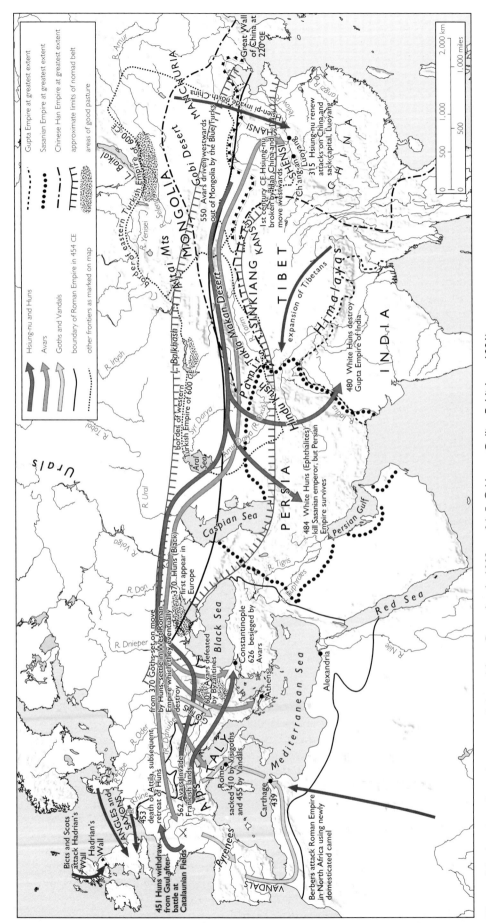

Figure 4.1 Invasions from the steppes (adapted from Barraclough, 1993, pp. 94–5; © HarperCollins Publishers 1991)

ravages have been immortalized in several European languages as a synonym for destructiveness: the Vandals. For the Western Roman Empire, already decaying, partly as a result of population decline, this was the death-blow. By the late fifth century, the map of western Europe had been transformed (see Figure 4.3, p.119). The provinces of the Roman Empire had been replaced by a series of successor Germanic kingdoms: Anglo-Saxon England, Frankish France, Ostrogothic Italy, Visigothic Spain and Vandal Africa. And these new kingdoms were Christian: most of the Germanic tribes had been converted by the influence of a now Christian Rome. These events mark the conventional end of the ancient world and the start of the Middle Ages.

Classical antiquity was a civilization based on the city. By contrast, the barbarian Germanic invaders dwelt in the countryside: there were no towns in their homelands north of the imperial frontiers. While some of the new Germanic kings had palaces in Roman cities (Ravenna, Pavia and Toledo are examples) and much of the former Roman administration was preserved, the barbarians continued to prefer rural life. A tentative explanation of their avoidance of cities comes from Hobley (1988, p.69). He draws on an ancient text which, commenting on the Germanic invaders of Roman Gaul and their dread of the Roman cities they encountered, states that they perceived city walls as a trap, and that they reminded them of the walled tombs of their ancestors. Hobley is thereby led to suppose that 'town walls represented a formidable psychological barrier, as well as a physical one' to the Saxon invaders of Britain. And so he can account for the 'total lack of archaeological evidence in London for Saxon occupation within the walls' in the sixth and seventh centuries, 'notwithstanding unprecedented programmes of excavation over the last twelve years'. To which he adds that 'the early Saxon settlers were farmers who preferred open settlements'.

So the great question arises of how much survived of classical urban life: what were the effects of the Germanic invasions on the cities, their populations, building types and public amenities? It is an important but difficult question, which continues to be debated. The difficulty arises from the sparsity of evidence. But it now seems clear that there was a wide variation in the degree of urban survival in the different regions of the former empire.

Urban continuity and decay in western Europe

Britain was distinctive. Situated at the periphery of the Roman Empire, it was the least Romanized of all the European provinces. Latin was hardly spoken, Roman law had not taken root, and the Roman legions had abandoned the island in the early fifth century (before the arrival of the Anglo-Saxons) to deal with a military crisis in Gaul. The invaders therefore confronted Celts, rather than Romans, and a weak urban culture which for centuries they did nothing to revive, because they disliked cities. The warrior chiefs lived in wooden halls in the countryside, their followers in wooden or mud huts. There was almost no demand for the Roman building skills of the stonemason: only some early monasteries were built of stone. Anglo-Saxon fortifications were earthworks with wooden upper structures. For a century or two, British towns almost entirely vanished. There is even some doubt about London in the fifth century; its continued functioning then is 'difficult to establish', though continuous settlement is considered probable. Not until the eighth century is there clear evidence of Saxon merchants taking advantage of its location at a river crossing near the Thames estuary (Ward-Perkins, 1992), and still outside the walls, on the Strand, an open-site shore that also facilitated the beaching of their trading boats (Hobley, 1988, p.73). The only Roman streets which have survived in English towns are those linking the city gates (the *cardo* and *decumanus* of the Roman city; see Section 3.2 of Chapter 3), which would be the only streets of a city 'likely to survive total abandonment' and to be later revived (Ward-Perkins, 1984, p.180).

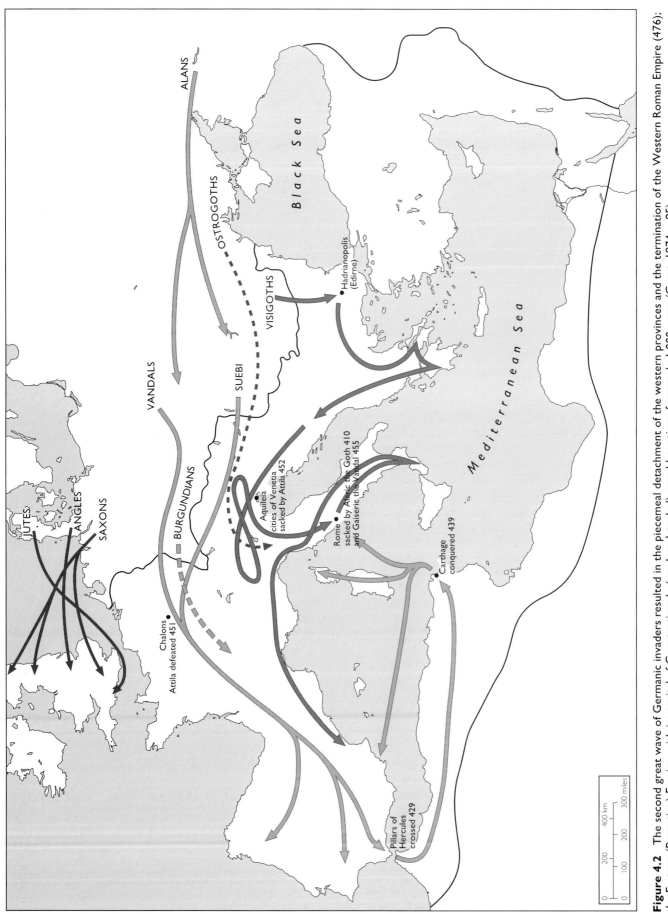

Figure 4.2 The second great wave of Germanic invaders resulted in the piecemeal detachment of the western provinces and the termination of the Western Roman Empire (476); the Eastern (Byzantine) Empire, with its capital of Constantinople (modern Istanbul), would continue for nearly 1,000 years (Grant, 1974, p.85)

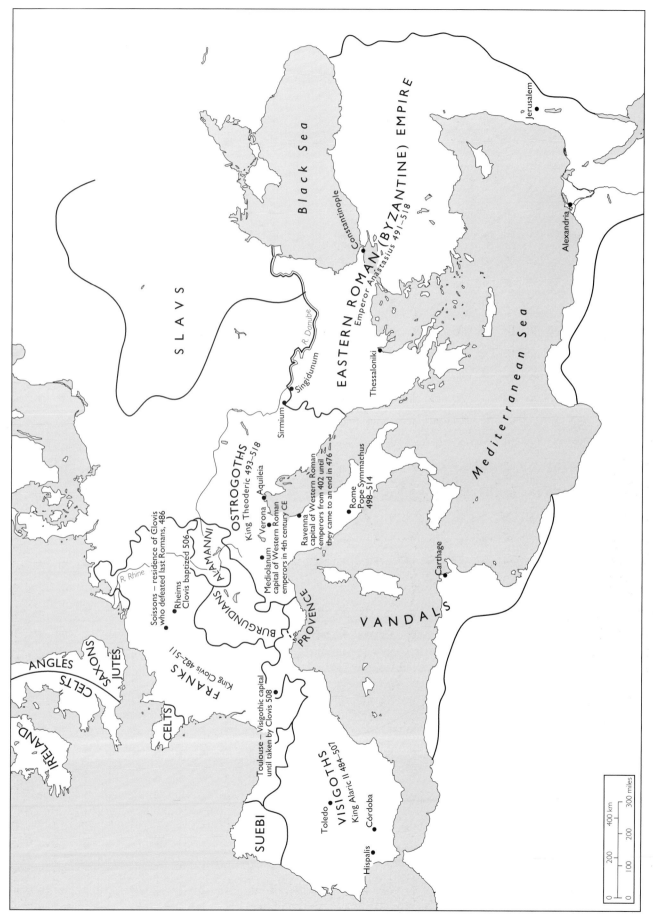

Figure 4.3 Europe in 500 CE (Grant, 1974, p.86)

Figure 4.4 Piazza del Mercato, Lucca (photograph: Alinari)

Some rare evidence of continuity of Roman urban life on the north-western periphery of the empire comes from Huy, a town on the Meuse (in modern Belgium). Archaeological research there has established not only continuous settlement into the seventh century but also a continuing technology: the remains of pottery kilns and metal workshops of the seventh century have been discovered on the very same site as their Roman predecessors.

In sharp contrast to the ruralizing of the periphery of the Western Empire, there are strong indications of urban survival at the centre, in Italy itself. But even here the evidence is patchy: the written records have obscurities, the archaeological work remains slight. The barbarian invasions had left a trail of urban destruction. When Milan was captured by the Ostrogoths in 539, its population was massacred (the women were sold as slaves) and the city walls destroyed. On the Adriatic coast, the city of Aquileia, a commercial centre, was ruined; other towns in the area never recovered. Where the Roman structures were too large to destroy or move, they survived the devastation. The extensive system of Roman sewers in Pavia continued to function throughout the Middle Ages and still does so today. In Rome, the colossal mausoleum of the emperor Hadrian, occupying a strategic site at the central Tiber crossing, was used by the Goths as a fortress; later renamed as Castel Sant'Angelo, it would become an impenetrable refuge for the pope. Roman amphitheatres did not always survive. Of the one at Lucca (in Tuscany) not a stone remains, but its outline is perfectly preserved by the oval of houses built on the site (see Figure 4.4), today called the Piazza del Mercato (market-place).

Aerial photography over northern Italy shows the near-perfect survival of Roman street plans in several towns, such as Lucca, Pavia and Piacenza. That suggests continuous settlement on the site after the collapse of the Roman Empire, and maintenance of the street layout. In the case of Verona, such suggestive evidence rises to the level of proof: archaeological excavations have revealed that, in the centuries after the barbarian invasions, there was continuous occupation in houses whose frontages carefully observed the existing Roman pattern of streets (Ward-Perkins, 1992). And the open space of the Roman forum remains today in Verona's Piazza delle Erbe, maintaining this ancient focus of urban bustle in the function of a fruit and vegetable market. Similarly, in Rimini, Florence and Brescia, the forum became the space for a market in the Middle Ages.

The Roman forum was, however, built over at Luni (in Tuscany). Excavations there have unearthed the remains of houses built in the sixth century on the area of the forum. These structures show a decline in building technology. No longer to be seen is the solid domestic architecture of the Romans, with its walls of brick and stone, strong floors and tiled roofs. Instead, Luni's post-Roman dwellings were built from weaker materials: wood, clay and dry-stone. Excavations at other sites in the region confirm this image of urban decline, revealing private houses with earthen floors and roofs of thatch. It all 'points to a drop of living standards, with people living in less durable, less weatherproof, and less hygienic homes' (Ward-Perkins, 1988, p.18). The same decline is discernible in post-Roman public buildings, which were built without bricks, apart from those taken ready-made from existing Roman structures. The superb paving of Roman streets and squares, the raised pavements that shielded pedestrians from mud, were no longer supplied. The surface of the urban fabric had deteriorated in the economic decline and political chaos that had followed the collapse of empire.

Urban water supply under the new regime

But what happened to that most essential of urban amenities, water supply? The Roman aqueducts, sewers and public baths had been among the greatest feats of civil engineering in the ancient world. Did these facilities also disappear from the cities of Italy in the aftermath of the barbarian invasions? The evidence from archaeology shows only small repairs to one or two of Rome's major aqueducts. But neither in Rome nor anywhere else in early medieval Italy is there evidence of a concerted programme to maintain or build aqueducts for the purpose of supplying the urban residents with water for drinking and washing. Drinking-water could, of course, be provided by means other than aqueducts, which were so expensive to maintain. An example is the eighth-century well, endowed by a priest and installed outside the church of San Marco in Rome, notable for the dedicatory inscription cursing anyone who dared to levy a charge on this drinking-water. This provision of water in the city also illustrates the general replacement of secular municipal officials by ecclesiastical authority that characterized the early Middle Ages. When, in the sixth century, Italy was again overrun by an invading Germanic tribe – this time the Lombards (hunters and pig-breeders) – the conquerors continued to live in the countryside. In Italy's cities there was therefore a power vacuum, and it was soon filled by the bishops, who by canon law were compelled to reside in towns. There, as Ward-Perkins (1984) has argued, the Christian interest in urban water supply seems to have been motivated less by concern for drinking-water than for hygiene and, above all, ritual.[1] Certainly, the Roman passion for bathing was seen by the church as a reprehensible indulgence, although the early popes seem to have enjoyed their own luxurious bath in the Lateran Palace. Ward-Perkins raises the interesting (and plausible) conjecture that early medieval popes may have been inspired by the image of Jesus washing his disciples to embark on plans to restore a Roman

[1]   See the extracts from Ward-Perkins's text in the Reader associated with this volume (Chant, 1999).

aqueduct to provide public baths for the clergy, pilgrims and the poor. In part this was out of concern for hygiene, like the sixth-century public lavatory installed by Pope Symmachus in front of St Peter's. But water supply for ritual was a dominant ecclesiastical concern. Hence the installation of fountains in atria, the porticoes in front of the principal doors of churches, so that the faithful might wash their hands before entering, as a spiritual preparation.

Ecclesiastical administration of urban water resources meant a sharp reduction in provision for bathing compared with the imperial age, when the Romans had brought bathing to the entire urban population, except for the poorest who could not afford even the low entrance fee. Now popes and bishops provided a bath for the poor and the clergy, but omitted the rest of the citizenry. Did that mean the sacrifice of urban hygiene? As in many historical questions of this nature, there is no simple answer. Ward-Perkins is unable to arrive at an unqualified conclusion. While the great Roman public baths were no longer in favour and the urban population must have bathed less, this does not mean they washed less: they could have used basins and tubs at home. The degree to which they did so is not open to historical inquiry and so all Ward-Perkins can do is express his 'suspicion' that the populations of early medieval Italian cities washed less thoroughly and less often. He is more confident of a fall in standards of urban hygiene, due to the decline of the Roman public lavatory and the general loss (Pavia excepted) of the system of flushing drains with waste aqueduct water. But he qualifies this with the reflection that hygiene may have been improved by a medieval decrease in the size and density of the urban population, creating extra open space for cesspits (Ward-Perkins, 1984).

From these contrasts between ancient and medieval Roman baths, it seems clear that urban water supply has not been merely a matter of technology and economics. Values and beliefs have also been powerfully influential. In ancient Rome, these huge public works were part of the propaganda of imperial munificence and power, intended to strengthen the loyalty of the urban populace. In early medieval Rome, the paramount purpose of water supply was the reinforcement of Christian belief through ritual washing.

A new building type

The ecclesiastical succession to municipal power had other powerful effects on the urban environment. The Christian religion, once persecuted by the Romans, had become the official religion of the late empire after the conversion of the emperor Constantine. Now it became possible to build places of public Christian worship. The urban landscape was changed by the building of numerous churches large and small. In Syracuse (in Sicily), one of the great cities of the ancient Greek world, a large church was built in the seventh century, its construction facilitated by the plundering of the nearby Temple of Minerva (which dated from the fifth century BCE); no fewer than twenty-one columns of that temple continue today to form part of the fabric of the cathedral at Syracuse (see Figure 4.5). But Christian churches could not be modelled on the temples of the Ancients. Their functions were different. A pagan temple was just a small shrine for the statue of a deity; processions and sacrifices occurred outside the temple. A Christian church had to accommodate an entire congregation within its walls. A type of assembly room was needed and, according to a widely held view of historians of art, it was supplied by the Roman basilica (see Chapter 3, Section 3.4), an urban, secular building type which served as a lawcourt or other place of assembly. ('Basilica' comes from the ancient Greek word meaning 'royal'; so it conveyed to the Roman building type a sense of pomp and ceremony.) The basilica was an oblong with a raised dais at one end, often in the recess of an apse, where a judge or other dignitary could be seated in a conspicuous position; the entrances were on opposite sides of the longer walls. This design had to be adapted for Christian use, so that processions could enter the main doors and move along the main length of the building. So the axis of the basilica was rotated through ninety degrees. The dais became the location of the altar. Basilicas had the further advantage of being easy to build; no special engineering skills were required. Pevsner, a distinguished historian of art and architecture, disagreed with this explanation of the origin of the early Christian churches. He objected that Roman basilicas, such as that at Pompeii, had features which bore no resemblance to the first churches – for example, an internal four-sided colonnade, which was also cut off from the apse (see Figure 4.6). Instead, he preferred an origin from the apsed hall found in Roman palaces and large houses (Pevsner, 1961, pp.33–5; see Figure 4.7).

In the wake of the barbarian invasions, Italy in the fifth and sixth centuries had experienced a general destruction of Roman urban institutions. In this heart of the empire, 'the ancient municipalities must have declined into sordid if picturesque decay, villages in a setting of classical ruins' (Waley, 1969, p.12). Yet nowhere in the Western Empire was the urban way of life so deeply rooted, and these roots, from the tenth century on, would make Italian cities the most precocious of all in the West's urban revival of the central Middle Ages.

Figure 4.5 Ancient Greek columns reused in the fabric of the early medieval cathedral of Syracuse, seventh century, rebuilt with additions in the seventeenth and eighteenth centuries (Wheeler, 1964, p.15; photograph: Hirmer)

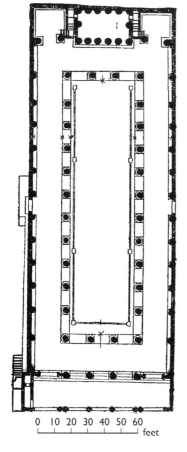

0 10 20 30 40 50 60 feet

Figure 4.6 Basilica, Pompeii, c.100 BCE (Pevsner, 1961, p.33)

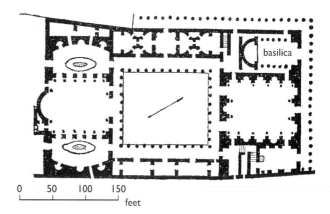

basilica

0 50 100 150 feet

Figure 4.7 Palace of the Flavian emperors, Rome, late first century CE (Pevsner, 1961, p.34)

Urban change in the Eastern Roman Empire

The barbarians had destroyed the Western Roman Empire, but the Eastern Roman (or Byzantine) Empire (which encompassed Greece, Asia Minor, Syria and Egypt) persisted. It too had experienced Germanic invasions: in 276 CE the Heruli destroyed Athens. But the Eastern Empire survived – the heart of it for another 1,000 years. Why it survived is not easily answered, but part of the explanation lies in its greater population and wealth compared with the West at the time.

Athens, despite the destruction, continued to be inhabited, but only in the contracted zone of the Acropolis and within the Roman fortifications. Excavations have revealed that the houses in medieval Athens were badly built and 'show evidence of desperate attempts to collect rainwater' (Bouras, 1981, p.628). The same determination to secure drinking-water is discernible in the remains of medieval Pergamum (in Asia Minor). Everywhere in the former cities of the ancient Greek world, the acropolis functioned as the strategic centre of the medieval city (a reversal of what had occurred in antiquity, when the acropolis, once all-important, was stripped of its political and economic functions and reduced to a religious sanctuary). But there seems to have been a general abandonment of town-planning in the Byzantine provinces. Miletus, once famed for its chequerboard design, was rebuilt over its ancient ruins in a wholly disorderly way. The medieval remains of Athens, Corinth and Ephesus all show the same decline: cheap building materials, streets of changing width, an absence of monumental buildings, and an apparently uncontrolled intrusion of cemeteries to the very centre of cities.

From all this, it has been concluded that from the early medieval centuries, the urban fabric of the Eastern Roman Empire had markedly deteriorated, evidence that 'the ancient order imposed on space had collapsed', and that there had been a loss of 'organizational ability' (Bouras, 1981). The force of this assessment (which was based only on Byzantine provincial cities) is greatly weakened as soon as we consider the Byzantine capital – Constantinople. It was exceptional in every way: in its size, topography, impregnability and in the extraordinary feats of technology achieved in ambitious projects intended to make this the world's greatest city.

A new capital

It all began when the emperor Constantine decided to create a new capital for the Roman Empire, shifting its focus towards the East. A myth later attributed the foundation to Constantine's desire to create a Christian capital to outshine pagan Rome. An explanation closer to the truth was the strategic value of a capital situated midway between the Danube and Euphrates frontiers, with good land and sea communications to both military zones. The new city, which came to be called 'New Rome' and the 'queen of cities' more frequently than 'Constantinople', was founded in 324 CE on the site of the ancient Greek colony of Byzantium. The new Rome was still under construction twelve years later. Constantine built wide colonnaded avenues; a forum; a great Roman circus (the Hippodrome), a vast open space around 440 metres long in which chariot races were held; a huge palace complex occupying 10,000 square metres – a city within a city; and several churches, including the church–mausoleum of the Holy Apostles built for the interment of emperors (an interesting example of the new Christianizing function of the ancient secular mausoleum). Building regulations attempted to prevent any use of wood, though that proved impossible beyond the monumental centre of the city. Constantine envisaged a giant city of some 200,000 inhabitants. To encourage settlement, he provided incentives for the wealthy to build houses by the grant of state lands and, to attract the less wealthy, he installed twenty public bakeries to supply free bread every day to sustain 80,000 citizens.

But the new city continued to lack essential amenities after Constantine's death; it was left to his successors to supply them. Port facilities were expanded, capacious buildings were erected for storing food imports, and the emperor Valens provided an adequate supply of drinking-water with the construction of a gigantic aqueduct (which continues to astonish the modern visitor to Istanbul) and a huge reservoir. The source was 100 kilometres away, near the present Bulgarian frontier. It was an enormous engineering project. Since the source lay close to territory threatened by continuing attacks of Ostrogoths and Huns, the city's supply was safeguarded by the construction of some 100 cisterns – three of these alone had a capacity of nearly one million cubic metres. Also in the fourth century, the city's area was doubled by the construction (by the emperor Theodosius) of an extensive line of masonry walls, which converted the city into a formidable fortress, protecting it from barbarian invaders for centuries. These fortifications were unmatched in the medieval West – a line of walls six kilometres in length and up to fifteen metres high, reinforced by towers at regular intervals, with a second wall and towers in front of the first. A broad moat served as a further line of defence. This elaborate defence system would not be breached until the advent of improved artillery, which enabled the Ottoman Turks to take the city in 1453.

It was under the emperor Justinian that Constantinople experienced one of the most dazzling technological achievements of the age – the building of the great church of Hagia Sophia (completed in 537 CE), whose astonishing structural form came to dominate the city skyline. For almost 1,000 years it would be the world's largest vaulted structure. Its design was a combination of a basilica and the central plan characteristic of Roman mausoleums. The most striking feature was to be its huge central dome, around thirty-five metres in diameter (the Pantheon's dome had a diameter of around forty-three metres) and rising some sixty metres above the floor. The materials used were brick and stone in alternate courses. To build it, Justinian employed two architects with strong mathematical interests: Anthemius of Tralles and Isidore of Miletus. The main problem was to design structures which could withstand the enormous thrusts of the dome. Their innovative solution made use not only of buttress towers but also a series of half-domes resting on other half-domes (see Figure 4.8). At the base were four enormous stone piers connected by arches (see Figure 4.9). Thousands of workmen were engaged on the vast project and the work progressed at an almost incredible rate – the building was completed in just six years. The view from within gave the illusion of weightlessness, provoking Procopius, a contemporary observer and the chronicler of Justinian's reign, to remark that the dome appeared to be suspended by a golden chain from heaven rather than supported from below. The structure proved a little too daring, and there was a partial collapse some twenty years after its completion; but the minor reconstruction carried out then guaranteed its survival to the present day. The great church served not only as a centre for worship – matins were held there – but was also used for great ceremonial occasions such as the crowning of the emperor. The city was designed to give the emperor direct access from his palace to the Hagia Sophia, as well as to his box seat in the Hippodrome.

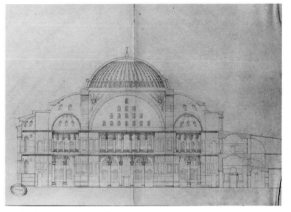

Figure 4.8 Hagia Sophia, Constantinople, 532–7; cross-section drawing by Charles Texier, 1834 (by courtesy of the British Architectural Library, RIBA, London)

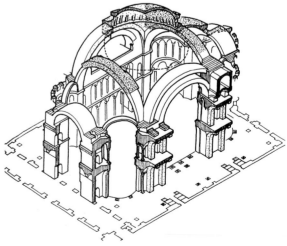

Figure 4.9 The supporting structure of the great dome of Hagia Sophia (Mainstone, 1969, figure 25)

Estimates of Constantinople's population in the sixth century vary from 300,000 to 500,000. The population fell sharply in the eighth century but recovered from the ninth; in the Early Modern period it would remain one of the world's largest cities.

Yet more invasions

Fresh waves of nomadic invasions from the eighth to the tenth century brought widespread pillage and destruction to settled Europeans, inhibiting the process of urban revival. The Vikings – Germanic sea-nomads and pirates – roamed far and wide, from Orkney to the Mediterranean and the Russian river system, burning and looting on the sea-shore and sailing up rivers accessible to their longboats.

They brought devastation to Cologne; the flourishing trading port of Dorestad, on the Rhine delta, disappeared forever from the map of Europe; and the raids on the city of Trier forced its inhabitants to erect a palisade defence which cut across the Roman street grid. These invasions did not have an entirely negative effect on future urban development. The new fortifications and rebuilt Roman walls (in Metz, for example) stimulated growth. And the Vikings themselves participated in trade, generating the development of ports: Dublin (a Viking entrepôt), Cork, Kiev and perhaps Novgorod – though that is vigorously contested by nationalist Russian historians who are loath to recognize any such Scandinavian origins of the Russian state (Riasanovsky, 1977, pp.25–30).

In the ninth century there was terror throughout Europe at the arrival of the Magyars, a Turkish nomadic tribe from the Asian steppes, who settled in Hungary and then proceeded to launch rapid attacks in all directions. Between 899 and 955 CE there were thirty-three Magyar raids on Italy, with widespread pillaging and deportation of the inhabitants into slavery; the result was a general flight from the plains to the safety of the hills. A battle at Lechfeld in 955 finally put a stop to these attacks.

But a far greater threat in these centuries was posed by an eruption from another quarter: invasions from the Arabian peninsula. In sharp contrast to the barbarian Germanic kingdoms, the Arab Empire was based from the start on cities. The consequences would be profound, leaving an indelible imprint on the urban fabric of the medieval and modern world.

4.2 Cities of Islam

A preference for cities

Visions of the archangel Gabriel were the sparks that triggered one of the greatest waves of military conquest in all history, and the rise of a new and eminently urban empire. Around 610 CE, Muhammad became convinced that the heavenly messenger had appeared before him, ordering him to spread the word of Allah and so rid the world of idolatry and polytheism. Muhammad had been a trader in the town of Mecca, a prosperous commercial centre which controlled the Arabian camel transport from Yemen to Damascus, the caravan route that carried much of the trade between the Indian Ocean and the Mediterranean. Now he turned from trade to prophecy and war.

His immediate task was the conversion of Arabia; the first step began when he was accepted by a small group of Meccan merchants. Most of his fellow citizens rejected him, however – perhaps fearing his rise to political power, certainly resenting his criticism of them for not giving some of their wealth to the poor – and Muhammad's dire warnings to the Meccans of the final Day of Judgement fell on deaf ears. Eventually, he moved his base north to the oasis town of Yathrib, soon to be renamed Madinat al-nabi ('city of the prophet'), or Medina, where he won support. Now, at the head of a small Muslim force, he conducted a jihad, or holy war, against the Meccans, routing them in battle and capturing Mecca in 630 CE. In the same year, Muhammad's forces defeated a tribal alliance of bedouin. For the first time in its history Arabia was united into a single community, based on faith in Allah and obedience to His prophet, Muhammad.

Muhammad's 'revelations' were collected together to form the holy book of Islam, the Koran. And that book contains the clearest signs of the new religion's commitment to promoting city life. Muhammad's religious struggle for Arabia had merged with the social tensions, deep and long-seated, between the settled populations of Arabia's coastal and oasis towns, and the warring nomadic tribes of the desert. Muhammad the citizen poured scorn on the bedouin (*ahl bedu* is Arabic for 'dwellers in the open land'), condemning their paganism, or indifference to religion. The Koran strongly criticizes their morality. Eventually, Muhammad, commanding his small army of converts, defeated a tribal alliance of bedouin in 630 CE, at the Battle of Hunayn. Islam subsequently poured praise on individuals who took the meritorious step of abandoning nomadic life to dwell in a city. Those who obstinately remained nomads suffered disabilities under the new religion: they were disqualified from leading prayers, and in judicial proceedings they could not testify against settled citizens. Under Islam the desert nomads would continue to be despised as lawless barbarians, infidel who pillaged caravans (Marçais, 1961).

Muhammad died in 632 CE, leaving no plans for the succession. But a series of caliphs (from the Arabic for 'successor') soon continued his great enterprise and, with extraordinary rapidity, Muslim forces burst out of the Arabian peninsula to conquer the richest provinces of the Byzantine Empire (Egypt and Syria) and totally destroy the Sasanian Empire of Iran. The ease of these victories is partly attributable to the weakening of those empires in their recent wars against one another, but also to the fanatical conviction of the Muslim forces that God was on their side. Families of Mecca's urban elite were the commanders, and they eventually established dynasties in Syria (the Ummayads) and Iraq (the Abbasids). Expanding still further, the armies of the Prophet reached the Indus and the borders of China; and westwards they surged along the entire coast of North Africa to reach the Atlantic (see Figure 4.10 overleaf). In 711 CE a Muslim army crossed the Straits to invade Spain. One of the first places captured by the Berber commander, Tariq ibn Ziyad, was named Jebel al-Tariq ('Tariq's mountain'), later corrupted to 'Gibraltar'. The irresistible Muslim forces swept through the Iberian peninsula, crossed the Pyrenees and temporarily occupied a few towns in France, until their progress was halted at Poitiers in 732 CE. Had their conquests continued further north, there would have been momentous consequences, imagined in the unforgettable words of the eighteenth-century historian Edward Gibbon:

> The Arabian fleet might have sailed without a naval combat into the mouth of the Thames. Perhaps the interpretation of the Koran would now be taught in the schools of Oxford, and her pulpits might demonstrate to a circumcized people the sanctity and truth of the revelation of Mahomet.
>
> (Bury, 1896–1900, p.15)

But the Muslim presence in Europe continued to be felt. The Arabs would remain in Spain for almost 800 years; they conquered Sicily, and their sack of Rome in 846 CE caused Pope Leo IV to build a massive defensive wall around St Peter's and the surrounding complex of monasteries and pilgrim hostels.

Wherever they went, the forces of Islam brought profound change: the religion of the Prophet, the Arabic language and the urban ideal. At first, in the early phases of their campaigns, garrison cities were created as springboards for further invasions: Fustat (in Egypt), Kairouan (in Tunisia), Rabat (in Morocco) and Basra (in Iraq). All soon became nuclei for flourishing new

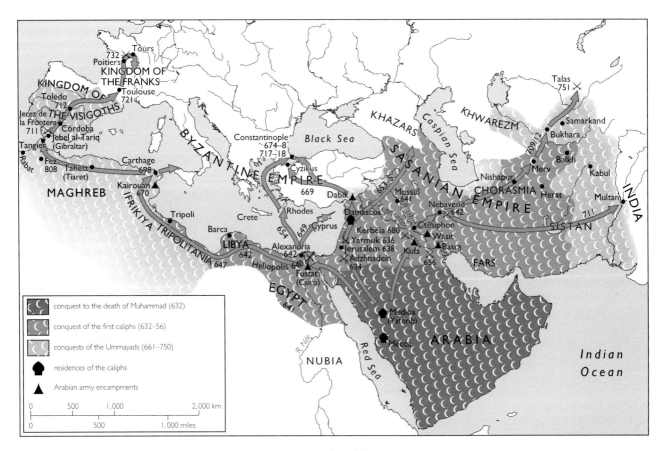

Figure 4.10 The outburst of Islam (Kinder and Hilgemann, 1974, p.134)

cities, once they had served their purpose of a temporary military base. Mud-bricks replaced tents. Elsewhere, the Arabs became the new masters of ancient cities still great (Damascus) or ruined and now revived (Córdoba, Seville).

Building a labyrinth, and why?

But can we speak of 'the Muslim city' as a distinctive type? That, today, is a burning question in Islamic studies. Torres Balbás, the historian of Spain's cities, was in no doubt as to the answer: the Muslims had imposed a new form on Iberian cities which survived into the twentieth century, half a millennium after the Christian reconquest of the peninsula. Take the city of Toledo: 'Visitors to Toledo, who do not know it well and wish to explore it, will frequently retrace their steps as they reach the end of a cul-de-sac' (Torres Balbás, 1968, p.83; see Figure 4.11). And elsewhere in Spain, in present-day Andalucia, vestiges of the same winding city roads are evident in Granada and Seville, where the main street is called Sierpes (Castilian for 'serpent') and another is Siete Revueltas ('seven twists'). Who gave Spain this tortuous street layout? Certainly not the Romans, with their rectangular grid system; nor is there any evidence to suppose it came from their successors, the Visigoths. Torres therefore concluded that it was brought from the Near East by the conquering Arabs, and in support of that he referred to the plan of an excavated section of the Muslim city of Fustat, on the Nile (see Figure 4.12). This was the typical city of Islam, with winding radial roads from which

> other narrow, secondary roads sprang, twisting and turning at each step; and from these in turn arose numerous blind alleys which penetrated deeply into the residential areas, ramifying into a labyrinth, like the veins in the body.
>
> (Torres Balbás, 1968, pp.79–80).

Why had the Arabs built cities like this? According to Torres, it was the result of 'a lack of municipal organization and building regulations', an absence of controls that also led to encroachments of buildings on the public way. Consequently, the streets of Muslim Spain had been narrow, dark, airless and dirty. And after the reconquest it was left to the rulers of Christian Spain to do what they could to open streets up, straightening them and clearing obstructions, restoring the tradition of the Western classical city – though much was to remain unchanged (Torres Balbás, 1985, pp.294, 424; see Figure 4.13 overleaf).

Other historians have similarly blamed the 'anarchy' of Muslim cities on a failure of authority. One historian of art could not otherwise explain this irregularity of design, 'which has always been alien to Islamic art … indeed, in [Islamic] architectural design there is usually an over-zealous desire for symmetry' (Jairazbhoy, 1964, pp.59–60). (He might have added that the same strongly geometrical planning is characteristic of the tradition of illuminated manuscripts of the Koran.) A different explanation attributes the irregular built environment to the religious principles of Islam, which 'do nothing to encourage urban display'. Hence – so the argument continues – the lack of urban pride, the lack of cohesion, the irregularity of plan, the rejection of luxurious and lofty houses, regarded as reprehensible arrogance, and the use of perishable materials (clay, mud or wood) to stress the unimportance of individuals and material possessions. And hence also the special features introduced in house-building to observe Islam's demands for family privacy and the concealment of women (Planhol, 1959, pp.7, 15, 23, 25). So private life turned away from the street to inner courtyards, Muslim houses had few or no external openings, and in Kairouan the houses were staggered to protect the inhabitants' privacy from neighbours on the opposite side of the road. In Muslim Spain, probably from the thirteenth century, *ajimeces* (shuttered windows on projecting balconies) were introduced to allow women to be in the open air and view the street without being seen (see Figure 4.14 overleaf).

Another, very different, explanation has come from Bulliet (1975; see Extract 4.1 at the end of this chapter). According to him, the irregular street layout is not a manifestation of oriental backwardness but a practice full of advantages. Second, Bulliet rejects the religious explanation, focusing instead on the technology of transport. The lands of Islam had abandoned wheeled vehicles for camels; wide, straight thoroughfares were therefore unnecessary, and urban life in Muslim cities was dominated by human needs, not those of vehicles.

While this explanation is plausible, there are some underlying difficulties in the argument which the author acknowledges and strives to overcome. Why did the camel replace the wheeled vehicle as the standard means of transport across the entire Islamic world, from Afghanistan to Morocco? Any explanation alleging that

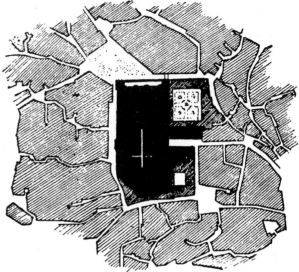

Figure 4.11 Central Toledo, showing the roads around the cathedral (Torres Balbás, 1968, p.84)

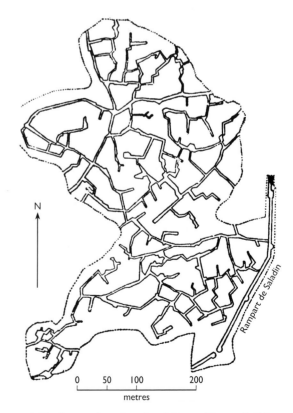

Figure 4.12 Street plan of medieval Fustat (Torres Balbás, 1968, p.79)

Figure 4.13 Blind alley in Córdoba (Torres Balbás, 1968, plate XIII)

Figure 4.14 Shutters for female seclusion – Muslim influence in Braga, Portugal (Torres Balbás, 1985, p.413)

camel transport was cheaper is 'difficult to establish' (Bulliet, 1975, p.21). Why did competition from the camel result not only in the replacement of wheeled commercial transport but in the total abandonment of all wheeled vehicles? And why did the camel replace the ox-cart when it did, 'sometime after the third and before the seventh century', and not hundreds of years earlier (Bulliet, 1975, pp.21, 27)? Why something did not occur at a particular time is a question that is hardly open to scholarly research. Nevertheless, intelligent speculative interpretation is still possible, and Bulliet argues that from the eighth century BCE a very slow eclipse of the military chariot began in the Near East until, by the first century BCE, it was considered improper for a person of status to ride in a vehicle instead of on horseback. More important, so the author argues, there was a decline in the manufacture of wagons and harness. A long, slow development of camel domestication occurred, and 'sometime between 500 BCE and 100 BCE a special camel saddle was invented which would transform the economic, political and social history of the Middle East' (Bulliet, 1975, p.87). This saddle, permitting the use of swords by mounted riders, shifted the balance of military power in the desert, with the camel-breeders securing control of the caravan trade. Now the wheel could be replaced by the pack-camel. But what of Spain, where neither camels nor wheeled vehicles were conspicuous in the first centuries of Muslim rule? Here Bulliet surmises that wheeled vehicles 'probably did not exist in Spain in large numbers' at the time of the Arab conquest, because Iberian roads had fallen into neglect after the barbarian invasions. Consequently, the Arabs had 'little incentive' to adopt wheeled vehicles there, with the result that irregular, 'non-vehicular city designs' also appear in medieval Spain. The inevitably imprecise chronology and the surmise prevents this from being a clinching interpretation; all the same, it is stimulating stuff, as is his later suggestion that since maximum loads in the Islamic world were limited to the carrying capacity of a camel, this affected the transport of heavy building materials and so 'probably helped bring about a brick-orientated architecture' (Bulliet, 1989, p.380). But the entire argument is fired by a highly questionable technological determinism. As the author puts it: 'the state of technology and of the economy is the prime determinant of people's attitudes and actions, at least within the sphere of transportation' (Bulliet, 1975, pp.228–9). And more than transport: he clearly reduces the choice of building materials to the same determinism.

Can we assume that the transformation into labyrinths of the regular street layout in cities of the former Roman Empire was due to the Arab conquerors? The question is anything but simple.[2] The Roman grid plan in Damascus and other Syrian towns had already lost much of its regular form before the Arabs arrived. But what most weakens the association of urban labyrinths with Muslim rule is the evidence of symmetrical city plans from various regions of the Islamic world. Eighth-century Anjar (in the Lebanon) had a rectangular street system, and so at first did Cairo, founded in the ninth century (Kennedy, 1985). But Muslim planning is nowhere stronger than in Iraq, where the Abbasid dynasty built the large mud-brick capital of Samarra – dividing it into rectangular blocks for houses and gardens – and before that, in the first capital of Baghdad.

The caliph al-Mansur himself selected the site of his new capital: the Persian hamlet of Baghdad, on the Tigris, situated in a fertile region with an existing canal network and near the hub of caravan routes. From a village Baghdad grew to become one of the largest cities of the medieval world. That had not been al-Mansur's intention; it was planned as a strategic military base and administrative centre. The details of his plan are uncertain because the site has never been excavated and scholars have to rely on literary evidence of varying reliability to reconstruct it (see Figure 4.15 overleaf). But what is undisputed is that the new city had a circular plan, with al-Mansur's palace at the centre. The circular form has subsequently been variously interpreted as a symbol of the cosmos, as an advantageous shape for defence, and as the mere imitation of a Persian circular military camp. It was a huge building project – the circle may have been as large as two and a half kilometres in diameter. The caliph brought in engineers and surveyors from all parts of his empire and engaged a large labour force – some sources speak of 100,000. For building materials, neighbouring ruins of Sasanian palaces were plundered. Work began in 762 and the building of the city was completed in four years. Two concentric walls and a moat supplied the fortifications. In each quarter of the circle was a gate opening to a wide arcaded street which led to the central palace and an adjacent mosque, built of sun-dried bricks and surmounted by a dome. This inner palace district could be sealed off by gates, isolating the caliph and protecting him from riot and assassination. The palace area was surrounded by a ring of houses occupied by the caliph's sons and army officers. Further away from this was another belt of houses, for other residents. Baghdad was therefore not an integrated city, but rather an administrative centre. There were shops: at first these were in the city quadrants, but soon – perhaps for security reasons, to avoid gatherings close to the centre at a time when the caliph had many enemies – they were banished beyond the city precincts. The dominant function of the capital was to serve as 'a symbol of total rule and universal legitimacy of the caliph' (Al-Sayyad, 1991, pp.115–26; Lassner, 1970). But after the caliph's death, prosperity from trade brought large-scale immigration, and Baghdad became a densely populated residential city and a centre of theological study.

---

[2]   See the text by Hugh Kennedy in the Reader associated with this volume (Chant, 1999).

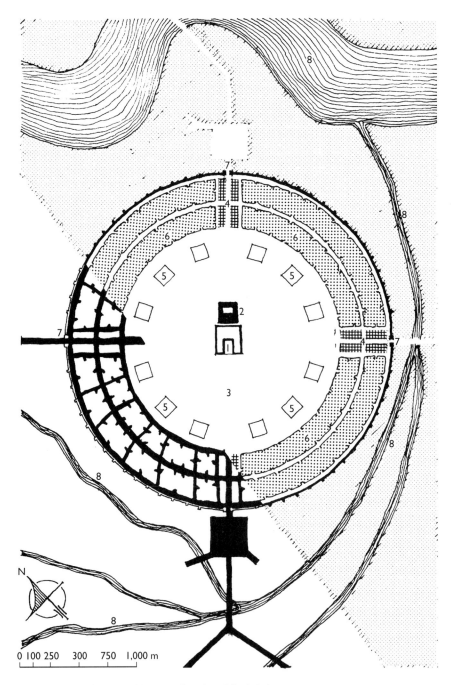

Figure 4.15 Suggested layout of medieval Baghdad

Key 1: caliph's palace; 2: mosque; 3: *rahbah*; 4: market; 5: palaces of the royal family;
6: restricted quarters; 7: gates; 8: rivers and canals (Al-Sayyad, 1991, p.121)

Recognition of the motives of individual caliphs in planning cities such as
Baghdad (and Cairo) for their own benefit, instead of for their subjects, is one
reason why the tendency of recent scholarship has been to reject the earlier
stereotype of the chaotic Muslim city, because it clearly shows that central
planning occurred in some Arab cities (Al-Sayyad, 1991).

Building types and urban amenities

Muslim cities therefore can no longer simply be categorized as labyrinths
without a qualifying acknowledgement that regular geometry was sometimes
imposed. Similarly, it is no longer accepted that the general absence of formal
municipal institutions meant a total absence of organization in urban life. An

alternative system was provided. Islam is not just theology but also law, regarded as being of divine origin and essential for the proper regulation of civic life in accordance with the will of Allah. Responsibility for supervision of the community rested with the *ulema* (from the Arabic for 'learned'), pious scholars who were experts in expounding the sacred law. A literate urban elite, they came from every class of society – they could be masons or pharmacists – and functioned as scribes, educators and market inspectors who sought to maintain high standards of honesty in weights and measures, as well as guaranteeing the quality of food merchandise (Lapidus, 1967, pp.108–9, 119). And in another way, some of the functions of municipal authority were supplied by the *waqf* ('pious foundation'), a public trust inspired by the Islamic emphasis on charity. These bequests generated the funds for essential urban services – hospitals, alms kitchens and waterworks – and for education, as in Muslim colleges, where the subsidies might endow the salary of a professor of law or provide students with lodgings and food (Makdisi, 1988, p.68; Powers, 1989).

For the French historian Maurice Lombard, the Muslim city was to be interpreted not as a religious but as an economic creation. According to this view, the motor for the expanding urban Muslim world was the flow of gold from the Sudan. It was the circulation of money that 'best characterizes' the creation and maintenance of the Muslim city. And it was most powerfully reflected in the *suqs* (bazaars) and caravanserais (combined merchant hostels and warehouses) that were so prominent a part of the urban environment (Lombard, 1957, pp.24–5). In sharp contrast to the classical city, in which Aristotle had advised keeping the central forum clear of merchandise, the Muslims generally placed their markets near the centre, next to the mosque. The truth in Lombard's analysis is evident in the hectic urban manufacturing that went on in medieval Islamic lands, immortalized in the vocabulary of European languages: 'fustian', a type of cloth, reflects the importance of textile-manufacture in Fustat; 'damask' and 'damascene' come from the thriving silk and steel manufacturers of Damascus; 'cordwainer' comes from the manufacture of punched and embossed leather, for which Córdoba was famous.

But the religious influence on Muslim cities was pervasive, reaching even into the market-place. Apart from the above-mentioned religious superintendents, the topography of the bazaar was such that trades supplying religious artefacts (incense, holy books, candles and mosque slippers) were located in lanes closest to the mosque, while baser trades were kept at a distance. And the dirty work of tanners, dyers and paper-makers was in Damascus confined to the very edge of the city.

Religion made it very difficult for Muslims to live anywhere but in cities. It has been argued that Islam's demand for the cloistering of women can only be satisfied in an urban society: women can be effectively enclosed in a mud or stone house but not in the mobile tent of nomadic bedouin (Marçais, 1961, p.61). But, above all, it was the central requirement for communal prayer that forced the faithful to become urban and led to the most characteristic building type of the Muslim city: the mosque. The *hadith*, part of the scriptural corpus of Islam, declares that communal prayer is twenty-seven times more effective than praying in solitude and 'actualizes the ideal of the perfect city', celebrating the Creator's glory in unison (Michon, 1980, p.25). Religious doctrine insisted that this could only be achieved within a permanent building; the principal rite of Friday communal prayer had no validity if performed by a group of nomads in a circle of tents.

It is alleged that, unlike cathedrals, mosques were not located at the hub of major roads and did not dominate the urban vista, a lack of prominence perhaps attributable to 'the absence of an established clerical hierarchy in Islam' (Al-Sayyad, 1991, p.4). The premise is questionable. The sheer size of the largest city mosques and their enormous domes makes them focal points of Jerusalem, Isfahan and Istanbul. In contrast to the classical separation of religious and political space, the mosque united the functions of a place for prayer and a public forum. In Aleppo (in Syria), the mosque was built on the Roman forum, retaining the expanse as a spacious courtyard capable of accommodating large numbers.

Technology, in the form of the manufacture and use of scientific instruments, had an important function in the erection of a mosque. From Córdoba to Samarkand, all mosques, on Muhammad's instructions, had to face Mecca. The *qibla* (direction of prayer) was established by astronomical observation, and fixed by the orientation of the *mihrab* (the prayer niche), the key feature of a mosque, and lavished with decoration. In their form, minarets may have been influenced by the ancient lighthouse of Alexandria; the word 'minaret' is derived from the Arabic for 'lighthouse'. From these conspicuous towers the muezzin (from the Arabic for 'ear') called the faithful to prayer. And still in mid-twentieth-century Morocco, the time of prayer was determined by the astrolabe, the main observational instrument of medieval astronomy.

Until the eleventh century, the main mosque provided higher education (theology, law), but afterwards this function was increasingly transferred to the *madrasah*, private collegial foundations with resident students (see Figure 4.16).

The cities of Islam were mostly situated in an arid geographical belt. Water was precious. Medina was a desert oasis; the very existence of the settlement depended on its wells. Throughout the Arab world, determined efforts were made to capture rain. In Algiers, the design of roofs and gutters was influenced by the need to conserve rainwater, so that the rain was directed into domestic cisterns instead of pouring on to the street. The same was achieved in Kairouan, where the huge mosque was given slightly sloping terraced roofing so that the rain gradually trickled down, flowing into gargoyles and finally into subterranean cisterns (see Figure 4.17). One of the most charitable works a Muslim can perform is to give water to the thirsty pilgrim or traveller. Islamic jurists debated the rights to water use, one tradition proclaiming the universal right to secure water, if necessary by force of arms, another interpretation permitting the owners of sources to refuse water to others (Mazloum, 1936, p.27).

Figure 4.16 The great fourteenth-century *madrasah* of Sultan Hasan, Cairo; watercolour by David Roberts, 1839. Each portico was allocated to the study of a particular rite; at the centre was an octagonal fountain (Fossier, 1997, p.265)

The largest Arab cities, such as Baghdad and Damascus, could not have survived without their elaborate networks of canals. Extensive irrigation of the surrounding countryside was essential to produce the rice, sugar-cane and oranges that the Arabs cultivated. The 'agro-city', an urban area with a rural hinterland, has been identified as the unit of Near Eastern society, and irrigation agriculture as the most distinctive characteristic of Islamic cities (Hourani, 1970, p.22). Hydraulic technology became a Muslim speciality. From Spain to Syria, imposing dams, water-wheels and aqueducts were built. Much was based on Roman techniques, as in twelfth-century Seville, where the unearthing of a buried Roman aqueduct was the occasion of jubilation; it was soon supplying drinking-water to the city, irrigating gardens and powering mills. Roman in principle also were the giant vertical water-wheels (*norias*) erected on the banks of rivers in Córdoba, Toledo and Homs (in Syria). These wheels – some were well over thirty metres in diameter – had earthenware pots attached to their rims, so that when the force of the river rotated the wheel, the pots were immersed and then rose to pour their contained water into a tank that fed an aqueduct (see Figure 4.18). Through mastery of these techniques and their extensive application, the Muslims permanently transformed the landscape around Valencia into an irrigated complex of gardens (*huertas*) yielding rice and citrus fruit.

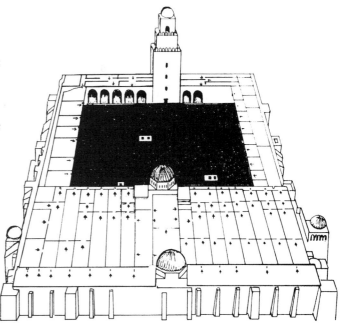

Figure 4.17 Sloping roof of the great mosque of Kairouan, designed to capture rain for the provision of drinking-water. The arrows show the direction of the flow (Pavón Maldonado, 1990, p.84)

Figure 4.18 *Noria*, still working in present-day Murcia, Spain (Pavón Maldonado, 1990, plate CXXII)

But it was the religious requirement that most distinguished Muslim
hydraulic technology in the medieval world. Water in huge quantities, far
greater than that needed for drinking, was supplied for ritual washing.
Ablutions were performed in the precincts of mosques and in *hammams*
(Arabic for 'hot') – steam or, as they later came to be known in the West,
'Turkish' baths. They were installed in considerable numbers: twelfth-century
Damascus had over fifty; Córdoba possessed hundreds; Baghdad was alleged
to have thousands. As though for some starred, medieval travel guide, Arab
writers rated the bathing establishments in various cities. Their multi-domed
structures were a conspicuous part of the urban environment and their remains
still catch the eye of today's traveller in Spain; some, in fine condition, have
recently been discovered there (see Figure 4.19). As in the Roman baths from
which they are thought to derive, the public entered a sequence of rooms
maintained at different temperatures, but with the difference that the Arab
baths were much smaller and dispensed with the sporting activities of the
classical *frigidarium* (see Figure 4.20). Water was supplied via pipes to a boiler
housed in brick and covered by a low vault to prevent the escape of steam.
The steam passed through perforations to the adjoining hot room; cooler
rooms were supplied with hot water by means of pipes. The use of paving
materials of marble or polished stone ensured a watertight floor. To conserve
heat and steam there were no windows and no ventilation; light penetrated
through thick pieces of glass, like bottle-ends, inlaid in the domed
superstructure. Heat losses were similarly prevented by the use of thick,
limestone walls and a corridor, kept closed by counter-weights, which led in a
zig-zag from the changing room to the steam bath. The basic form of this
design has remained unchanged, and some of Damascus' medieval baths still
function (Ecochard and Le Coeur, 1942–3).

Figure 4.19   Baths of Muslim Spain, Jaén, tenth to eleventh century (Pavón Maldonado, 1990, plate CXXXI)

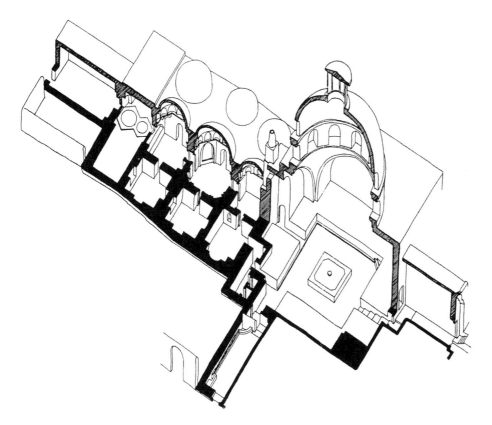

Figure 4.20 Cross-section of fourteenth-century baths, Damascus (Ecochard and Le Coeur, 1942–3, p.57)

Complementary to urban water supply was the system of waste disposal, and here archaeologists have been particularly impressed with what the Arabs achieved. Muslim Córdoba had a network of sewers which descended from the upper city to the river, passing through the principal streets and at the same time collecting the waste waters of secondary sewers. Built of limestone blocks and lined with concrete, these sewers were huge – up to two metres wide. The drainage system of medieval Fez was so efficient that the French colonizers of early twentieth-century Morocco continued to use it. Fez is situated in a deep valley, on a river with a rapid current. When the city needed cleaning, lids in conduits were opened so that fast-flowing water swept through the streets.

The city of Fustat, the seventh-century capital of Muslim Egypt, no longer exists, but excavation of its ruins, three kilometres south of the later capital of Cairo, has revealed elaborate systems of water supply and sewage disposal, designed to fit local conditions. The first permanent buildings were of mud-mortar, too weak a material to support anything above a single storey. This housing was no longer adequate once the city became a bustling, populous centre of textile-manufacture and commerce. So a crucial change in building materials and technology was introduced – the use of lime-mortar and the reinforcement of the house walls by the insertion of columns made of the local sandstone (see Figure 4.21). This permitted the construction of houses with five and even six storeys; such houses, built in complexes around a courtyard, became the typical city structure – tenements housing up to 100 occupants. It all required a sanitation system far more developed than the simple cesspool of the days of the one-storey house. Flues, now introduced on each floor, carried the waste to deep channels (covered and running beneath the floors and courtyards), to be deposited in large cesspools that were covered daily with sand. To flush the latrines,

Figure 4.21 Sandstone column reinforcing a multi-storey house in Fustat, about twelfth century (Scanlon, 1970, plate 1a; by courtesy of Fustat Expedition/American Research Center in Egypt)

a laborious process of water supply was implemented. Vast pits were dug in the sandstone rock on the outskirts of the city to form reservoirs. Teams of pack-animals continuously brought water from the Nile in skin containers, emptying them into the pits. From there the water was carried by donkeys, or on the backs of water-carriers, to supply the needs of the domestic sanitation system (as well as mosques and *hammams*). The same Nile water was used for drinking, carefully kept separate from the waste in containers cut into the rock of the dwelling and plastered over (Scanlon, 1970, pp.186–92; see Figure 4.22).

Figure 4.22 Internal water supply for houses in Fustat (Scanlon, 1970, plate 1b; by courtesy of Fustat Expedition/American Research Center in Egypt)

4.3　Urban revival of the Latin West

Interpretations of the revival and their critics

In a famous thesis developed in the 1920s and 1930s by the great Belgian historian Henri Pirenne, it was argued that the Muslim conquests represented a crucial turning-point in the history of Europe; those conquests, not the earlier Germanic invasions, were presented as the real beginning of the Middle Ages. Until the Muslim expansion, the Christian West had maintained its essential, long-established trading links with the Levant, importing Eastern luxuries (spices, papyrus, silks, purple-dyed cloth). But then, so Pirenne continued, the Arab conquests cut off this trade and the Mediterranean became a 'Muslim lake'. The West turned in on itself, the circulation of money shrank to a trickle, merchants and urban life disappeared, towns and cities dwindled, and western Europe was reduced to a subsistence economy, 'a civilization which had regressed to the purely agricultural stage' (Pirenne, 1939, p.242). Pirenne believed that the Arabs had established an iron curtain separating the Faithful from the Infidel, so isolating western Europe into an economic and cultural backwater, and causing a revolutionary change in its outlook: 'The West was blockaded and forced to live upon its own resources. For the first time in history the axis of life was shifted northwards from the Mediterranean' (Pirenne, 1939, p.284).

The Pirenne thesis, as it came to be called, would excite debate for decades. It was never wholly accepted and is now a thing of the past. Some questioned its details, denying that trade between East and West had been cut off. More interesting, Lombard was moved to propose a diametrically opposed interpretation, which portrayed the march of Islam as highly beneficial for the West and the wider world because it resulted in an upsurge of cities. According to this grandiose explanation, the urban revival began with an increased flow of gold from the Sudan, which fired Muslim commercial activity, and this in turn stimulated the urbanization of Muslim Spain (including the creation of the great port of Almería), North Africa and Muslim Sicily. Next, the demands of cities throughout the lands of Islam for certain merchandise (naval timber, iron, tin, fur) could only be satisfied by the barbarian West in exchange for gold. Hence the rise of Italian cities (Amalfi, Naples and, above all, Venice) from the eighth century, and a general urban revival in the West in the ninth and tenth centuries as the wave of commerce stimulated by Muslim money was propagated to the Rhine, to Flanders, to England. The same causes led to the economic revival of Byzantine towns, the creation of towns in black Africa (Sudan, the east coast and Timbuktu), and the rise of riverine towns in Russia, where the ferment of Muslim wealth was also felt. It all added up to a new era and 'a major epoch in urban history' (Lombard, 1957).

A very different interpretation of urban revival in the Latin West, based on technological innovation rather than the flow of bullion, was developed from 1940 by an American medievalist, Lynn White, Jr.[3] Like Pirenne, White set out to explain the transfer of the centre of European civilization from the Mediterranean northwards.

For White, the great change was attributable to a set of agricultural innovations which, gradually introduced between the eighth and tenth centuries, provided the necessary base for rapid urbanization. First, a new and more complex type of plough permitted the cultivation of the heavy, but potentially very fertile, wet soil of northern Europe. The much simpler Mediterranean plough, the ard (with a simple point of wood or iron), which merely scratched the surface of that region's light and dry soil, was inadequate for cultivation in temperate Europe. A heavier plough, fitted with additional components (iron coulter, iron ploughshare and wooden mould-board), could simultaneously cut the soil vertically and horizontally, and turn it over (see Figure 4.23). This efficient agricultural instrument not only saved peasant labour; extensive areas of hitherto uncultivated soil now became available. A marked food surplus was therefore created in the Latin West, a condition that stimulated population growth: releasing people from the soil permitted urbanization and leisure.

But the heavier plough also required more efficient traction. And so two associated fundamental innovations appeared: the collar horse-harness and nailed horseshoes. These improved the horse's pulling power, protected it from injury and, so White alleges, made overland transport of agricultural produce economically attractive for the first time, encouraging peasants to consider sales of cash crops.

The final piece of this revolutionary system of agricultural technology was a new technique of crop rotation, the three-field system, introduced in the eighth century. The benefit was not merely more food but better food. The produce included oats to sustain horses labouring at the plough, and for humans nutritious legumes, which provided the population with a new vigour, leading to growth, the multiplication of cities, and an unprecedented flourishing of

Figure 4.23 The fully developed medieval plough, with ox-team, ploughshare and mould-board; an illumination from the Luttrell Psalter, England, c.1340 (British Library Add. MSS 42130, folio 170; by permission of the British Library Board)

---

[3] Extracts from White's book are reproduced in the Reader associated with this volume (Chant, 1999).

commerce and industry. In fact, so White concludes, the new food surplus and improved diet created a new class of artisans and merchants, and generated capitalism. This bubbling environment reached its climax with the 'medieval industrial revolution' based on wind- and water-driven machinery, that exploitation of power technology which has ever since characterized the modern world. And it all came ultimately from ploughs, horse-collars, horseshoes and crop rotation (White, 1962).

Can we accept White's interpretation? A caustic review of his book accused him of indulging in naive technological determinism and adventurous speculation, 'particularly attractive to those who like to have complex developments explained by simple causes' (Hilton and Sawyer, 1963). For these reviewers, the argument amounted to 'a chain of obscure and dubious deductions from scanty evidence about the progress of technology'. They objected that the chronology was all wrong, arguing that the heavy plough had come to England much earlier – as early as the first century BCE – and that oats were not introduced in the early Middle Ages but were known in northern Europe even in prehistoric times, as archaeology demonstrated. White's enthusiasm over the substitution of horse for ox was misplaced because, among other evidence, records showed that in late thirteenth-century Bedfordshire the ox continued to be preferred. And as for the alleged bumper crops, the scanty evidence suggested very poor yields in these medieval centuries – hardly an indication of an agricultural revolution. Rejecting also White's supposed large-scale transformation of the tenth-century diet by legumes, they argued instead for an increase in cultivation of bread grains; peas and beans were not important in England until the fourteenth century. Hilton and Sawyer concluded that White's interpretation muddled cause and effect. Was it increased food supplies that caused a population increase? Or was it (as they preferred) the other way round: that an already growing population provided for its subsistence by an extension of the cultivated area, as much as by any increase in productivity?

This did not spark off a debate, probably because the evidence on medieval crop yields is slight, and that on ploughs ambiguous. The vocabulary used to denote ploughs, for example, in documents of the twelfth and thirteenth centuries relating to eastern Europe makes it very difficult to discern exactly what type of implement was being used in that region (Bartlett, 1993, pp.149–52). The heavy plough contained more iron parts and – although iron ore was becoming increasingly mined – until the metal became cheaper the new tool could only be adopted slowly. Some historians have remained sceptical about attributing great importance to the replacement of the ox by the horse, or to the three-field system (for example, Lopez, 1971, pp.214, 216). Yet several historians today seem to accept that there must have been some connection between improving agriculture and the striking urban growth of the medieval West from the eleventh century. But they avoid White's linkages of single cause to effects, instead considering progressive agricultural technology to be just one of the conditions that stimulated urban development in the medieval West.

Population explosion and calamity

What is not in dispute is the occurrence of a sharp rise in population in the West from the eleventh century, boosting the size of both towns and rural settlements. This period was far removed from the modern practice of accurate censuses and exact population statistics. No such medieval records exist, and so precise data on population are not to be expected. The degree of uncertainty is evident from the widely varying figures given by present-day historians for the population of fourteenth-century Paris, probably the largest city of the medieval West: estimates range from 80,000 to 200,000. Yet in other cases there is more agreement on approximate population size, based on thirteenth-century counts of householders for the purpose of tax collection; the problem here for the historian is to decide on the average size of households and so determine the multiplier for deducing the population, allowing also for the sizeable sectors of the tax-exempt population such as the nobility and clergy. Population trends have also been deduced from the prices of cereals, a rise in prices suggesting the rising demand of an expanding population. And strong evidence of growing city populations comes from the progressive expansion of their surrounding walls.

Current estimates of the population of medieval Europe in 800 indicate that it was quite low – some twenty-five million. At this time, most of the land still consisted of forests and marshes, and the population lived in scattered, closely knit rural settlements. Such was the continuous extent of forests in early medieval Europe that one historian has reflected that a squirrel could have travelled from Moscow all the way to Europe's Atlantic seaboard, 'passing from tree to tree without ever touching the ground' (Heaton, 1948, p.63). Then, from around the early eleventh century, came Europe's great population surge. Encouraged by the cessation of the Viking and Magyar raids, and (some suggest) an improving climate, Europeans ventured widely to clear, reclaim and colonize new lands and found new cities. This expansion stimulated population growth and also supported it by the produce of wider cultivation. Droves of German peasants pushed eastwards across the great river systems of the Elbe, Oder and Vistula, reaching the Slav lands and founding hundreds of new towns, including Lübeck (1159), Riga (*c.*1200) and Berlin (1242); and refounding others, such as Kraków (1257), recently destroyed by the steppe Tartars, and Danzig (1263). At the same time, the Christian reconquest of Muslim Spain began, accompanied by the foundation of new cities to repopulate deserted regions. It is thought that by 1300 Europe's population had soared to a peak of around 100 million, and that Flanders and Italy, the most heavily urbanized regions, had as much as one-quarter of their population living in towns. By this time, the population of several Italian cities (Siena, Pisa, Padua) exceeded 30,000; rapidly expanding Florence probably amounted to 120,000; Venice and Milan may have reached 180,000 (Herlihy, 1984). It is alleged that the urban environment discouraged early marriage and therefore reduced reproduction. Urban males (so the argument goes) could not afford to marry until they were established in a craft or profession. By contrast, in the countryside, a peasant acquiring a farm needed the immediate help of a wife and offspring. In the towns the apprenticeship of male children learning a craft was long and expensive, again a discouragement to reproduction (Herlihy, 1984, p.144). And the death-rate was higher in towns. Medieval cities grew only because of immigration from the countryside.

A disaster of staggering dimensions struck Europe in the mid-fourteenth century. Between 1347 and 1351 the Black Death, a terrifying epidemic, spread from the Crimea throughout Europe, killing one-third of Europe's entire population. From a population of some 30,000 in 1335, the city of Toulouse had by 1430 contracted to around 8,000. This was bubonic plague, propagated by black rats whose fleas carried the plague bacillus. The continental spread of the lethal infection depended on the existence of an established shipping network between the Mediterranean, the Atlantic and the North Sea, providing long-distance transport for infected rats and their fleas to a multitude of European ports. A plausible explanation of the astonishing mortality levels is that the population had exceeded the available resources at a time when technology was not sufficiently developed to boost food supplies: there was inadequate transport for the rapid transit of food, a lack of chemical fertilizers and, of course, the absence of modern techniques of refrigeration and canning. And so Europe's overstretched population in the fourteenth century succumbed to famine and the malnutrition that generally precedes the onset of epidemics. Not until the sixteenth century did Europe's population fully recover from the devastation of the Black Death.

Genesis of towns, urban forms and building types

New medieval cities were generated by a variety of stimuli, the reason for the foundation often influencing both the urban form and the building types. Religious fervour was one motivation. The city of Santiago de Compostela, on the north-west tip of the Iberian peninsula, was no different from many others in France, Italy and England, in its construction, during the central Middle Ages, of an imposing and architecturally magnificent ecclesiastical complex of cathedral and neighbouring bishop's palace. But what made Santiago unique was its genesis from the supposed discovery of the tomb of one of the apostles, St James ('Santiago'), by a ninth-century monk who was guided to the holy spot by a bright star (hence, *campus stellae*, 'field of the star', which became corrupted to 'Compostela'). First a chapel, then a massive cathedral, was built over the tomb. A wave of excitement brought a swarm of pilgrims from all over Europe to this remote region. From then on, Santiago became the patron saint of Spain and the added inspiration for the Christian reconquest of the peninsula from the Muslims. A crop of new towns sprang up on the pilgrim route from the Pyrenees, providing hospices, shops and churches for the travellers. Particularly interesting is the common form of these urban creations on the road to Santiago: they are generally longitudinal, reflecting their function as a passageway to the pilgrimage centre. A good example is Santo Domingo de la Calzada (on the border of Castile and Navarre), founded in the eleventh century after the building of a bridge across the neighbouring River Oja facilitated the westward passage of pilgrims. The nucleus of the town was a long street running east to west, continuing to the bridge (see Figure 4.24). Accommodation, shops, churches and convents lined the street.

Other medieval cities were born as strategic strongholds. When, in the eleventh century, the long Christian reconquest of Spain began, the most important centres captured from the Muslims were already populous cities, such as Toledo, Seville and Córdoba. But in a few cases some wholly new fortified towns were built by the conquering Christian monarchs, to a characteristic plan. In the east of Spain it was the monarchs of Aragon who, with their armies, progressed south, eventually capturing Valencia. Tens of thousands of Muslim peasants dwelling in villages in the hills now came under Christian overlordship, and to subjugate them and maintain their cultivation of the land, several royal towns were built in the plains, fortified like some Roman military camp. The type is well illustrated by Villarreal (some fifty kilometres north of Valencia), founded in 1271 by the conquering James I of Aragon. It consisted of a rectangular walled enclosure, protected by towers at each of the four corners, and four fortified gates, symmetrically placed in the middle of each side of the rectangle. The two pairs of facing gates formed the extremities of the two

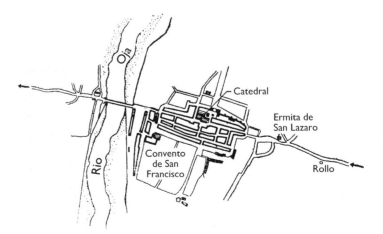

Figure 4.24 Santo Domingo de la Calzada, northern Spain (Torres Balbás, 1968, p.108)

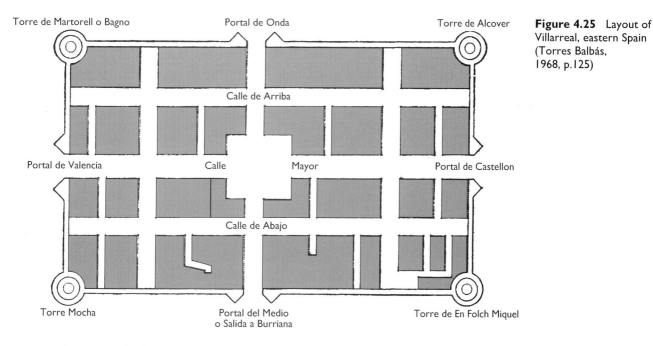

Figure 4.25 Layout of Villarreal, eastern Spain (Torres Balbás, 1968, p.125)

principal streets, which intersected at right angles. The point of intersection was the centre of a spacious central square, or plaza, the site of the town hall, the jail and the principal church. Secondary thoroughfares were laid parallel to the main streets to form a perfect grid (see Figure 4.25). A later engraving of the neighbouring city of Castellón de la Plana, founded by the same monarch in 1251 and for the same reason, gives some idea of the appearance of these fortified cities (see Figure 4.26). When, at the end of the fifteenth century, Ferdinand and Isabella finally completed the reconquest by besieging and capturing the last Moorish stronghold of Granada, they built (in 1491) for the long siege a square camp some eight kilometres from Granada. It had gates at the centres of each side and a grid street plan. At the centre was a large open space in which the army could assemble. This was Santa Fé, which became a permanent small town and still preserves much of its original form of a military camp. Elsewhere in medieval Europe towns grew up around a castle, acquiring a circular form – for example Barnstaple in Devon.

Lübeck represents another type of medieval city, where neither religion nor war was the principal influence on the built environment. Lübeck's form was irregular, defined by a circuit of walls which, as in many other cities, was dictated by topography – in this case the area of land between two winding rivers (see Figure 4.27a overleaf). The town has the usual medieval cathedral and monasteries, but what dominates here is commerce and the crafts. The centre of the city was the market, consisting of shops of a whole range of specialized merchants and craftspeople, concentrated together (as in many other cities) in their own areas: a street of goldsmiths, a street of herring merchants, and so on (see Figure 4.27b). A mint, that new urban engine of the medieval money economy, is also conspicuous. And there is a weighbeam, or steelyard, a characteristically large balance for commercial use. Everywhere there are signs of a town under commercial control. Even the large Marienkirche in the central market was the merchants' church. On the periphery are the water-mills with their associated crafts.

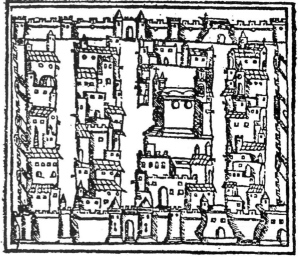

Figure 4.26 Castellón de la Plana, eastern Spain; from a sixteenth-century engraving (Torres Balbás, 1968, p.123)

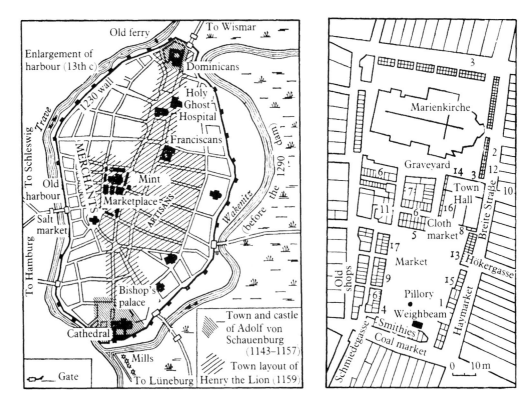

Figure 4.27a Lübeck   **Figure 4.27b** Lübeck's market-place

Key 1: armourers; 2: butchers; 3: bakers; 4: beltmakers; 5: money-changers; 6: shoemakers; 7: spice merchants; 8: needlemakers; 9: fellers; 10: herring merchants; 11: grocers; 12: minters; 13: goldsmiths; 14: cookshops; 15: saddlers; 16: tailors; 17: tanners (Le Goff, 1988, p.77)

Lübeck had been founded in 1159 by Henry the Lion, Duke of Saxony, with a view to attracting merchants from neighbouring lands and prosperity. Its subsequent growth exceeded his expectations and greatly reduced the ducal power. For Lübeck was not just another merchant town. Situated on the western edge of the Baltic, its rapidly developing sea-borne trade was accompanied by expansion of its harbour (as Figure 4.27a indicates). The town eventually became the nucleus of a whole system of German Baltic trade pursued by a confederation of towns known as the Hanseatic League (including Hamburg, Lüneburg, Cologne, Danzig and Riga). Their ships established regular trading routes to Novgorod, bringing back furs and forest products which, along with Lüneburg salt for preserving fish and meat, were distributed throughout northern Europe and exchanged for cloth and other manufactures. This was a new economic power, which reached a peak in the fourteenth century. The German middlemen acquired wealth and political power. Merchant elites governed Lübeck and the other Hanse cities. And Lübeck purchased considerable areas of land in surrounding rural Saxony, controlling well over 200 villages there.

The ferment of long-distance trade also helps to explain the location of the most urbanized zones of all in medieval Europe. Flanders was the arrival point of much international trade from the North Sea, the Baltic and the Mediterranean. Northern Italy was a terminal for the traffic across the great Alpine passes and Mediterranean sea routes. In the case of Italy, increasing urbanization was usually not a matter of building wholly new cities but of developing sites formerly populated in antiquity. Their development in the Middle Ages transformed these cities into world centres of commerce. It had all started with the use of ships by Pisa, Genoa and Venice to transport troops in the crusades. Trading privileges and colonial possessions in the Holy Land soon followed. Genoa gradually acquired a vast trading network stretching

from the Black Sea to southern Spain. Venice formed a considerable empire of islands and fortified capes from the Adriatic to the eastern Mediterranean. As in Lübeck, merchant oligarchies in Italy secured control of urban government and built flamboyant town halls, symbols of urban independence, power and prestige. By the twelfth century, Florence, Siena, Milan and other Italian cities constituted autonomous city states, which dominated their surrounding countryside.

Technology, in the form of mining, generated the birth of a few medieval towns on sites that had previously been unpopulated. Perhaps the best example is Kutná Hora, some eighty kilometres east of Prague. Important silver mines were discovered there in the thirteenth century, and a new city grew up rapidly, second in Bohemia only to Prague. The prominent buildings included a municipal weigh-house (a public building for weighing silver); and all around there were stores of timber and coal-fuel, lead stocks for processing the silver ore, and buildings in which to store the precious silver yield. This was a boom town and the mining brought an influx of immigrant workers – Hungarians and Poles. Several of central Europe's new mining towns were in mountainous areas, such as Annaberg (Saxony). Yet, because of the hectic technological activity, a majority of the population in this part of Saxony, remote and of difficult access, became town-dwellers (Molenda, 1976).

The form of medieval cities had a pronounced vertical emphasis, not merely because of the tall spires of cathedrals pointing to the heavens. With their confining walls and narrow streets, cities were crowded places, and lack of space in the centre could force buildings up to four or five storeys, as in Paris. Genoa was a packed place, squeezed by topography between the mountains and the sea; therefore the citizens (some 100,000 by the fifteenth century) lived in a densely populated environment of tall houses (see Figure 4.28). There was no space here for large squares, just narrow streets. Elsewhere in Italy very tall buildings arose for another reason – the urban feuding of the nobility, as well as their determination to outdo rivals by erecting a taller residence. The face of medieval Florence and of Bologna was powerfully influenced by this social and political rivalry, but nowhere have its traces survived as much as in the Tuscan hill-town of San Gimignano (see Figure 4.29).

The retail shop was one of the products of the new commercial environment of medieval cities. Typically, the shop was combined with the living quarters (just as it had been in ancient cities). In Lübeck's merchant houses there was a ground floor for business, a middle floor for living accommodation and an attic for storage. In Rome, money-changers operated in the Via del Banchi ('road of the counters'), named after the tables of the dealers' stalls – from which source comes the English word 'bank'. In Florence and Paris, the money-changers had shops on the city bridges. Medieval bridges had more functions than the mere communication or fortification of later periods. Bridges were provided with shops and houses (as can still be seen on the Ponte Vecchio in Florence), chapels, and water-mills for grinding corn; in twelfth-century Paris, the bridge was even the site for some university lectures. Cities became the principal sources of education. In the Italian cities, elementary schools were founded to teach commercial arithmetic. And everywhere in the medieval West, universities were founded, bringing new and sometimes magnificent building types to Salamanca, Valladolid, Padua, Oxford, Cambridge and Prague. The growth of twelfth-century Paris was partly due to the foundation and expansion of its university on the left bank of the Seine, and to the creation of numerous monasteries.

One of the most characteristic features of the medieval city was the town wall – so much so, that it was frequently the symbol on a town's seal. It was not quite ubiquitous – most of the towns of the Tyrol were without walls – but

Figure 4.28 Crammed
between mountains and the sea,
Genoa was a city dense with tall
buildings, as this fifteenth-
century painting shows
(Fossier, 1997, p.175)

Figure 4.29 Medieval fortified
dwellings of the nobility of San
Gimignano, Tuscany, mostly
thirteenth century
(Waley, 1969, pp.180–81;
photograph: courtesy of
Weidenfeld Archive)

the majority of medieval citizens lived in a place that was felt to be special because of its walled boundary. Indeed, the community itself was bound together by the labour (it was a compulsory obligation) expended in the erection and maintenance of these stone walls. In Italy, a large part of the budget of the commune (the principal jurisdictional authority in self-governing medieval cities) was spent on this public work, which served for both defence and the control of tolls on goods entering and leaving the city.

4.4 The urban stimulus to medieval technology

New building techniques

The town much more than the countryside stimulated the development of building techniques. Building upwards in crowded areas called for advanced techniques in construction and the provision of a sound framework. At first timber was the principal building material of northern Europe, so plentifully supplied with forests. In a temperate or cold climate, timber (if thick enough) had the advantage of good thermal insulation. It was also easy to work with simple tools, and even easier when, from the eleventh century, iron saws and chisels became increasingly available as a result of the expanding iron output of medieval mines. And by the early thirteenth century, water-powered saws made woodworking more efficient. One result of these developments was the widespread introduction of the mortise-and-tenon carpentry-joint (already used by the ancient Romans in the building of their ships). That joint strengthened the wooden framework of domestic architecture, permitting the construction of the multi-storey house, as well as the upper projection that sheltered the lower floors from the weather and provided extra floor space in the crowded cities (Bechmann, 1989). But wooden cities were vulnerable to fire. Conflagrations destroyed much of London in 1136, destroyed Rouen six times in 1200–25, and in Lübeck, too, in 1276, the wooden houses and shops went up in smoke. In each case the result was the same: a turning away from timber to stone and brick. And for greater protection, a widespread municipal practice was adopted of ringing a bell in the evening to order householders to cover all open fires – *couvre-feu*, as it was called in France, whence the English word 'curfew'.

In much of southern Europe, the more readily available stone was the common building material. Better suited to a hotter climate, stone houses were cooler. Again the medieval harnessing of water power produced a saw that could cut some types of stone rapidly and more efficiently than a chisel. Granite was in use in medieval masonry construction, but would have been too difficult to work by saw. But soft sandstone and limestone, common building materials of the age, would have been easily workable by this technique. Those tall domestic fortresses of Italy's urban nobility were built of stone. Traces of the technique used to erect them are still clearly visible in the regular array of apertures scattered over their pitted fabric (see Figure 4.30 overleaf). These were the progressively higher slots to take the horizontal scaffolding used by masons on the rising structure (see Figure 4.31, p.149). But the greatest advances in masonry building occurred in the great cathedrals. Stability here was achieved only by a revolution in building techniques. To support the enormous outward thrusts of these massive stone structures, the flying buttress was introduced in Paris in the late twelfth century for the Notre-Dame and for the still taller cathedrals of Chartres and Bourges. An unprecedented, breathtaking structure arose: a skeleton of stone with walls pierced by large areas of stained glass.

Figure 4.30 Medieval fortified dwellings, Pavia, Lombardy, showing apertures for builders' scaffolding (Waley, 1969, p.174; photograph: courtesy of Weidenfeld Archive)

Figure 4.31 Building labourers at work on the construction of a tower, showing scaffolding in use; detail from Ambrogio Lorenzetti, *Allegory of Good and Bad Government*, Palazzo Pubblico, Siena (photograph: Lensini)

The transport of huge quantities of stone from quarries to building sites was expensive and, in an age of primitive roads and road transport, difficult. As in classical antiquity, water transport was used wherever possible. And any explanation of the rise of Gothic cathedrals should consider that their sites in northern France, the Netherlands and England were close to navigable rivers that flowed near quarries of high-quality stone (Bechmann, 1989, p.562). For the great Gothic cathedral of Milan, an irrigation canal was used to transport stone from the distant quarry to the central city site. The final stretch was uphill and, at the end of the fifteenth century, occasioned Leonardo da Vinci's invention of a special lock to adjust the water-level.

Advances in hydraulic engineering

From the early twelfth century, the Italian city of Pistoia directed much of its efforts to diverting the course of local rivers that were bringing down torrents of debris and mud from the mountains; soon its surrounding plain became a productive agricultural area instead of a quagmire. In the next century, the French Alpine city of Grenoble, threatened by floods from the River Drac, commissioned experts to investigate ways of altering the river-bed. They used a series of wooden, watertight enclosures filled with stones to form an embankment, and the city was saved. It was all part of the hydraulic engineering that was one of the great growth areas of medieval technology from the eleventh century. Everywhere in the medieval West, marshes were drained, rivers diverted, and canals dug to drive the host of urban water-mills. Those mills not only ground flour but were now provided with transmission gears (cams) which converted rotatory motion to the up-and-down movements needed to operate hammers for a variety of industrial processes, such as the fulling stage (pounding cloth to clean it and remove gaps between the fibres) in textile-manufacture. In the words of one historian of these developments, 'water was the economic nerve centre of preindustrial urbanization; without water, there would have been neither millers nor weavers, neither dyers nor tanners, nor would communities have existed' (Goubert, 1988, p.52). According to Goubert, there were two distinct phases. First, in the eleventh century, a network of new canals was constructed and water-mills installed. At Caen (in Normandy), where William the Conqueror created a new city and port by an impressive drainage scheme, the labour of canal-digging was facilitated by the local, soft, peat soil. In Beauvais, where the subsoil was chalk, the excavation was harder, and required 'extremely good knowledge of levelling and surveying', but the labour was reduced by unearthing the underlying ancient Roman streets, lowering the road-bed, and using it as a canal (Goubert, 1988, pp.56, 58–9). In this first phase of development the aim was to provide flour mills, which Goubert sees as unimportant for urbanization, stimulating 'only a slight increase in employment' (Goubert, 1988, p.59). The much bigger change occurred in the subsequent phase, when a variety of artisans settled in towns, attracted by the new resources of stationary water. Several of the processes of textile-manufacture called for still rather than running water: the degreasing of wool by immersion in water baths; the dyeing process, which again required the cloth to be immersed – first in boiling water containing the mordant, which fixed the dye, and then into a bath of the dye; and tanners, too, relied on standing water to prepare leather.

For the Tuscan hill city of Siena, artificial water supply was a matter of survival in an environment devoid of streams and lakes. Water was needed, in much larger quantities than for drinking, to maintain the local industries of wool and leather. From at least the thirteenth century, Siena's water supply became an urgent government responsibility. The commune offered subsidies and supplies of limestone to encourage citizens to build wells, and it commissioned engineers such as Lando di Pietro to provide fountains, in addition to his other contracts to work on city defences, streets and bridges. A hectic spate of difficult projects on Siena's aqueducts and fountains culminated in the successful conduction of water up to the height of the city centre, where it issued from an elaborate fountain – a masterpiece of medieval sculpture, which embellished the city as well as supplying it with an essential resource.

Siena's commune proclaimed severe penalties for pollution of the water-supply system. It was a problem felt in many cities. In medieval London, attempts were made to control the disposal of butchers' waste: entrails were still allowed to be dumped in the Thames, though at a point judged to be innocuous. In fourteenth-century Paris, butchers were moved out of the city centre, but only to slaughterhouses on other parts of the Seine, which they polluted with their waste materials. And everywhere in France, cities had streets which functioned as public toilets, the excrement seeping into the water table and so into the drinking-water. Archaeologists have begun to take an interest in medieval urban pollution. Samples of fourteenth-century sediment have been taken from up to a depth of ten metres from London's River Fleet. They show a high concentration of a diatom (unicellular algae) usually associated with polluted water, modern evidence to support Edward III's repeated calls to the Mayor of London to clean up that area of the city (Schofield and Vince, 1994, p.179).

Developments in transport

Notable progress in transport technology – intimately connected with the rise of medieval cities – occurred from the twelfth century. Much light on this has come from research on Pisa (Herlihy, 1958).[4]

Pisa's flourishing commerce was the main stimulus for improving the city's communications. As in the rest of Europe, the state of Italian roads made land

---

[4]   See the text by David Herlihy in the Reader associated with this volume (Chant, 1999).

transport slow, expensive and subject to total disruption by heavy rain. The flat terrain surrounding Pisa was marshy and impassable for half the year, forcing travellers by land to make wide diversions up into the mountains and down again, even to reach neighbouring villages.

The main technical difficulty in constructing Pisa's new roads was the problem of drainage. Gradually this was solved by the widespread introduction of techniques that had been used in antiquity – elevated roads, also functioning as dikes, and flint paving. By the late thirteenth century, Pisa's once swampy valleys were traversed by wide, all-weather roads that carried increased cart traffic, a transport system dependent on the construction of three new bridges. These technological developments were facilitated by the ruthless expropriation of resources and draconian organization of the labour force, measures implemented by the city's tyrannical government (Herlihy, 1958, pp.90–105).

One of Pisa's aims was to colonize and control its rural surroundings, partly for economic gain. Those aspirations for territorial expansion were foiled in the long term by the rising power of Florence, just sixty-five kilometres further up the River Arno, on which Pisa stood. (In the early fifteenth century Pisa would be conquered and made part of an expanding Florentine state.) But in the twelfth and thirteenth centuries, Pisa was the main outlet for Florence's wool trade – the source of that city's opulence and power. The Arno tended to dry up, so building an all-weather road between Pisa and Florence was important. Pisa's harbour, badly affected by silting, was totally inadequate to serve as the port of Florence. Instead, Porto Pisano, some eight or nine kilometres south of the mouth of the Arno, became the main port for Pisa and for the whole of Tuscany. This in turn made it necessary to build a good road to the port from Pisa.

The port was the centre of Pisa's own considerable maritime trading network. Pisa had first become prominent in the eleventh century as a centre for piracy. This aggressive naval action was the response to Arab raids up the Arno and the destruction of Pisa in 1004. Pisan ships were soon participating in revenge attacks on Muslim cities in North Africa, some of the booty being used to embellish the new cathedral – the nucleus of what was to be one of Europe's outstanding architectural complexes, including the famous campanile, or leaning tower, and the imposing circular baptistery. Pisa's fleets conquered the neighbouring islands of Corsica and Sardinia, participated in the crusades and, with her competitors, Genoa and Venice, gained control of the Mediterranean sea routes, acquiring an important trading post in Alexandria in 1173. Pisa's naval power and associated overseas commercial empire were destroyed in a naval battle with Genoa in 1284. But before that, Pisa was a flourishing Mediterranean power with numerous wharves, a thriving shipbuilding industry, and an animated port. And a canal and locks facilitated the transport of massive marble by barges, for delivery to the site of the cathedral complex.

For much of the Middle Ages, the Rhine remained the main highway of Europe; traffic proceeded more surely on rivers than on the generally poor roads. But then, from the twelfth century, land transport began to improve: along with the gradual improvement of the surface of medieval roads there were important innovations in the design of horse-drawn vehicles. The horse-collar, first introduced in Europe around the ninth century and widely adopted by the twelfth, enabled a horse to pull a cart loaded with several passengers, to back up, and also to brake on a downhill journey. The shaft-cart, already invented by late antiquity, was widely adopted in the early Middle Ages. Greater power came from the practice of harnessing several horses in tandem. And greater manoeuvrability was achieved by the invention (by the twelfth century) of the swingletree – a wooden bar, connected at its centre to the front of the vehicle and hooked at its ends to the straps fixed to the horse-collar. During the fourteenth century, a suspension system made land travel more comfortable for the elite: luxury four-wheeled chariots were designed so that the carriages hung from chains or leather straps.

Figure 4.32 Reconstruction of a thirteenth-century Italian merchantman (Landström, 1961, p.87)

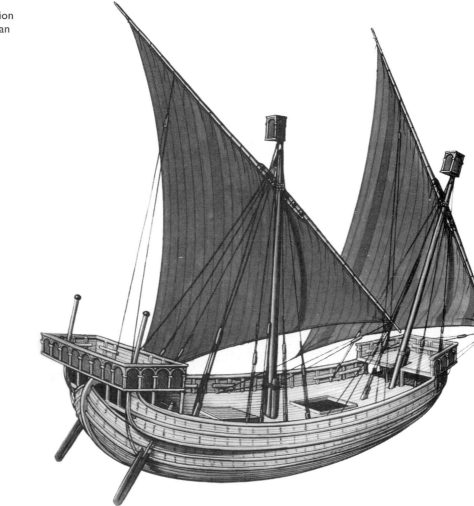

Expanding long-distance commerce stimulated the development of port facilities and modifications to ships – urban products. The Italian city republics of Pisa, Genoa and Venice, which dominated the trading routes to Africa, the Levant and the Black Sea, developed two types of craft. An enlarged oared galley served as a warship as well as for the transport of small-volume, high-cost luxury goods, such as the sought-after Eastern spices. These galleys were built in large numbers at Venice's arsenal. But the carrying of cheaper bulk cargoes, such as grain and wine, called for more capacious hulls, and so the Italian republics produced their characteristic roundships. There was nothing completely new about their triangular lateen sails, edge-to-edge planking or beamy form.[5] But the exaggerated curvature of the hull gave these ships a distinctive shape (see Figure 4.32) and made them effective, but slow, cargo-carriers. Some were large – with up to three decks. These were the vessels used to transport crusaders and horses to the Holy Land, and for the increasingly lucrative trade in conveying pilgrims to the same destination. The Mediterranean is a tideless sea. Lack of tides meant that arriving vessels had to be hauled out of the water and beached before cargoes could be unloaded. This was a laborious, time-consuming operation, unacceptable to the profit-conscious Italian maritime

[5] 'Lateen' is a corruption of 'Latin', the adjective used to describe these sails by medieval northern Europeans who first saw them in the Mediterranean (see Chapter 3, Section 3.1). Of triangular shape and suspended at an angle of about forty-five degrees to the mast, the lateen sail has the great advantage of being able to be moved according to the wind direction. Its origin is uncertain – some attribute the invention to the eastern Mediterranean, others to Arabia – but its antiquity is clear from its representation in an illustration of the second century CE. The lateen sail is still to be seen today on craft in parts of the Mediterranean, upper Nile, Red Sea, Persian Gulf and Indian Ocean.

Figure 4.33 The Hanse cog
(Landström, 1961, p.71)

republics with their increasing frequency of sailings and soaring quantities of merchandise in their ships' holds. So wharves, not employed since Roman times, reappeared in medieval Pisa and Genoa, facilitating the loading and unloading of cargoes.

A particularly frustrating aspect of maritime history is the impossibility of knowing when and where new ship types, new types of sail, rudders and other fundamental parts of a vessel were first introduced. All of these things remain obscure and elusive. Nevertheless, it is established that by the twelfth century the shipbuilders of the Hanse ports developed an emerging new carrier, the cog, which became the characteristic vessel of the northern merchants plying the Baltic and the North Sea. It was slow and hardly elegant, but it was sturdy and its hull of straight lines and a flat bottom made it a capacious and economic bulk carrier of grain and barrels of wine. By the thirteenth century it was fitted with a sternpost rudder and bowsprit (a large wooden support, projecting over the bows and serving to fasten the rigging of the foretopmast), both features facilitating the handling of the ship. Castles were added for defence (see Figure 4.33). The cog stood low in the water, favouring port and urban development at the mouths of rivers rather than the shallower inland ports. The cog was a natural partner to the wooden wharves that had also been sprouting in northern ports in response to the surging new trade. Wharves made it possible to cope with the loading and unloading of ships, now larger and sailing much more frequently. In 1368, as many as 700 ships sailed in and out of Lübeck's harbour, some of them repeatedly. Barrels of wine were rolled off the deck along a plank on to the wharf. And from the thirteenth century, the urban environment of several ports of the Hanse trading

Figure 4.34 Medieval crane, Bruges (Bayerische Staatsbibliothek, Munich, Cod. Lat. 23638, folio 11v.)

network was altered by the appearance of giant cranes for lifting heavy merchandise on to or off ships. There was one in Bruges (see Figure 4.34) operated by a treadmill, a hoisting technique already well established in ancient building practice (see Chapter 3, pp.94–5). But the huge, medieval crane of Danzig (which survived right up to its destruction in the Second World War) was water-powered, and that had no precedent in antiquity.

Clocks and books

Everywhere in the medieval world, life and work were regulated by the seasons and the rising and setting of the sun. Nature set the pace of life – the crowing cock signalled that it was time to rise. The times of prayer of monks, priests and their flocks were similarly determined by the sun and seasons: matins, vespers and the other 'canonical hours' divided up the day, to the sound of bells. But these were not invariable hours; as in ancient Rome, the changing length of daylight was divided into twelve, similarly the night. The length of the monastic hour therefore varied considerably throughout the year. Yet for the early and central medieval centuries this was the variable unit of time throughout society.

Then came the great change in time measurement which the leading French medievalist, Jacques Le Goff, saw as the most profound of all influences of urban culture on the medieval West (Le Goff, 1972, pp.86–7). It was in the commercial cities that rising demand for a more precise determination of time first occurred. In the world of business it has become a cliché to say that 'time is money'. But in the commercial revival of the medieval West such expressions were a novelty, revealing a changing mental attitude among urban merchants. The great fourteenth-century Tuscan banker and entrepreneur Francisco Datini, the founder of an extensive trading company with branches in Pisa, Florence, Genoa, Barcelona, Valencia and agencies throughout north-west Europe, spoke from the experience of competitive success when he advised that 'he who knows how to spend his time better will outstrip the others' (quoted in Gurevich, 1990, p.279). A more exact measure of hours was felt to be particularly necessary for the regulation of labour in textile-manufacturing, the principal industry of the medieval city, and one in which the merchant usually supplied the materials and employed the workers.

This demand was satisfied by the invention, probably in the late thirteenth century, of the mechanical clock, perhaps the most impressive of medieval technological achievements. Now the day was divided into twenty-four invariable hours, signalled by the regular chimes of the new public clocks erected on towers in the centres of towns. In 1355, a new belfry was erected in Aire-sur-la-Lys, a town (south-west of Calais) then said to be 'governed by the cloth trade' and in need of a mechanical clock to ensure that the workers 'come and go to their work at fixed hours'. For Le Goff, this communal clock represented 'an instrument for the merchants' economic, social and political domination of the commune' (Le Goff, 1960, p.425). In Beauvais (another cloth town of northern France), in 1390, a dispute occurred with the guild of weavers over the length of the working day and the timing of breaks. It was settled by the Parliament of Paris, which ordered the installation of a mechanical clock to establish reliable time and fix the hours to be observed by the workers (van Rossum, 1987, p.40).

The spread of the mechanical clock was pronounced: recent research has shown that by 1380 every city of the medieval West with 10,000 or more inhabitants had a public clock; by 1450 there were 460 recorded public clocks (van Rossum, 1987, pp.32, 38; see Figure 4.35). From now on, Genoese notaries recorded the hour as well as the date of the preparation of their documents. And, in general, historians have seen the coming of the mechanical clock as setting a new framework for city life, as the decisive step in 'the evolution of the machine-like rhythm of collective life in modern society' (van Rossum, 1987, p.29). In the towns at least, ecclesiastical time was no longer dominant; it had been replaced by lay, municipal, merchant, technological time. This established a sharp divide between the town, whose commercial environment had led to the generation of the mechanical clock, and the countryside, where the older practices continued.

Towns vied with each other in the installation of the new machinery. Although quarter-hours were chimed, clocks were not yet accurate. Yet they had become objects of urban pride and an addition to the urban environment, sometimes displaying arresting movements of automata at the striking of hours.

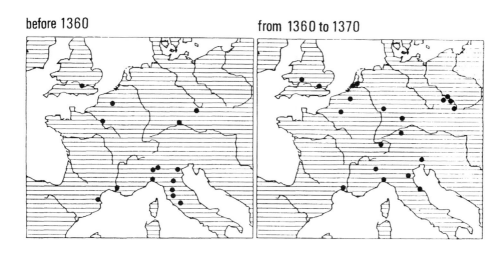

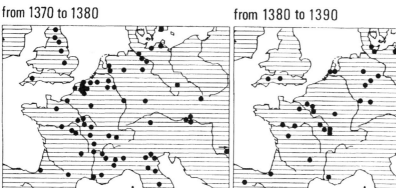

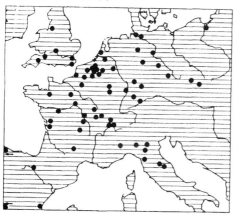

Figure 4.35 The diffusion of public clocks in medieval European cities (van Rossum, 1987, p.33)

These clocks – made of iron and bronze – were the products also of an intensifying concentration of urban metal-workshops. Nuremberg became one of the great centres of clock-manufacturing. All the conditions favoured it. The city was situated at the nexus of a well-developed road system, with good land communications to the iron, copper and silver mines of the neighbouring Upper Palatinate region, Saxony, and Bohemia. It was Europe's main producer of metal objects, a prime centre for the manufacture and wide distribution of armour, cannon, swords, church bells, household pots and pans, and luxury silverware for the dining-table. By the late fifteenth century, the mechanical clock was added to this long list of Nuremberg's metallurgical output.

At the close of the Middle Ages there occurred another technological innovation in metallurgy, which for ever after would have an essential role in modern society: printing. In the city of Mainz, on the Rhine, in the 1430s, Johann Gutenberg, who came from a family of goldsmiths, began a series of experiments that culminated around 1453 in the production of the first book to be printed with movable metal type. The type was made by pouring molten lead into moulds cut into the shapes of letters of the alphabet, in forms then current in the handwriting of urban scriveners. The higher literacy of cities had created a demand that could not be met by copyists of manuscripts. Printing had originated in an urban environment and from Mainz it spread with remarkable rapidity in all directions to other cities: Cologne (1465); Rome (1467); Venice (1469); Nuremberg (1470); Paris (1470); Florence (1471); Naples (1471); Zaragoza (1473); Seville (1476); Barcelona (1480); London (1480) – the list can easily be extended. This was urban culture and it encouraged a still higher literacy. Entrepreneur printer-publishers and scholars congregated in the hectic atmosphere of the printing-houses. Venice soon became the most important of Europe's printing centres, using its developed commercial network to export books far and wide, by this time chiefly Latin and Greek classics. This was now a different age: the Renaissance.

4.5 Town and country

In view of the outstanding characteristics of medieval cities as centres of trade, technological innovation and political development, and taking into account the normal physical separation of cities from their surrounding countryside by that ring of walls which gave citizens a shared space and a sense of common identity, should we draw a sharp dividing-line between the medieval urban and rural worlds?

That conclusion is firmly rejected in Elizabeth Ewan's paper on medieval Aberdeen, in which she draws on archaeological evidence.[6] She finds that Aberdeen was, in various ways, inextricably linked to its countryside. Its citizens depended on the countryside for their food supplies, building materials, peat-fuel and raw materials for a variety of urban crafts. The city's population was maintained only by the immigration of rural folk. The citizens exploited the resources of the countryside through royal grants of economic privileges and by investment in croftlands. And some of the city's craft specialists – tanners and perhaps potters – normally operated on the outskirts and beyond. Residents of town and country were joined by bonds of family or friendship, just as they were treated indistinguishably when received by the religious houses that cared for the sick and the poor.

Scottish medieval towns were unusual in having no surrounding walls. But Ewan suggests that, even where such walls existed, they were no barrier to the type of intimate relationship between town and country that had developed in Aberdeen. However, she exaggerates in alleging that historians have only just

[6]   See the extracts reproduced in the Reader associated with this volume (Chant, 1999).

begun to appreciate this symbiosis of walled cities and their rural surrounds (Ewan, 1992) – it has been well known for decades.

According to Le Goff, the real distinction in the medieval world was not between 'town' and 'country' but between the inhabited, cultivated, ordered world of 'town–countryside' and 'wilderness' – uncultivated and largely uninhabited – such as the forests. Town and country were 'a pair that formed a single whole' (Le Goff, 1972, pp.72, 93). There is ample evidence to support this view. Medieval towns, with their poor hygiene and higher mortality than the countryside, could not be sustained by the natural increase of their populations; towns survived and grew only because of the influx of rural immigrants seeking greater opportunities for employment. The majority of a town's inhabitants were peasants; rural folklore became an integral part of the town's culture. And rural aristocracy also moved into the city, as in eleventh-century Barcelona, where they maintained houses and secured for their relatives positions as urban clergy. Cities – Florence, Venice and Nuremberg are typical examples – annexed large areas of their surrounding countryside to guarantee food supplies, as well as to exert growing political and economic power. Venice drew on a rural hinterland within a radius of about fifty kilometres to satisfy its basic food requirements. Siena's government in the early fourteenth century strove to provide sufficient food for its citizens first by building urban fishponds, and when, by the 1460s, that measure proved inadequate, undertook an elaborate project to create an artificial lake in the countryside by the construction of a masonry dam on the River Bruna. Europe's flourishing textile industry, the source of much urban wealth and political power, similarly displays the inseparability of town and countryside. In Florence, one of Europe's principal producers of fine cloth, the textile industry, in its various stages of processing, was scattered over an extensive area stretching from the city centre to the surrounding countryside. The raw material – English and Spanish fleeces – arrived in the Florentine entrepreneur's urban shop, where it was sorted into grades, followed by a series of preliminary processes: washing in the city river, drying and carding. The wool was then sent to the countryside to be spun by women who were paid a pittance. The resulting yarn was sent back to the city, where it fed weavers' looms in urban shops and private dwellings. The woven cloth was fulled in mills on outlying streams. The brilliant dyes were added at workshops on Florence's Corso dei Tintori ('dyers' street'). And the city was also the site of the finishing processes of stretching and shearing. The cloth, a luxury product, was then ready to be packed and sealed in the central office, for export throughout Europe and the Levant. Its production had depended on the integration of work in the town and its countryside – an inseparable industrial complex.

Yet, despite this symbiosis of town and countryside, there was also something which really did distinguish the city of the medieval West as a separate entity – its legal status. The fact was that a citizen enjoyed legal privileges denied to the peasants in the surrounding countryside. Within the city walls there was equality as well as freedom from serfdom and other subjections imposed on the countryside by princes and feudal magnates. 'Town air makes free' was a dictum proudly proclaimed by medieval cities, especially the new towns whose foundation was often accompanied by royal grants of such legal privileges. A very short period of residence by a rural immigrant was enough to obliterate the shackles of bondage and servitude. That did not apply to Russian towns, however, where there were no such privileges to distinguish citizens from peasants, and where a sizeable section of the population consisted of serfs or even slaves. Despite the prosperity of the trading city of Novgorod, it never became politically autonomous and had no protecting urban law. Nor was this urban freedom possessed by other medieval cities such as Muslim Córdoba and Baghdad, Constantinople, Delhi or Beijing. Only in the medieval West did cities acquire this special legal status, which more than anything else distinguishes them from the cities of both the ancient and the modern worlds.

The evidence surveyed in this chapter again points to a strong and intimate connection between cities and technology. Wherever one looks in the medieval world, at the rising cities of the Latin West or those of Islam, one finds that building techniques, transport technology and hydraulic engineering have all been stimulated by urban genesis and growth.

Extract

4.1   Bulliet, R.W. (1975) *The Camel and the Wheel*, Cambridge, Mass., Harvard University Press, pp.224–9

Whoever has attempted to characterize medieval Middle Eastern and North African cities has sooner or later commented upon the narrow streets, the blind corners, the encroachment of buildings upon the public way, and in general upon the labyrinthine quality that strikes so forcibly the Western visitor. Many scholars have attributed this quality in some way to the Islamic religion and have implied that it is a universal feature of Islamic cities. None has seen it as a characteristic of a society without wheels.

A rectilinear layout with streets of uniform width lined with buildings of similar height and design has until recently represented in Western thinking good order and intelligence of design. Even individuals not normally given to admiring this type of design are apt to feel the adjective 'oriental' come to mind when confronted with a city composed of winding streets and narrow alleys. Dirt, darkness, and crowding are thought of as inevitable and evil conditions of this 'oriental' type of city, while parallel Western conditions such as motorized danger and the isolating quality of broad avenues are taken in stride. Islamic society is often described as turning its back on the street, hiding inside walled courtyards and behind windowless walls to shut out the cheerlessness and contagion of the public way. Private life is supposedly given precedence over communal life, which is seen by many to be deficient in the Islamic religion. To an earlier generation such streets might be regarded as visible evidence of the supposed inadequacy of Islam.

This entire conception of the 'oriental' city plan being generated by Islamic social principles runs counter both to logic and to fact. A great many factors, some deliberately planned and others unconscious and incremental, come into play in the development of a particular urban environment. Religious principles undoubtedly have their place among them. When it comes to the layout of streets, however, what cries out most for explanation is the rigid application of abstract geometrical forms. A particular shape or compass orientation might be dictated by religious belief, astrology, legal principle, or the caprice of a ruler; but whatever the motive, it is reasonable to expect it to be ascertainable. The same cannot be said of a city whose streets are not laid out in a formal pattern but according to the lay of the land and the inclination of the builder. Disorder requires explanation only if order is taken to be normative.

Since from the Western viewpoint regular patterns in urban topography are generally considered to be good and the absence of such patterns bad, it is not difficult to understand why the feeling arose that the 'oriental' city plan had to have an explanation, particularly in view of the fact that Roman cities throughout the Middle East and North Africa are known to have exhibited a uniform rectilinearity. But if narrow, winding streets are not inherently bad, the rationale for seeking an explanation for this alleged falling away from perfection evaporates. And in fact, narrow, winding streets have much to recommend them. They easily follow the lay of the land; in hot countries they provide shade; they diminish winds; they permit a higher density of habitation which in turn makes a sizable city accessible to pedestrians; they facilitate social relationships; and they are easily defensible. As for enclosed courtyards with windowless exteriors, they provide secluded open spaces where many household tasks are carried out, as is desirable in a warm climate; and they allow for careful regulation of water consumption, as is desirable in an arid climate …

Since, then, the transition from Roman rectilinearity to medieval disorder was not necessarily inherently bad, the need to explain it in moral or ideological terms is greatly diminished and the way is open for a more prosaic physical explanation. Wheeled vehicles – and this can come as no surprise to today's city dwellers – are inflexible in the restraints they put on city life. Streets must be flat, without stairsteps or precipitous grades, and, if possible, paved. Moreover, they must be maintained in this state if circulation is not to be interrupted. They must always be as wide as a single axle – as wide as two if the citizens are to be spared immoderate language. Corners must not be too sharp or narrow to be maneuvered; dead ends must be eschewed. Encroachments on the public way either by buildings or by merchants displaying goods cannot be tolerated. And on top of all of these burdens is the fact that wheeled vehicles are noisy and dangerous.

Freed from this vehicular straitjacket by the disappearance of the wheel, it is scarcely a matter for wonder that Middle Eastern and North African cities gradually evolved types and arrangements of streets suited more closely to human needs. With only pedestrians and pack animals to accommodate, the street could become an open market or a narrow cul-de-sac giving access to residences. In the absence of any ideological sanction of constant widths and right angled turns, only enough legislation was needed to keep the streets passable … planning was not the rule because without wheeled vehicles the necessity for plans was negligible. As late as 1845 the width of a major new street in Cairo was determined by measuring the combined width of two loaded camels.

… Islam came into being in a society that had recently abandoned the wheel, and hence it incorporated in its growth no ideological bias in favor of vehicular traffic. The evolution from a geometric to an organic urban design within the zone of the wheel's disappearance followed as a natural consequence. Outside that zone other, equally Islamic – or, rather, equally non-Islamic – urban patterns arose. The dispersed cities of Indonesia made up of discrete, villagelike *kampongs*; the precise rectilinear design of Jaipur in India; and the striking 'skyscraper' cities of southern Arabia illustrate the variety of urban design to be found in Islamic lands and testify to the irrelevancy of Islamic religious principles in this domain. It is the absence of the wheel that goes furthest toward explaining this characteristic feature of Middle Eastern and North African urban environments …

Camels, donkeys, and pedestrians do not need paved roads. Given that throughout the zone of the wheel's disappearance the climate is dry during most of the year, it is more comfortable to walk on dirt. Furthermore, natural obstacles, such as boulders, do not have to be removed to provide for a constant minimum width, nor do ruts have to be filled in. Cost of maintenance is as negligible as cost of construction. In a non-vehicular economy the most important physical features of a road are its bridges. One bridge in place of a ford or ferry can make an enormous difference in the ease and cost of transportation. After bridges, the most important features are accommodations for travelers. A regular daily stage of travel for a caravan does not exceed twenty miles [thirty-two kilometres], and a good road will afford a stopping place at the end of every stage, whether it be a town, a village, or a caravanserai. Beyond these two things, bridges and caravanserais, the physical upkeep of roads is insignificant; but bridges and caravanserais themselves can be very costly.

The reflection of this state of affairs is everywhere apparent in the history of the Islamic Middle East. References to the upkeep of roads are almost nonexistent, but powerful dynasties frequently show their interest in promoting trade by building bridges and caravanserais. Investment in these two things is functionally equivalent to roadbuilding in a wheelless society. There is no need to search for an ideological explanation for a nonexistent neglect of public ways. Middle Eastern governments acted with complete rationality in investing in bridges and caravanserais instead of in useless grading and paving …

Whether attitudinal or physical, the effects of the disappearance of the wheel as analyzed above imply the operation of a principle that should be more clearly stated. This is the principle that the state of technology and of the economy is the prime determinant of people's attitudes and actions, at least within the sphere of transportation. The counterargument to this has already been noted in the discussion of urban topography. According to this counterargument Arabs, or Muslims in general, can be viewed as constructing about them a certain type of world wherever they go; the transport economy can either be seen as a function of a positive ideological viewpoint, a pro-camelline bias, or a physical consequence of negative viewpoints such as absence of communal concern for roads and urban orderliness.

References

AL-SAYYAD, N. (1991) *Cities and Caliphs: on the genesis of Arab Muslim urbanism*, Westport, Conn., Greenwood Press.

BARRACLOUGH, G. (ed.) (1993) *The Times Atlas of World History*, London, HarperCollins.

BARTLETT, R. (1993) *The Making of Europe: conquest, colonization and cultural change 950–1350*, London, Allen Lane.

BECHMANN, R. (1989) 'Construction: building materials' in J.R. Strayer (ed.) *Dictionary of the Middle Ages*, New York, Charles Scribner's Sons, vol.3, pp.558–64.

BOURAS, C. (1981) 'City and village: urban design and architecture', *Jahrbuch der Osterreichischen Byzantinistik*, vol.31, pp.611–53.

BULLIET, R.W. (1975) *The Camel and the Wheel*, Cambridge, Mass., Harvard University Press.

BULLIET, R.W. (1989) 'Vehicles: Islamic' in J.R. Strayer (ed.) *Dictionary of the Middle Ages*, New York, Charles Scribner's Sons, vol.12, pp.379–80.

BURY, J.B. (ed.) (1896–1900) *E. Gibbon: the history of the decline and fall of the Roman Empire [1776–88]*, London, Methuen, vol.6.

CHANT, C. (ed.) (1999) *The Pre-industrial Cities and Technology Reader*, London, Routledge, in association with The Open University.

ECOCHARD, M. and LE COEUR, C. (1942–3) *Les Bains de Damas*, Beirut, Institut Français de Damas, vol.2.

EWAN, E. (1992) 'Town and hinterland in medieval Scotland', *Medieval Europe 1992*, Conference on Medieval Archaeology in Europe, York, University of York, vol.1, pp.113–21.

FOSSIER, R. (1997, rev. edn) *Le Moyen Age*, Paris, Armand Colin, vol. 3 (first published 1983).

GOUBERT, A. (1988) *The Age of Water: the urban environment in the north of France AD 300–1800*, College Station, Texas A. and M. University Press.

GRANT, M. (1974) *Ancient History Atlas*, London, Weidenfeld and Nicolson.

GUREVICH, A. (1990) 'The merchant' in J. Le Goff (ed.) *The Medieval World* (trans. L. Cochrane), London, Collins and Brown, pp.243–83.

HEATON, H. (1948, rev. edn) *Economic History of Europe*, New York, Harper.

HERLIHY, D. (1958) *Pisa in the Early Renaissance: a study of urban growth*, New Haven, Yale University Press.

HERLIHY, D. (1984) 'Demography' in J.R. Strayer (ed.) *Dictionary of the Middle Ages*, New York, Charles Scribner's Sons, vol.4, pp.136–48.

HILTON, R.H. and SAWYER, P.H. (1963) 'Technical determinism: the stirrup and the plough', *Past and Present*, vol.24, pp.90–100.

HOBLEY, B. (1988) 'Saxon London. Lundenwic and Lundenburgh: two cities rediscovered' in B. Hodges and B. Hobley (eds) *The Rebirth of Towns in the West AD 700–1050*, London, Council for British Archaeology, pp.69–82.

HOURANI, A.H. (1970) 'The Islamic city in the light of recent research' in A.H. Hourani and S.M. Stern (eds) *The Islamic City: a colloquium*, Oxford, Bruno Cassirer, pp.11–24.

JAIRAZBHOY, R.A. (1964) *Art of the Cities of Islam*, New York, Asia Publishing House.

KENNEDY, H. (1985) 'From *Polis* to *Madina*: urban change in late antique and early Islamic Syria', *Past and Present*, vol.106, pp.3–27.

KINDER, H. and HILGEMANN, W. (1974) *The Penguin Atlas of World History*, Harmondsworth, Penguin Books, vol.1.

LANDSTRÖM, B. (1961) *The Ship*, London, Allen and Unwin.

LAPIDUS, I. (1967) *Muslim Cities in the Later Middle Ages*, Cambridge, Mass., Harvard University Press.

LASSNER, J. (1970) 'The caliph's personal domain: the city plan of Baghdad re-examined' in A.H. Hourani and S.M. Stern (eds) *The Islamic City: a colloquium*, Oxford, Bruno Cassirer, pp.103–18.

LE GOFF, J. (1960) 'Temps de l'église et temps du marchand', *Annales: économies, sociétés, civilisations*, vol.15, pp.417–33.

LE GOFF, J. (1972) 'The town as an agent of civilisation 1200–1500' in C.M. Cipolla (ed.) *The Fontana Economic History of Europe*, London, Collins-Fontana, vol.1.

LE GOFF, J. (1988) *Medieval Civilisation 400–1500*, Paris, Éditions Aubier/Flammarion.

LOMBARD, M. (1957) 'L'évolution urbaine pendant le haut moyen age', *Annales: économies, sociétés, civilisations*, vol.12, pp.7–28.

LOPEZ, R.S. (1971) 'Medieval and Renaissance economy and society' in N.F. Cantor (ed.) *Perspectives on the European Past: conversations with historians*, New York, Macmillan, pp.208–27.

MAINSTONE, R. (1969) 'Justinian's church of St Sophia, Istanbul: recent studies of its construction and first partial reconstruction', *Architectural History*, vol.12, pp.39–49.

MAKDISI, G. (1988) 'Schools: Islamic' in J.R. Strayer (ed.) *Dictionary of the Middle Ages*, New York, Charles Scribner's Sons, vol.11, pp.64–9.

MARÇAIS, W. (1961) 'L'islamisme et la vie urbane' in *Articles et conférences*, Paris, pp.59–67 (first published 1928).

MAZLOUM, S. (1936) *L'Ancienne Canalisation d'eau d'Alep*, Damascus, Institut Français de Damas.

MICHON, J.-L. (1980) 'Religious institutions' in R.B. Serjeant (ed.) *The Islamic City*, Paris, UNESCO, pp.13–40.

MOLENDA, D. (1976) 'Mining towns in central-eastern Europe in feudal times', *Acta Poloniae Historica*, vol.34, pp.165–88.

PAVÓN MALDONADO, B. (1990) *Tratado de arquitectura hispanomusulmana*, Madrid, Consejo Superior de Investigaciones Científicas, vol.1.

PEVSNER, N. (1961) *An Outline of European Architecture*, Harmondsworth, Penguin Books.

PIRENNE, H. (1939) *Mohammed and Charlemagne* (trans. B. Miall), London, Allen and Unwin.

PLANHOL, X. DE (1959) *The World of Islam*, Ithaca, NY, Cornell University Press.

POWERS, D.S. (1989) 'Waqf' in J.R. Strayer (ed.) *Dictionary of the Middle Ages*, New York, Charles Scribner's Sons, vol.12, pp.543–4.

RIASANOVSKY, N.V. (1977, 3rd edn) *A History of Russia*, Oxford, Oxford University Press.

SCANLON, G.T. (1970) 'Housing and sanitation: some aspects of medieval Islamic public service' in A.H. Hourani and S.M. Stern (eds) *The Islamic City*, Oxford, Bruno Cassirer, pp.179–94.

SCHOFIELD, J. and VINCE, A. (1994) *Medieval Towns*, Leicester, Leicester University Press.

TORRES BALBÁS, L. (1968, 2nd edn) 'La Edad Media' in A. García y Bellido (ed.) *Resumen histórico del urbanismo en España*, Madrid, Instituto de Estudios de Administración Local, pp.65–170.

TORRES BALBÁS, L. (1985) *Ciudades Hispano-Musulmanas*, Madrid, Instituto Hispano-Arabe de Cultura.

VAN ROSSUM, G. (1987) 'The diffusion of the public clocks in the cities of late medieval Europe 1300–1500' in B. Lepetit and J. Hoock (eds) *La Ville et l'innovation en Europe*, Paris, Éditions de l'EHSS, pp.29–43.

WALEY, D. (1969) *The Italian City-Republics*, London, Weidenfeld and Nicolson.

WARD-PERKINS, B. (1984) *From Classical Antiquity to the Middle Ages: urban public building in northern and central Italy AD 300–850*, Oxford, Oxford University Press.

WARD-PERKINS, B. (1988) 'The towns of northern Italy: rebirth or renewal?' in R. Hodges and B. Hobley (eds) *The Rebirth of Towns in the West AD 700–1050*, London, Council for British Archaeology, pp.16–27.

WARD-PERKINS, B. (1992) 'Roman "continuity"', *Medieval Europe 1992*, Conference on Medieval Archaeology in Europe, York, University of York, vol.1, pp.33–8.

WHEELER, M. (1964) *Roman Art and Architecture*, London, Thames and Hudson.

WHITE, L., Jr (1962) *Medieval Technology and Social Change*, Oxford, Clarendon Press.

Chapter 5: RENAISSANCE CITIES

by David Goodman

5.1 Clarifying the period

The introduction of the term 'Early Modern' to designate a distinctive period in European history is fairly recent; originating in the 1960s, it has come to be widely adopted. It signifies the period from the fifteenth to the late eighteenth century. At its beginning it overlaps with another division of history, much older, and now firmly established: the Renaissance, a term of very wide currency but notoriously problematic, because it has long defied precise definition. Indeed, the whole idea of the Renaissance has been highly controversial, and there are still eminent medievalists who will have nothing to do with it. (See, for example, Fossier, 1986, pp.494–523, where the Renaissance is rejected in order to present the Middle Ages as continuing well into the sixteenth century.) But for historians of art everywhere, 'Renaissance' is indispensable terminology, expressing that astonishing development of painting, sculpture and architecture in Italy from the fourteenth to the sixteenth century. And that is just what the Renaissance meant to fifteenth-century Florentines, conscious that they were living in the midst of a cultural revival in the visual arts and literature, a revival that had brought them out of the alleged darkness of the Middle Ages, ushering in a new age in which the long-lost artistic skills of the Ancients were recovered and even enhanced. It was a sixteenth-century Florentine painter, Giorgio Vasari, who coined the word *rinascita* (Italian for 'rebirth') to express this powerful upsurge of culture, which has ever since been acknowledged by the general recognition of the genius of Leonardo da Vinci, Raphael and Michelangelo. But it is the French word for 'rebirth', *renaissance*, which now has the greater currency. That is due to Jules Michelet, a French historian, who in 1855 extended the sense of rebirth from an Italian cultural development to a historical epoch applicable to the whole of Europe. For him, 'Renaissance' meant the discovery of 'the world and man', the age of Columbus, Galileo and Vesalius, an age filled with a vigorous inspiration which he could not detect in the later Middle Ages.

Michelet's understanding of the Renaissance was in one respect defective: he confined it to the sixteenth century and ignored the precocious developments in Italy in the previous two centuries. That was soon rectified by the Swiss historian Jacob Burckhardt, whose great *Civilization of the Renaissance in Italy* (1878; first published, in German, in 1860) has remained strongly influential in portraying the Renaissance as a historical period centred on Italy of the fourteenth to sixteenth centuries. But still today there is no general agreement on the time-span covered by the Renaissance. Some historians argue for an even earlier beginning, in the thirteenth century (Herlihy, 1986; Holmes, 1986). So there is something of a tug-of-war between medievalists and Renaissance scholars for the territory of the thirteenth, fourteenth and fifteenth centuries. It is generally sensible not to look for a sudden, sharp separation from the Middle Ages. Renaissance scholars frequently recognize that the people they study have a foot in both periods; certainly Leonardo da Vinci is best interpreted in that way. While many Renaissance scholars continue to focus on Italy, others – like Michelet before them – insist that an adequate understanding of the Renaissance requires a much wider geographical compass. The introduction to a collection of interpretative essays by distinguished scholars declared that 'we apply the term Renaissance freely, perhaps more outside Italy than inside that country, to everything that happened between about 1350 and 1600 – politics, science, and of course the fine arts, literature and learning' (Hay *et al.*, 1982, p.8). Similar sentiments are expressed in Porter and Teich (1992, pp.1–2, 68, 70).

Among the compelling reasons for applying 'Renaissance' to urban history in fifteenth-century Italy, there is the new approach to town-planning, the revival of classical architectural forms, and the innovative technology – all of which altered the appearance of Italian cities. Here again Florence, commonly regarded in so many other ways as the cradle of the Renaissance, stands out. (For an explanation of why Florence became the engine of the Italian Renaissance, see Hay, 1989, pp.375ff.) It was a fifteenth-century Florentine, Matteo Palmieri, who extolled the virtues of living in cities, drawing on an ancient quotation from Cicero:

> For there is indeed nothing more acceptable on earth to that great God who rules all the world, than those assemblies and gatherings of men in social bond, which we call cities.
>
> (quoted in Whitfield, 1981, p.84)

Other Florentines were conspicuous in designing ideal cities and applying inventions, both to improve amenities and impart heightened beauty to cities – now seen again, as the ancient Greeks and Romans had once regarded them, as the essential centres for civilized living.

5.2   Building technology in Renaissance Florence

Since the early fourteenth century Florence had been successfully pursuing an aggressive policy of territorial expansion, eventually ruling the whole of Tuscany. It had become one of the great powers in Italy, along with Milan, Venice and the Papal States, each of which had expanded in the same way. Florence's political power derived from a flourishing economy based on the export of woollen and silk cloth, and its role as one of the greatest banking centres in Europe. The Medici, the dynasty who dominated the government of the city for three centuries, had originated as highly successful entrepreneurs in the cloth industry and the owners of an extensive network of banks; eventually they rose to become dukes and grand dukes of Tuscany. The accumulation of wealth stimulated a building boom in Renaissance Florence, while the city's territorial conquests and artistic achievements engendered a deep sense of civic pride. Abundance of money and the burning desire for magnificent display together created a strong demand for public buildings to celebrate and beautify the city, and for private palaces to establish the fame of the most powerful citizens for posterity. Such was the intensity of interest in construction in Renaissance Florence that

> while intellectuals worked out a rationale for building, many more ordinary citizens learned about architecture by participating directly in the management of construction projects, not only in the private sphere but in the public one as well. The characteristic institution that involved them in this activity was the *opera* – the works, or building committee ... The permanent, autonomous building committee was a common institution in Florence ... Members of these committees gained considerable experience in building practice.
>
> (Goldthwaite, 1980, p.90)

Richard Goldthwaite's pioneering study has thrown much new light on the building of Renaissance Florence – the materials, techniques and industrial organization.[1] Florence was unusually well endowed with the essential raw

[1]   Extracts from Goldthwaite's work are reproduced in the Reader associated with this volume (Chant, 1999).

materials needed to satisfy the city's rising demand for building. Along the banks of its River Arno, there were abundant supplies of sand for making mortar, and gravel for foundations. Municipal statutes of the fourteenth and fifteenth centuries gave builders privileged rights to exploit these natural resources, irrespective of private property claims. There was high-quality limestone in the hills surrounding the city, and readily available clay and wood-fuel to make bricks – the most widely used of all materials in the buildings of Florence. The brick-making of medieval Florence had gradually developed into something of an organized industry, and by the early fourteenth century municipal regulations were announced to ensure that there were enough kilns in operation to satisfy urban demand. In this respect Florence was not unique; elsewhere in Italy, cities such as Venice and Parma were introducing similar regulations, and so was that other great centre of brick-production: medieval Holland.

By the fifteenth century, Florence was much more solidly built because of the widespread substitution of brick and stone for the timber structures which had repeatedly been destroyed by medieval conflagrations. Balconies and projections were banned. The streets, too, were improved by carefully laid stone-paving, and kept cleaner by a complex system of sewers. And amenities were notably improved by the revival of interest in a water-supply system, based on the use of elaborate earthenware piping which delivered water to nobles' palaces and some other individual properties, where latrines were also installed. A new concern with domestic plumbing is evident in the allowance, at the design stage of construction, for apertures in building foundations for the reception of pipes.

Florence was distinctive in its techniques for the manufacture of brick and lime. Its numerous kilns were almost unique in their permanence: in other centres, where brick-making was less continuous, it was normal to dismantle the apparatus used for the firings. Florentine kilns were unusual also in serving to produce both brick and lime. Some were occasionally situated within the city itself, in Via delle Fornaci ('street of the kilns'), well away from the areas of denser population because of the smoke and smell. But to avoid this pollution, and for greater proximity to clay and limestone, the kilns were generally located in the countryside beyond the city walls. (For these paragraphs I have drawn heavily on Goldthwaite, 1980.)

As a result of these generations of hectic building, the city of Florence possessed a reservoir of artisans highly proficient in the building crafts. The number of stonemasons and carvers – they catered to the Renaissance taste for classical columns and cornices – is thought by Goldthwaite to have been unequalled in Europe. These skills were nothing new: they signified a recovery of the levels achieved in ancient Rome. The vast building project of the cathedral at Florence drew on these skills to supply the enormous quantities of brickwork and stone for the fabric. But it was the achievement of the crowning feature, the cupola, which revealed that, in their building techniques, the Florentines had surpassed the Romans.

The great dome of Florence

Santa Maria del Fiore, the cathedral of Florence, like most other Italian cathedrals, was begun as a medieval public-building project. The foundation stones were laid in 1294, but progress on the main building was slowed when attention was concentrated on the detached bell-tower, the famous campanile designed by Giotto and completed in the 1350s. Although the city was devastated by the Black Death in 1348 – its population is thought to have been reduced by a half or even three-quarters – the Florentines, foreseeing recovery by immigration from the countryside, modified their original plans and enlarged the nave to such an extent that one of the biggest churches in the world was now projected. It was assumed that the enormous nave would be covered by a huge dome, though at the time no one knew how that was to be achieved.

The solution to this technical problem, one of the milestones in the history of architectural engineering, was provided by Filippo Brunelleschi (1377–1446). Trained in the city as an apprentice goldsmith, he had extraordinary talents for experiments in design and the application of elementary mathematics to artistic space, and this elevated him to a position of central influence and inspiration in Renaissance art. He was one of the pioneers of linear perspective, a geometrical method of representing the three-dimensional world on a two-dimensional surface, so creating the illusion of depth in paintings and, of still wider importance at the time, a powerful artistic means of controlling the spectator's attention when looking at paintings, architecture and urban space. Perspective viewpoints would become a dominant concern of Renaissance architects and their patrons in the urban layout of rebuilt cities such as sixteenth-century Rome.

In 1420 Brunelleschi was appointed to superintend the construction of the cathedral dome. His daunting task was to provide the technology for the erection of a massive and stable superstructure. There was no question here of a rigorous mathematical analysis of the stresses and strains in planned domes: neither Brunelleschi nor anyone else for centuries would have at their disposal the necessary advanced mathematics which involves differentiation and integration. Indeed it is thought highly unlikely that Brunelleschi had any understanding of structural stress in the abstract, scientific sense (Mainstone, 1969–70, pp.125–6). He relied instead on empirical means – the building of small-scale models.

It used to be thought that Brunelleschi's solution was inspired by his observation of that other great domed edifice, the Pantheon of ancient Rome. But now that view is rejected, because the Pantheon dome is a single shell; Brunelleschi's is double, consisting of an inner and an outer dome. Also in sharp contrast to the concrete Pantheon, the principal material of the cathedral dome is brick. Brunelleschi's design has some similarities to earlier and smaller brick domes built in Persia, but it is unclear whether these had any influence on him. The same uncertainty applies to the influence of the brick dome of the mausoleum of the Roman emperor Diocletian at Spalato (Split, Croatia). This too was built without a central supporting structure, the most distinctive feature of Brunelleschi's building. The citizens of Florence were astonished to see the huge cupola rising without scaffolding, buttresses or any other central support – temporary or permanent. Instead Brunelleschi was using a self-supporting masonry system which, on this scale, had never been used before anywhere in the world. The advantages were that the cost of erecting scaffolding and other temporary supporting structures was eliminated, and at the same time building was speeded up by avoiding this preparatory phase. Furthermore, building operations could proceed from the interior, unencumbered by the presence of scaffolding.

The secret lay in his use of large bricks arranged in a pattern of interlocking herring-bone units, secured by quick-drying mortar (see Figure 5.1). The repeating unit of this brickwork consists of two vertical bricks (such as A and B in Figure 5.2) separated by a slightly inclined band of nearly horizontal bricks, their number varying from five (as between A and B) to one (as at the top of the figure), depending on their position in the dome. And so, in effect, the dome was built up by a succession of almost-flat, horizontal arches of diminishing lengths, the constituent arches being divided into short sections defined by two vertical bricks. Each of these short sections could have been completed to form self-supporting courses by 'teams of two or three

Figure 5.1 Herring-bone brickwork in the dome of the cathedral at Florence (Mainstone, 1969–70, plate XIV; photograph: by permission of the Newcomen Society)

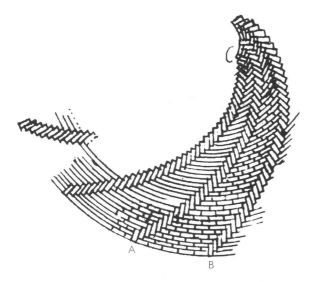

Figure 5.2 Detail of sixteenth-century drawing by Antonio da Sangallo the Younger, showing sectional units of brickwork making up the dome (adapted from Mainstone, 1969–70, p.114; courtesy of the Soprintendenza alle Gallerie)

bricklayers spaced around the dome on light working platforms and holding the cementing bricks in place until they had set' (Mainstone, 1969–70, p.114). The curvature of the rising brickwork dome was controlled by the use of templates, made of pine and lined with lead.

But how could the great quantity of building materials be raised some sixty metres above the cathedral floor to the base of the dome? In addition to masses of large bricks, giant blocks of marble had to be put in place to form the cornice and drains. It was here that Brunelleschi again showed his inventive genius, designing two special machines essential for the completion of the great project. They had to be big. One was a hoist, over seven metres high, with a pulley system operated by the pulling force of oxen. But the machine was required alternately to lift and to lower materials. That could have meant laborious, repeated removal of the yoke from the oxen; and then turning the animals around for each sequence of the operation. It was avoided by Brunelleschi's invention of a screw-actuated reversing clutch – an alternating change of gear (see Figure 5.3). The other invention was a large crane,

Figure 5.3 Modern working-model of Brunelleschi's great hoist (Saalman, 1980; photograph: courtesy of Professor Salvatore di Pasquale, Faculty of Architecture, University of Florence)

Figure 5.4 Modern working-model of Brunelleschi's giant crane (Saalman, 1980; photograph: courtesy of Professor Salvatore di Pasquale, Faculty of Architecture, University of Florence)

operating high up on a loading platform at the base of the dome, and functioning as a load-positioner, carrying material vertically and horizontally to its destined position (see Figures 5.4 and 5.5). (Elsewhere, as at Salisbury, medieval cathedral-building relied on treadmill cranes.) It took sixteen years to complete the dome, which measures over forty-three metres in diameter and thirty-two metres in height. Ever since the 1430s it has continued to dominate the skyline of Florence, and it remains the largest masonry dome in existence (see Figure 5.6 on p.171).

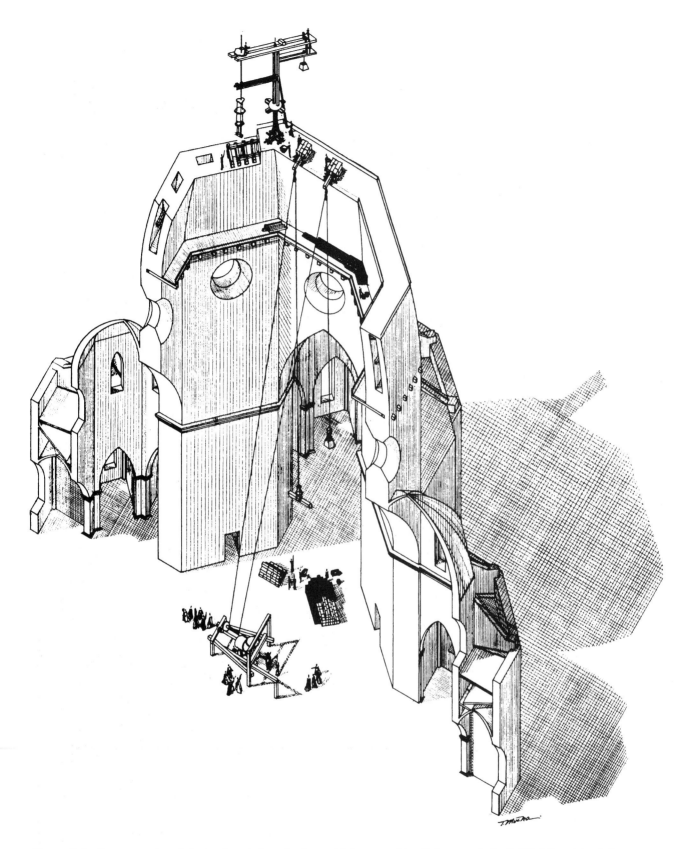

Figure 5.5 Reconstruction of the loading platform in place, with (at ground-level) the great hoist (1420–21) and (top) the great crane (1423) in operation (Saalman, 1980; reproduced by kind permission of Philip Wilson Publishers)

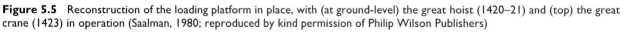

Figure 5.6 Florence, dominated by its cathedral (Black *et al.*, 1993; photograph: J. Allan Cash)

5.3 Geometrical cities, imaginary and real

The cult of beauty was a feature of intellectual life in the Renaissance. That was particularly inspired by the influence of Neoplatonic philosophy, which sought fundamental truths in the closely related ideas of symmetry, proportion and harmony – that foundation of music, long known to be associated with simple arithmetical ratios. In fifteenth- and sixteenth-century Italy (Florentines are again conspicuous here), these ideas were manifest in architectural treatises whose authors set out their views on the ideal city. Some were written to satisfy a prince's thirst for power and prestige. Frequently these town plans were sheer fantasy, caprices that never went beyond the sketch stage; but even these were mingled with practicable elements that responded to genuine needs.

Leon Battista Alberti (1404–72), architect and polymath, the illegitimate offspring of an exiled Florentine, wrote the first and most influential of these Renaissance treatises. His *De re aedificatoria*, which he began writing around 1450 and was posthumously published in 1485, was intended as a complete manual for the practising architect. Its pervasive theme is the identification of beauty with harmonious proportions, numerical ratios provided by music. Geometrical symmetry is also apparent in his recommended regular layout of houses on streets, and in his wholly symmetrical design for a market-place (see Figure 5.7).

But it was the Florentine goldsmith and architect Antonio Averlino (*c.*1400–69), better known by his nickname 'Filarete', who most indulged in ideal geometrical cities. His widely read treatise of the early 1460s took the form of a dialogue between himself, as the architect, and Francesco Sforza, the powerful Duke of Milan and patron of Filarete; the duke was then employing Filarete to build the Ospedale Maggiore, the first city hospital to have symmetrical cross-shaped wards. Filarete's treatise described a utopian city, Sforzinda, a name to gratify his patron. This dream city had the shape of an eight-pointed star. Much of the space within was divided up into a series of rectangular piazzas, each associated with various public functions. The central

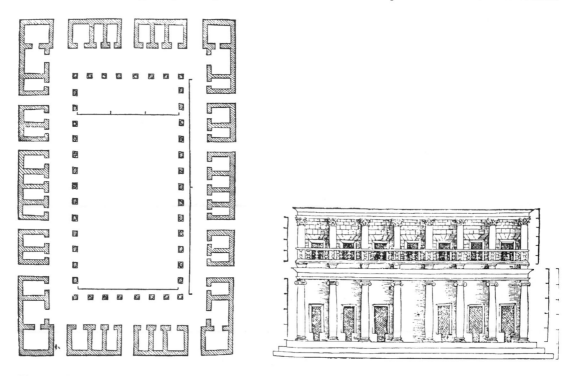

Figure 5.7 Design for a market-place; from Alberti's *De re aedificatoria*, 1485 (Blunt, 1940)

square was reserved for the cathedral and royal court, seats of the principal religious and secular authorities. In other, outlying squares, kept away from the centre, were the merchants, the market, the inns and brothels. A geometrical street-network connected squares with city gates. And, intent on avoiding the noise and congestion of vehicular traffic, he commented that:

> I would prefer few carts to pass through the city. So I arrange for all the water to go down the main streets. One could enter by boat and go all around the piazza on water.
>
> (Filarete, 1965, Book VI, folio 43v.)

Inspired by ideas very similar to this call for piazzas and symmetrical town-planning, Pope Pius II (in close co-operation with the Florentine architect Bernardo Rossellino) in 1459–64 rebuilt his birthplace. This was Corsignano, a Tuscan village, which he renamed Pienza – after himself. It still stands today as an early example of Renaissance planning: a small square with a cathedral, bishop's palace and town hall.

Geometry for urban defence

But it was technology, not aesthetic theory, which provided the main stimulus for the geometrical outline that became increasingly conspicuous in Italian cities from the late fifteenth century, and subsequently spread far beyond the peninsula – even far beyond Europe. A revolution in weapons technology had occurred: new artillery had been developed, which demanded a correspondingly great innovation in defence, and that defensive response was very much a matter of geometry. The shape of things to come had already been glimpsed by Alberti who, in a section of his treatise on architecture, advised that defensive fortifications would be all the more effective if they were built in a geometrical form resembling the teeth of a saw.

Urban defence was an obligatory concern for architects in Renaissance Italy because Italy had become the cockpit of Europe. Since the fourteenth century the city states of Florence, Milan and Venice, and the other peninsular powers of Naples and the Papal States, had been warring with one another to achieve yet more territorial expansion. Alliances were sought with foreign powers, and invitations extended by the Duke of Milan for French intervention, which occurred in 1494. The result was a large-scale international conflict, a contest between the French Valois monarchs and the Spanish Habsburgs for mastery of Italy. It culminated in a treaty of the mid-sixteenth century, recognizing Spanish possession of about one-third of the peninsula. Renaissance Italy was therefore the scene of endemic warfare, and it is not at all surprising that Italian architects became so heavily involved in military engineering. Leonardo da Vinci, in his famous letter of 1482 seeking employment from Ludovico Sforza, Duke of Milan, promised inventions of decisive value in warfare. It was the ability of architects to supply princes with effective fortifications which accounts for their rising demand and status in Italy. And their long experience made Italians the world's greatest experts in military engineering. Rich rewards were offered to induce them to emigrate and deliver their secrets to princes throughout Europe.

Over the medieval centuries, fortifications had become increasingly effective in defending cities from assault on their walls. The tactics of the enemy had been: to scale the walls by bringing up wheeled towers provided with scaling-ladders; to try to break down sections of a wall by repeated battering with rams or heavy stones hurled by torsional, catapult-like weapons; and to sling incendiary materials over the walls. The defensive response was to build thicker walls, and to modify the numerous tall towers, set at regular intervals along the walls, by introducing projecting platforms ('machicolations') extending beyond the ramparts, so that the defenders could drop boiling oil or

Figure 5.8 Defensive walls of late fifteenth-century Florence: accent on the vertical (Argan, 1969; photograph: Alinari)

missiles on the heads of the assailants. This was therefore warfare with an emphasis on vertical defence, and still at the end of the fifteenth century cities everywhere continued to display a peripheral form whose shape was dictated by this medieval strategy (see Figure 5.8). But even as cities, such as Siena, were spending huge sums on extensions to their medieval fortifications, developing weapons technology was rendering them obsolete.

Artillery, firing stone balls by the ignition of gunpowder, first appeared in Europe in the early fourteenth century. These were primitive weapons, liable to burst in firing, and well below the accuracy of the catapults in use. But gradually, a series of technical improvements led to the development of formidable cannon. By the mid-fifteenth century the techniques of church-bell manufacture were being applied to produce cast-bronze, muzzle-loading cannon. The results were spectacular. The massive walls of Constantinople, which had resisted Muslim onslaught for eight centuries, were finally breached in 1453 by the artillery of the Ottoman Turks. And the kingdom of Granada, the last Moorish stronghold in Spain which had stood for centuries, was at last conquered by the power of the artillery trains of Ferdinand and Isabella in 1481–92. This Spanish cannon included light artillery, firing small cast-iron shot of greater penetration, recently introduced in France. But mostly they were heavy guns firing stone balls, cumbersome weapons which had to be carried on horse-drawn wagons. These transport difficulties were already being overcome by the invention of the first gun-carriages. And the barrel of the cannon, permanently mounted on a carriage, was now cast with two small projections, the trunnions, which enabled the gun to be pivoted on its carriage for more rapid adjustment in targeting. The effectiveness of this mobile, penetrating new weaponry was demonstrated in the rapid progress of the French army invading Italy in 1494; the defences of cities and fortresses crumbled helplessly before its firepower.

A drastic change in traditional defence strategy became a matter of urgent urban concern. The first step taken was to thicken the walls and lower the towers. The superstructures of medieval battlements – those crenellations and machicolations, so conspicuous and characteristic a feature of the outer face of medieval cities – were now coming down, from artillery fire or precautionary design. These were the most vulnerable of all targets for artillery. But deprived of these tall towers, the urban defenders' range of vision was reduced, with increasing risk of a successful surprise assault. Nor was the conventional round or square firing-platform an efficient means of ground cover. Figure 5.9 shows how the lines of fire are seriously restricted, leaving much unprotected, 'dead' ground (indicated by the shading). Those areas could not be defended by guns firing almost vertically down because, apart from the greater difficulty of accurate targeting, the shot would tend to fall out of muzzle-loaded cannon. So a new system of flanking fire was indispensable if every part of the perimeter walls was to be protected from assault. And to achieve that, the precision of geometry was required to ensure that the fire cover would leave not the slightest gap. As an Italian military engineer, Girolamo Maggi, would later put it: 'Strength resides in form rather than in materials' (quoted in Hale, 1977, p.28). The solution was found in a regular polygon, a many-sided figure with re-entrant angles (i.e. angles pointing inwards). Figure 5.10 depicts one of the replicating units of a polygonal pattern of fortification. It shows how pointed forms eliminated the unprotected space in front of a tower, opening clear fields of fire to prevent sappers reaching the walls and digging gunpowder mines beneath them. There was now total cover for the perimeter wall. The essential new feature was the projecting, squat and pointed 'angled bastions' (A, B and C in Figure 5.10). They served also as spacious, solid platforms for artillery firing outwards on a besieging enemy. They first appeared in Italy. One of the earliest bastioned town walls was begun by the Florentines in the 1480s at Poggio Imperiale, strategically situated on one of the valley marching routes into Florentine territory from hostile Siena (see Figure 5.11).

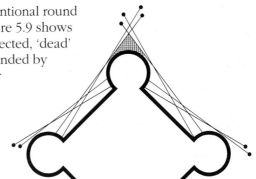

Figure 5.9 Round firing-platforms: the lines of fire are seriously restricted (adapted from Pepper and Adams, 1986; by permission of University of Chicago Press)

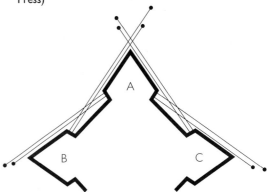

Figure 5.10 Polygonal fortifications: the lines of fire are unimpeded (adapted from Pepper and Adams, 1986; by permission of University of Chicago Press)

Figure 5.11 Angled bastion at Poggio Imperiale, Tuscany, late fifteenth century (Hale, 1965; photograph: courtesy of Professor John Hale)

Figure 5.12 Siena, c.1570, with fortifications old and new; from Braun and Hogenberg, *Civitates orbis terrarum*, 1572–1618 (Smith, 1967; photograph: Alinari)

The angled bastion, which has been seen as 'the most radically effective architectural element since the arch' (Hale, 1965, p.467), gradually spread to several Italian cities. In the 1520s Michelangelo was sketching bastions, and was appointed superintendent of Florence's fortification. A late sixteenth-century painting of Siena vividly shows the changing face of the city's perimeter, a visual representation of a new epoch: in the middle distance, a host of medieval nobles' towered homes surrounded by the city's medieval curtain wall; in the foreground, a complex geometry of Renaissance bastions projects into the countryside from the medieval wall (see Figure 5.12). And in the 1590s, to defend the Venetian frontier from invasions of Austrians and Turks, a wholly geometric new city, Palmanova, appeared: from a hexagonal city centre, radial roads led to nine bastions, forming a nine-pointed star whose beauty was only perceptible in a much later age of aviation (see Figure 5.13).

In their perceptive study of the development and spread of the new fortifications, Pepper and Adams have corrected the 'technological determinism' of some historians who see the adoption of bastions as an 'inevitable response' to the new artillery. 'Technological developments', they continue, 'only open doors; they compel no one to pass through them.' And, as an example of a Renaissance city which managed to survive well enough without the new bastions, they cite the strategically important papal possession of Bologna. There it was decided to adopt a different defence strategy – taking forces out of the city to face the enemy in the field (Pepper and Adams, 1986, pp.xxii, 29). That was certainly a cheaper way: building a network of bastions was enormously expensive – so expensive that it emptied city treasuries. And the cost was another reason why the new military technology was not automatically adopted, like some supposedly irresistible march of progress. A project to surround Rome with a protective belt of eighteen bastions had to be abandoned in 1542 when it was found that the cost of constructing just one had come to 44,000 ducats (about £10,000 then – equivalent to millions today). Palmanova's nine bastions represented a reduction of an originally envisaged project for twelve, judged too expensive to implement. When in the 1550s Siena proceeded recklessly with a huge defence project, involving the building of new bastions for the city and sixteen

Figure 5.13 Palmanova, the wholly symmetrical Venetian fortress-city, from the air (Smith, 1967; photograph: Aerofilms)

of its subject towns, the money ran out before the scheme could be completed. Siena had no funds left to pay for a relief army. When the Florentines attacked, Siena was unable to withstand the siege; it lost its independence and was annexed by Florence. Siena had fallen, a victim to the cost of the new technology. (These examples are from Parker, 1988, p.12.)

Yet despite the heavy costs, other cities and states in the sixteenth and seventeenth centuries managed by increased taxation or loans to pay for bastions, another item in the soaring expenditure on warfare. Italian engineers were busy in France and the Netherlands supplying what the French called *la trace italienne*, the Italian polygonal ground-plan of bastioned fortifications. They changed the face of Nancy, Antwerp and the Maltese fortress-city of Valletta. And a family of Italian engineers, the Antonelli, were employed by Spain's monarchs in the sixteenth and seventeenth centuries to strengthen defences in the Iberian peninsula and the Indies, installing bastions for Havana and other strategic centres. From the sixteenth century on, the frenzied building of fortifications was inseparable from the instruments of geometry (compass, protractor) and surveying (see Figure 5.14). The Renaissance bastion trace would remain a feature of urban defence into the twentieth century (for the notable seventeenth-century developments in the Netherlands, see Chapter 6, Section 6.3).

Figure 5.14 Planning a fortress; a seventeenth-century engraving from Mallet, *Les Travaux de Mars*, Paris, 1673 (Duffy, 1975)

5.4   Rome, the eternal city

The heightened interest in classical antiquity during the Renaissance was bound to give a special, new importance to the city of Rome. It had been the world capital of a classical civilization which was now the centre of attention and admiration in the cultural revival that was in full swing. Educated Florentines of the fifteenth century, nurtured on classical texts, liked to think of their city state as imbued with republican virtues inherited from ancient Rome in its republican phase – when, they supposed, the city of Florence was first created.

But while, in thought, the cultural achievements of ancient Rome were held in high esteem, the visible, material remains of the ancient city no longer declared the glory that had once been there. The splendour of the ancient forum was hidden beneath a wilderness of vegetation, a pasture for cows and horses, and a quarry for lime-burners, whose numerous kilns continued to be fed with whatever ancient marble could be uncovered. The great aqueducts, once the life blood of an expansive metropolis, had long collapsed; and the population, forced to concentrate around the River Tiber, now the main source of drinking-water, had shrunk to perhaps one-fortieth of its peak in antiquity. The people lived in shabby huts, and the narrow, winding streets were obstructed by projecting buildings. The city had largely reverted to rusticity. Great barns, filled with fodder for horses, occupied parts of the city centre; flocks of sheep, brought down from the hills to winter within the city walls, were driven at night through the streets; and herds of buffalo, kept on the islands of the Tiber, were sometimes used to tow boats upstream (Partner, 1976).

Yet this fifteenth-century Rome was the capital of one of Italy's largest territorial states, the Papal States which included the city of Bologna, some 300 kilometres to the north (see Figure 5.15). And Rome was also the spiritual capital of the Western world, the seat of the popes, who were the acknowledged successors to St Peter and the vicars of Christ on earth. Papal rule from Rome had only recently been re-established. Incessant war and endemic violence in Rome had forced the popes to abandon the city; for much of the fourteenth century they had resided in Avignon, an enclave in French territory, ceded and eventually sold to the popes by the Angevin princes of Naples. Although Gregory XI returned to Rome in 1378, it took more than half a century for papal authority to be regained, amidst challenges from rival popes and general councils of the church. Not until the reign of Nicholas V (1446–55) did Rome once again become the continuous residence of the popes, the administrative centre of a vast geographical network of ecclesiastical appointments, and the recipient of considerable revenue from the taxation of European clergy and the donations of the laity.

Renaissance popes and the planning of Rome

It was Nicholas V who first devised grandiose town-planning projects to restore the magnificence the city had once displayed through the public works of the ancient emperors. Here was a true pope of the Renaissance, an avid reader of classical literature (his collection of books was the germ of the great Vatican Library), a patron of artists and classical learning. He was convinced that architects and artists, inspired by classical models, held the key to the transformation of Rome into the world's finest city. Rome must again be given a face worthy of its status as the centre of universal Catholicism, a resplendent urban image that would serve to strengthen the faith and consolidate papal authority. Psychology mingled with practicality in this programme, which would be fostered by his successors. Nicholas insisted on the importance of

Figure 5.15 Italy, c.1500 (adapted from Brady *et al.*, 1994; by permission of Brill)

architecture for implanting 'solid and stable convictions in the minds of the uncultured masses', because the faith of the populace,

> [if] sustained only on doctrines, will never be anything but feeble and vacillating. But if the authority of the Holy See were visibly displayed in majestic buildings, imperishable memorials and witnesses seemingly planted by the hand of God Himself, belief would grow and strengthen from one generation to another …
>
> (quoted in Partner, 1976, p.16)

That must be the impression given to the hordes of visiting pilgrims who would then transmit it throughout Europe on their return.

But there would be no sudden urban face-lift in Nicholas V's reign. His vision of a rebuilt St Peter's would not be realized until the seventeenth century, achieved by cannibalized architecture – the papal destruction of

Figure 5.16 Terminus of the Acqua Felice aqueduct, the Moses fountain, Rome (Lavedan *et al.*, 1982)

mountains of ancient marble structures for the building material. Under Nicholas V's regime, only one of the ancient aqueducts, the Aqua Virgo, was restored, by Alberti, the drinking-water issuing in the city fountain of Trevi; but it was not maintained, and by 1475 it had ceased to function. Progress was slow, and interrupted in 1527 by the Sack of Rome – widespread destruction wrought by the invasion of the unpaid army of Charles V, a recurrence of the centuries-old rivalry between emperor and pope. But a succession of popes remained committed to the programme of creating a magnificent Rome, and notable steps were taken to accomplish it. Sixtus IV (1471–84) encouraged private building by proclaiming exemption of property from state confiscation after the owner's death. And he introduced compulsory purchase orders to demolish private property which stood in the way of road improvements – the creation of broad thoroughfares, essential for the free movement of crowds on the route through the city to the seven pilgrimage churches.

By 1520 two new avenues, straight and broad, crossed the city – Via Giulia and Via Leonina. But the rest of the city remained unimproved. Town-planning requires a powerful central authority with vision, determination and an abundance of money. All of these conditions were met by Sixtus V (1585–90), whose sweeping reforms still dominate the face of Rome today. They were paid for by the imposition of customs duties, the sale of offices, new taxes on the owners of coaches and houses (to fund road-building), and large loans. Sixtus V created a state commission to supervise road-building and water supply, and he increased the numbers and powers of municipal officials responsible for street maintenance. But his most elaborate project was to repopulate the hills surrounding Rome, relieving the crowded and unhealthy centre. And to provide the essential water supply for this, he engaged a specialist in hydraulic engineering, Giovanni Fontana. A papal monopoly was declared on the use of limestone for the repair and extension of the ancient aqueducts, culminating in the restoration of the ancient aqueduct of Emperor Severus, now to be called the Acqua Felice (from the pope's first name). By the end of 1586 the Acqua Felice, running underground for nearly twenty kilometres and over arches for another ten, brought water to the hills as well as to the lower city, where it issued at an ornate fountain (see Figure 5.16), one of many to add beauty to Rome, improve hygiene, and introduce the soothing sounds of rippling water. Sixtus V ensured maintenance of the supply by appointing experts to inspect the whole length of the aqueduct annually and to detect illegal tapping of the public supply by farmers irrigating crops. This extension of water supply contributed to the marked increase in Rome's population in the sixteenth century: between 1527 and 1600 it doubled to around 100,000. By the early seventeenth century, Romans were being supplied with no less than 180,000 cubic metres per day, a substantial volume which still proved adequate for the city in the mid-nineteenth century (Delumeau, 1957, pp.327–39). This can be compared with the figure, given in Chapter 3 of this book, of nearly one million cubic metres – the daily supply for a million ancient Romans.

Moving monoliths for the pope

Sixtus V employed his architect, Domenico Fontana (the brother of Giovanni), to build numerous monuments which greatly altered the physical appearance of the city. His work included the construction of the Vatican Library and the erection of giant monoliths – Egyptian obelisks. For the ancient Egyptians, obelisks had been potent, religious objects; for the ancient Romans who had transported them to their capital, they symbolized world domination; for Sixtus V, their principal function was to control urban vistas, guiding the eye along rectilinear streets to a climax at some monumental building. Fontana's reputation reached its peak with his successful re-erection of an obelisk in the square in front of St Peter's (see Figure 5.17). It was one of several which he would position to satisfy papal demands for Renaissance perspective on the city scale.

Figure 5.17 Egyptian obelisk in front of St Peter's, Rome (photograph: Alinari)

The tone of Sixtus V's letter appointing Fontana displays a confident assertion of power. It could not have been written by popes of the early fifteenth century, at least not with any real basis. But by this time, the steadily increasing power of the popes had reached a point where it was little less than absolute; no other monarch in Europe had greater power over his subjects. The checks on papal power had dwindled to insignificance: Rome's medieval commune, the municipal authority, was defunct; and the once-powerful college of cardinals was now little more than a rubber stamp ratifying papal decisions. Sixtus wanted the great obelisk moved from the position it had been set in by the ancient Romans and re-erected in front of St Peter's basilica, some 700 metres away. That is what Sixtus wanted and there was nothing that could stand in his way, even if objections arose. The necessary labour would be compelled, tools and materials requisitioned, houses demolished (here the edict of his predecessor, Sixtus IV, was exploited), and severe penalties announced for any interference with the operation – a heavy fine or even, we learn from Fontana's record, capital punishment. And yet these formidable powers fell short of what the despots of the ancient world had been able to deploy in their giant technological projects. The Pharaohs could draw on multitudes of slaves to build their enormous pyramids, and Roman emperors on gangs of waged labour for their monumental construction. But in the medieval West, slavery and serfdom had gradually disappeared, and while Renaissance princes could order their soldiers to participate in building fortifications, labour generally had now to be paid for at a fair price, which extra cost helps to explain the reduced scale of projects compared with those in the ancient world. Human rights had increased. Sixtus V here compels labourers and craft specialists but pays them well; he also declares a good price for tools taken, compensation for the commandeering of beasts of burden, and compensation to the owners of houses demolished.

Fontana left a vivid account of the technology he employed to re-position the obelisk. (Klemm, 1959, pp.198–205 contains translated extracts along with the pope's letter of appointment.[2]) Simple hoisting machines were used – the pulley, complemented by the capstan (a rotating drum fitted with levers). One end of each of the numerous long ropes was tied around the huge stone obelisk, the other end passed over a pulley to be wound around a capstan. The arduous work of turning the capstans was performed by the muscular effort of harnessed horses and gangs of men. There was no new technology here. Pulleys and capstans (in the form of a winch) were in regular use for building work in ancient Rome, and even earlier the ancient Greeks had used them to lift heavy stage scenery and timber for shipbuilding. And the rollers Fontana employed were of the same type that had been used to move obelisks in ancient Egypt.

Some preliminary calculations were necessary. To determine the weight of the obelisk, a small sample of the stone (a measured proportion of the whole) was cut out and weighed; the weight of the whole obelisk – well over 400 tonnes – was then calculated from a simple sum involving ratios. Another simple sum told Fontana how many capstans he would need to lift that weight. Some elementary surveying ('levelling') was essential to determine whether the new site was on the same level as the old. It wasn't, and to facilitate the transport of the obelisk, a large embankment of earth had to be made between the two sites. In the hoisting operations, considerable organization was indispensable for co-ordinating the pulling forces of the hundreds of men and scores of horses. That was achieved by visual and aural signalling, to synchronize the operation of individually numbered capstans. Here was a project on a grand scale, taking almost a year to execute. The work fascinated the citizens and there was general celebration at its successful completion.

---

[2]   These are reproduced in the Reader associated with this volume (Chant, 1999).

While Sixtus V was engaged in bestowing a new magnificence on the city of Rome, momentous changes were already beginning to transfer Europe's principal foci of economic and political importance away from their long-established Mediterranean centre. There would be a long-term shift in the centres of power, northwards from the Mediterranean to the Atlantic seaboard. Southern Europe (the states of Italy and Habsburg Spain) would be overtaken by regions of new prosperity and political power in north-west Europe, states that would mature into the dominant powers of the seventeenth century. This was accompanied by striking developments in their largest cities – London, Paris and (a focus of the next chapter) Amsterdam.

References

ARGAN, G.C. (1969) *The Renaissance City*, New York, George Braziller.

BLACK, C., GREENGRASS, M., HOWARTH, D., LAWRENCE, J., MACKENNEY, R., RADY, M. and WELCH, E. (1993) *Atlas of the Renaissance*, London, Cassell.

BLUNT, A. (1940) *Artistic Theory in Italy, 1450–1600*, Oxford, Oxford University Press.

BRADY, T.A. Jr, OBERMAN, H.A. and TRACY, J.D. (eds) (1994) *Handbook of European History, 1400–1600*, Leiden, Brill, vol.1.

BURCKHARDT, J. (1878) *Civilization of the Renaissance in Italy* (trans. S.G.C. Middlemore), London, Kegan Paul (first published 1860).

CHANT, C. (ed.) (1999) *The Pre-industrial Cities and Technology Reader*, London, Routledge, in association with The Open University.

DELUMEAU, J. (1957) *Vie économique et sociale de Rome dans la seconde moitié du XVIe siècle*, Paris, de Boccard, vol.1.

DUFFY, C. (1975) *Fire and Stone: the science of fortress warfare, 1660–1860*, Newton Abbot, David and Charles.

FILARETE (1965) *Filarete's Treatise on Architecture* (trans. J.R. Spencer), New Haven and London, Yale University Press.

FOSSIER, R. (ed.) (1986) *The Cambridge Illustrated History of the Middle Ages, 1250–1520* (trans. S.H. Tenison), Cambridge, Cambridge University Press, vol.3.

GOLDTHWAITE, R.A. (1980) *The Building of Renaissance Florence: an economic and social history*, Baltimore and London, Johns Hopkins University Press.

HALE, J.R. (1965) 'The early development of the bastion: an Italian chronology, *c*.1450–*c*.1534' in J.R. Hale, R. Highfield and B. Smalley (eds) *Europe in the Late Middle Ages*, London, Faber and Faber, pp.466–94.

HALE, J.R. (1977) *Renaissance Fortification: art or engineering?*, London, Thames and Hudson.

HAY, D. (1989, 2nd edn) *Europe in the Fourteenth and Fifteenth Centuries*, London and New York, Longman.

HAY, D., CHASTEL, A., GRAYSON, C., BOAS HALL, M., HAY, D., KRISTELLER, P.O., RUBINSTEIN, N., SCHMITT, C.B., TRINKHAUS, C. and ULLMANN, W. (1982) *The Renaissance: essays in interpretation*, London and New York, Methuen.

HERLIHY, D. (1986) *Pisa in the Early Renaissance: a study of urban growth*, New Haven and London, Yale University Press.

HOLMES, G. (1986) *Florence, Rome and the Origins of the Renaissance*, Oxford, Oxford University Press.

KLEMM, F. (ed.) (1959) *A History of Western Technology* (trans. D.W. Singer), London, Allen and Unwin.

LAVEDAN, P., HUGUENEY, J. and HENRAT, P. (1982) *L'Urbanisme à l'époque moderne, XVIe–XVIIIe siècles*, Geneva, Droz, vol.3.

MAINSTONE, R.S. (1969–70) 'Brunelleschi's dome of S. Maria del Fiore and some related structures', *Transactions of the Newcomen Society*, vol.42, pp.107–26.

PARKER, G. (1988) *The Military Revolution: military innovation and the rise of the West, 1500–1800*, Cambridge, Cambridge University Press.

PARTNER, P. (1976) *Renaissance Rome, 1500–1559: a portrait of a society*, Berkeley and London, University of California Press.

PEPPER, S. and ADAMS, N. (1986) *Firearms and Fortifications: military architecture and siege warfare in sixteenth-century Siena*, Chicago and London, University of Chicago Press.

PORTER, R. and TEICH, M. (eds) (1992) *The Renaissance in National Context*, Cambridge, Cambridge University Press.

SAALMAN, H. (1980) *Filippo Brunelleschi: the cupola of Santa Maria del Fiore*, London, Zwemmer.

SMITH, C.T. (1967) *An Historical Geography of Western Europe before 1800*, Harlow, Longman.

WHITFIELD, J.H. (1981) 'Cicero Marcus Tullius' in J.R. Hale (ed.) *A Concise Encyclopaedia of the Italian Renaissance*, London, Thames and Hudson, pp.84–5.

Chapter 6: THE EARLY MODERN CITY

by Peter Elmer

6.1 Introduction

From the point of view of the urban historian, the Early Modern era in Europe – roughly the period from 1450 to 1750 – is highly problematic. On the one hand, it was a period of considerable change, witnessing major upheavals in the social, economic, political, religious and cultural life of the continent. Yet on the other, it was an era characterized by a great deal of stability and continuity, nowhere more so than in the fabric and shape of its cities. This paradox has generated much debate among historians and informs a great deal of the writing on the nature of cities in the period, particularly those located in northern Europe where the forces of innovation and tradition were most starkly juxtaposed. Consequently, in what follows, I should like to focus on the rapidly expanding urban communities of northern Europe – in particular those to be found in France, the British Isles and the Low Countries – in order to explore further the complex nature of the relationship between technology and urban development in the period immediately before the onset of the Industrial Revolution.

Though historians may disagree about whether the Early Modern era should be categorized as either a footnote to the Middle Ages or a forerunner of the modern world, most would probably agree with Friedrichs' conclusion that Early Modern life was fundamentally 'burdened by technical and scientific limitations of enormous consequence' (Friedrichs, 1995, p.12; Gutkind, 1971).[1] This overwhelmingly negative view of the technical capabilities of Early Modern Europeans permeates urban studies of this period – to such an extent, in fact, that discussion of technology in the burgeoning literature on the Early Modern city has been negligible. As a result, where it is acknowledged that change did occur in the cities of this period, this is rarely attributed to technological advance. A well-founded example of this approach would be the explanation routinely offered to account for the rapid growth in population of many of northern Europe's towns and cities in the sixteenth and early seventeenth centuries. Case-studies consistently demonstrate that urban death-rates exceeded birth-rates, largely as a result of the high incidence of epidemics and infant mortality, with population growth over time largely to be accounted for by mass rural migration to the cities. Such demographic studies have thus helped to underline the point that low levels of technical knowledge – in this case medicine and hygiene – were a feature of the Early Modern scene (Palliser, 1974, p.59).

While Friedrichs' general approach is well founded, it is fair to raise again the wider issue of the impact of technology on the development of the city in the Early Modern era. In particular, two recent lines of historical thinking about the period vindicate such a response. First, it is becoming increasingly acceptable to view the period from about 1600 to 1750 as one which witnessed a remarkable transition in the physical appearance of the city – a transition not simply manifest in the long-acknowledged aesthetic revolution in architecture, but also grounded upon a fundamental reassessment of the role and place of the city in all aspects of Early Modern life (Borsay, 1989). And second,

---

[1] A fuller exposition of Friedrichs' views is given in the Reader associated with this volume (Chant, 1999).

historians generally have recently become much more open to the idea that technology and technical know-how in the sixteenth and seventeenth centuries should be granted a far higher profile than in previous accounts (Webster, 1975). Taken together, these two developments offer a compelling reason to reappraise the role which technology played in urban history in the period leading up to the Industrial Revolution. In the process, they also provide a convenient opportunity to put to the test the view of Friedrichs and others that the pre-industrial city was seriously constrained by the limitations imposed upon it by technological conservatism.

In order to reappraise the relationship between the city and technology in this period, we shall need to focus on a number of key technological developments which one might reasonably assume to have influenced the physical fabric and structure of the Early Modern city. Developments in three areas will be considered: warfare; transport and communication; and the wider urban environment, including discussion of water supply, hygiene, street lighting, fire-fighting and a number of other issues directly impinging on living conditions in the city. In each case, particular attention will be paid to the notion of technological determinism, and we shall need to assess the extent to which a technological imperative, as opposed to a social, religious or political imperative, is responsible for changes in the urban landscape of this period. This study will conclude with three case-studies – of Amsterdam, Paris and London – which should allow us not only to test some of the hypotheses outlined above, but also to make comparisons between cities with contrasting political, as well as geographical, contexts.

6.2   Technological innovation, the built environment and the Early Modern city

Introduction

A major plank in the argument for a limited role for technology in the evolution of the Early Modern city is the evidence of the various physical and geographical constraints under which urban development took place at this time.[2] Foremost among these constraints were the ubiquitous city walls which most urban historians of the period have seen as *the* defining feature of the Early Modern city. According to Friedrichs, for example, the city wall was probably 'a city's dominant architectonic feature' which, with its series of manned gates, must have posed a formidable obstacle to urban development and growth (1995, p.22). These barriers to expansion may also have severely constrained the introduction of technological solutions to the problems of Europe's cities, particularly those experiencing rapid population growth. One by-product of the reluctance of civic authorities and governments in general to authorize the growth and extension of cities beyond their traditional walled boundaries was the emergence of large, uncontrolled districts of suburban growth outside the official perimeters of the city. This was a problem common to the great metropolises of Early Modern Europe, and one which by and large compounded the technological deficiencies of urban planning. Indeed, sometimes, the introduction of new technology exacerbated the problem. The advent of new and more powerful forms of siege artillery in the late sixteenth and seventeenth centuries had necessitated a major reappraisal of the way in which vulnerable cities should be defended in the event of war. One response, originally pioneered by the Italians, but subsequently developed to its fullest extent by the Dutch and French, was to create ever larger and more

[2]   For a general discussion of these constraints and the way in which they inhibited the development of new urban building types, see the commentary of Friedrichs given in the Reader associated with this volume (Chant, 1999).

sophisticated series of walled defences or bastion rings which owed much to the burgeoning interest among scholars and others in mathematics and military engineering (discussed further below, pp.189–193). In the Low Countries and northern Europe, the result was the creation of large garrison towns and cities in which the needs of the civilian population were subordinated to the interests of defence. In such cases urban expansion, not surprisingly, was curtailed.

This reminds us that technology in itself – which today we so frequently associate with providing all the solutions to the ills of society – could often create more problems than it appeared to solve. Another example of this process is provided by the advent of coaches and carriages (discussed further below, pp.195–8). However, it was also the case that innovations in technology – such as those referred to above with respect to the demands of urban defence in the sixteenth and seventeenth centuries – might provide valuable opportunities for municipal authorities to introduce change and innovation to the cityscape. This was particularly evident in the case of cities which were far removed from the static war zones of Early Modern Europe. In Paris, for instance, during the second half of the seventeenth century, the old, confining ramparts of the city were either demolished or remodelled in order to promote new schemes of urban improvement in line with contemporary ideas on leisure (see below, p.194).

Even if, as was increasingly the case by the end of our period, city walls were less of an obstacle to urban growth and development, other features of the built environment continued to restrict plans for urban regeneration. A powerful example of this process is provided by the problem of pre-existing street patterns which often inhibited the successful introduction of new technology. Although the Renaissance had promoted a great deal of interest in new and more rational schemes of urban street planning and design (see Chapter 5), few of these schemes were ever put into effect. In an era when few new towns or cities were built, the opportunity to innovate in this way was highly limited. Exceptions to this rule were cities destroyed or damaged by human and natural disasters such as fires and earthquakes which provided municipal governments with the perfect excuse to build from scratch. Indeed, the example of fire-damaged cities – a frequent phenomenon at this time – provides us with an excellent opportunity to test the hypothesis that one cause of the lack of innovatory practice in the design and building of the Early Modern city lay in the low level of technological ability which some have seen as characteristic of the age. It is for this reason that much of the later section on London is devoted to its reconstruction after the Great Fire of 1666 (below, pp.228–33).

Experimentation in the physical layout of the Early Modern city, however, was not restricted to the opportunities provided by natural or human calamities. One aspect of the so-called urban renaissance of the late seventeenth century was the trend towards schemes of urban regeneration which emphasized the creation of public spaces and squares in Europe's cities. Though not confined by any means to the large metropolises, the most dramatic transformations were found in the capital cities of northern Europe. In Paris, for example, under the stimulus of two powerful reforming monarchs, Henri IV and Louis XIV, elaborate plans were carried out, often with clear technological implications, for the redevelopment of the city's squares, bridges, streets and public spaces. Even more dramatic was the planned development and expansion of the city of Amsterdam in response to the startling growth of the city's economy and population in the seventeenth century. Both demand careful study, not least because the political authority incorporated in the ruling bodies of both cities was sufficient to overcome the power of vested interest and individual property rights, both of which taken together arguably constituted as great a barrier to urban innovation in this period as the lack of technological creativity.

This final point leads us to the consideration of a vital issue – namely the extent to which technological innovation was itself circumscribed by the political and administrative limitations of the age. In this respect, the examples cited above may well represent exceptional instances of the ability of governments to implement urban innovations. If this was the case, then it is quite feasible that the main impediment to improvement in most European cities of the period was not the poor state of contemporary technology but rather the lack of political will or administrative ability on the part of the rulers of Europe's cities – itself a reflection of the low levels of resource available to staff large and effective bureaucracies. As many historians of technology in this period immediately before the Industrial Revolution are now beginning to show, technological acumen and know-how may well have *preceded* the ability of Early Modern municipal and national administrations to encourage and implement innovation in the fabric of their cities. This view, in turn, seeks to rehabilitate the place of technological innovation in the pre-industrial age – a view which will need to be borne in mind as we progress through the remainder of this chapter.

Warfare

Changes in the technology of warfare are widely considered by historians today to represent one of the prime influences on the social, economic and political evolution of Europe in the sixteenth and seventeenth centuries (Parker, 1988). Although many of the changes associated with this military revolution pre-date our period (for example, the discovery of gunpowder), it is now widely believed that their full effect was not felt until the sixteenth century with the development of ever more effective forms of heavy artillery and cannon. The consequences for cities under siege were momentous. As you read in Chapter 5, medieval-style fortifications and walls which ringed every city of any size in this period were no match for the new weaponry, thus prompting a revolution in urban defences which had clear implications for the fabric and design of cities in the centuries ahead. The result was a surge of interest in military technology, much of it widely disseminated thanks to the invention of the printing press, in which field the Italians of the Renaissance were the first to excel. The application of this new military technology to siege warfare is of particular interest since it clearly called for a major reappraisal of the way in which Early Modern cities functioned, as well as how they were constructed and defended. City defences, as we have seen, became ever more important, particularly in those parts of northern and central Europe where warfare on the grand scale was endemic throughout the sixteenth and seventeenth centuries (see Figure 6.1 overleaf).

The most obvious difference between medieval walled fortifications and those constructed by the urban theorists and military engineers of our period was the sheer size and scale of the new-style defences. As siege artillery grew in volume and accuracy, so it became necessary to build ever larger and more complex bastion rings and defensive earthworks around cities (as discussed in Chapter 5). The result was to create a vast cordon sanitaire around the urban core – a large area of land where once existed either suburbs or open space, but which was now given over exclusively to the needs of defence. This in turn cut off the city from its rural hinterland, and isolated it in a way which was quite unimaginable in the medieval period, fostering the first stirrings of that sense of urban and rural distinctiveness which was to become such a prominent feature of the industrial age.

It was not only the boundaries of the Early Modern city, however, that were affected by the introduction of the new military technology. Internally, cities in the front line of siege warfare were quite often forced to adapt their traditional street layouts and to undergo change to the urban fabric. Thus, the drawing

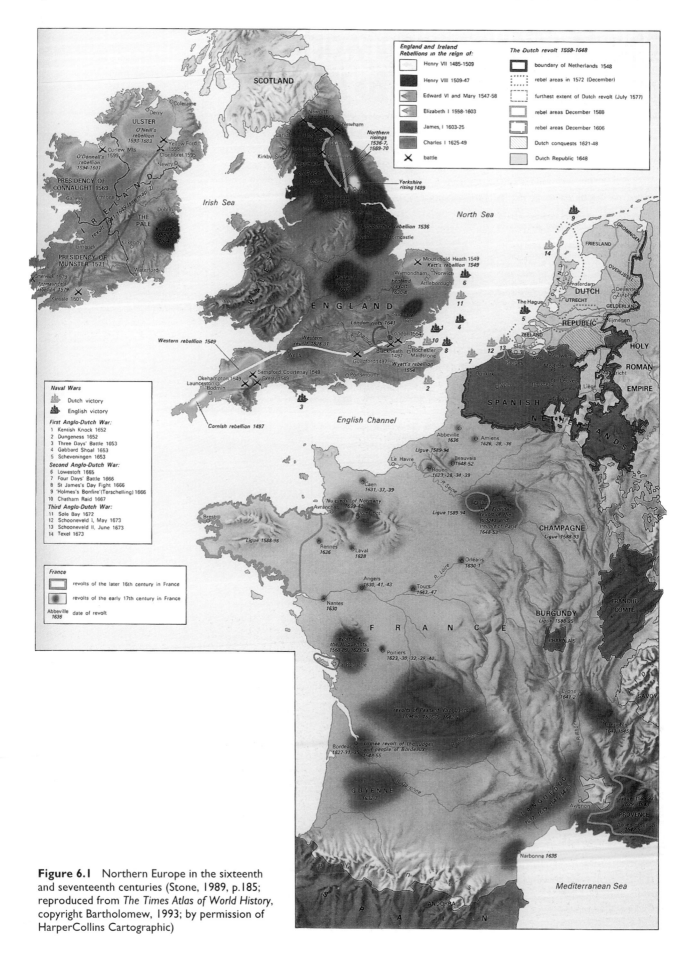

Figure 6.1 Northern Europe in the sixteenth and seventeenth centuries (Stone, 1989, p.185; reproduced from *The Times Atlas of World History*, copyright Bartholomew, 1993; by permission of HarperCollins Cartographic)

board sketches and plans of the urban and the military theorist (often one and the same person) quite frequently moved towards straighter street lines and wider public spaces, both crucial if the men and artillery of a besieged city were to be rapidly deployed in the event of an attack. Public squares were accordingly built, not simply for use in civic ritual, but primarily as large spaces conducive to parading and mustering large numbers of troops and horses. In addition, as defending armies, with all their military paraphernalia, grew in size, so space had to be found to house men, horses and equipment. Hitherto, the problem of accommodating garrison soldiers had largely been resolved by the highly unpopular ploy of billeting troops on the civilian population. From the early seventeenth century, however, it became increasingly normal for military governors to construct barracks which, like artillery magazines and arsenals, were strategically located to facilitate ease of deployment (see Figures 6.2 and 6.3).

Figure 6.2 Engraving of the artillery magazine in Metz, France, by Claude de Chastillon. The gunpowder was confined to the vaulted cellar where it was stored in barrels of different size according to its nature: fine for muskets and coarser for heavy artillery. Armour and other equipment was stored on the upper floor (Buisseret, 1984, plate 11; by permission of The British Library)

Figure 6.3 Standard gunpowder magazine, built to the specifications of the celebrated French soldier and military engineer Sébastien Le Prestre, Marquis de Vauban (1633–1707). Note the unusual number of pier buttresses, a borrowing from medieval architecture, which helped to brace the building on all sides to withstand the force of an explosion. The magazine itself was recessed in an excavated area and surrounded by another wall to provide further safety. The cases of powder were stacked on a paved floor to protect against damp, and small openings provided ventilation. Even nails, hinges and locks were made of bronze instead of iron to prevent sparks (Berger, 1994, p.176; photograph: Robert W. Berger)

The United Provinces (modern-day Netherlands) provide some interesting examples of the effect of these innovations upon the fortress and garrison towns which sprang up during the course of the Eighty Years' War with Spain (1567–1648), and which were consolidated during the struggle against the French in the late seventeenth century. Dutch society was probably the most militarized in seventeenth-century Europe (Duffy, 1996, pp.58–105; Israel, 1995a, pp.262–7). Its survival was dependent upon the adaptation of the majority of its cities and citizens to the thoroughgoing demands of military defence. Consequently, it comes as no great surprise to learn that Dutch writers and military engineers displaced the Italians as the masters of military urban planning in the early seventeenth century. Among the most celebrated was the mathematician and engineer Simon Stevin (1548–1620) (see Figure 6.4),

Figure 6.4   The Dutch Defensive Ring during the Twelve Years' Truce (1609–21). Among Stevin's many accomplishments, his work in surveying the Dutch frontier between 1605 and 1608, and constructing the most sophisticated system of defences yet seen in Europe, was of critical importance in securing the fragile independence of the Dutch (Israel, 1995, p.263; by permission of Oxford University Press)

whose *Art of Fortification* (*Stercktenbouwing*) (1594) incorporated many of
the principles of Renaissance town-planning. There can be little doubt that for
Stevin, military defence and technological innovation went hand in hand. This
was particularly evident in Stevin's advocacy of a number of new techniques in
the construction of canal locks and sluices which were an integral feature of
Dutch urban defences. They included the development of the swivel-gate lock
(see Figure 6.5), a multipurpose device for scouring the canals which encircled
most Dutch cities, as well as numerous refinements to sluices and water-mills
(windmills for draining land). At the same time, his theoretical work on urban
planning (largely derived from Italian sources) contained much on the
practical side of urban design including heating, water supply, sewerage and
sanitation as integral to the 'modern' city (Dijksterhuis, 1970, pp.93–102; Struik,
1981, pp.52–60; see Figure 6.6).

In some cases, the influence of this 'theoretical' planning on the actual
construction of cities can be clearly seen. The best example in the Dutch context
is provided by the city of Coevorden, an ideal city with a radial, concentric
street layout (see Figure 6.7 overleaf) in which civilian life was fully
subordinated to the demands of defence. A good example of what this meant
in practice is the strategic fortress town of 's Hertogenbosch where civilian
space was severely restricted by the need to erect barracks, magazines and
guardhouses, as well as the need to set aside some urban land for the cultivation
of crops. In the event, the municipal authorities issued an edict which imposed
draconian restrictions on the size of houses in the city (Gutkind, 1971, p.107).
Elsewhere in Europe, urban concentration of this kind frequently forced
civilians to build vertically rather than horizontally, leading to the construction
of buildings as high as ten storeys in some cities (Braudel, 1981, p.497;
Montagu, 1887, p.112). As for the highly complex series of military defences
designed to withstand lengthy sieges, these are best viewed (even today) in
the town of Naarden where there were constructed in the 1670s and 1680s
'mighty installations in the Vauban style [Vauban was a celebrated French
military engineer of the late seventeenth century], comprising six great bastions
and ravelins set in wide moats, protected gun emplacements and enfilade
firing points and a vast network of covered routes and passages connecting
armouries and ammunition stores to firing positions' (Burke, 1956, p.60).

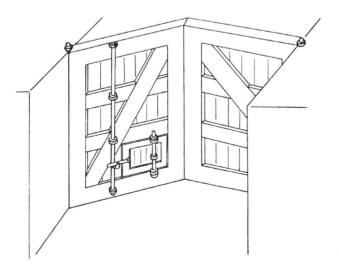

Figure 6.5 Stevin's swivel-gate lock (Dijksterhuis, 1970,
p.100, figure 31)

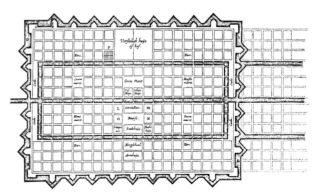

Figure 6.6 Ideal port city plan by Stevin (1590). The
Renaissance preoccupation with geometrical forms is self-
evident. But note also Stevin's pragmatism in allocating a vital
role, both economic and defensive, to an integrated canal
system (Dijksterhuis, 1970, p.114)

Figure 6.7 Coevorden. The reconstruction of the city was completed in 1605 by Prince Maurice of Nassau and incorporated recent Dutch innovations in fortress construction. The right-hand side of the plan contains a useful profile of the defensive earthworks which incorporated a wet ditch more than fifty metres wide, and extensive *glacis* (area of open terrain which provided the defenders with a convenient field of fire) (Duffy, 1996, p.92, figure 32)

It should not be assumed that all European cities were equally affected in this period by new methods of warfare. Indeed, for many, the new military technology inadvertently created new opportunities for urban planning and design in which the demands of warfare were minimal. By the second half of the seventeenth century, military and political developments were tending to concentrate warfare, particularly siege warfare, in the rather narrow, contested borderlands between countries, thus creating large zones in which the needs of defence were increasingly insignificant. This was particularly common in large and powerful military nations such as France where, under Louis XIV, cities such as Paris could begin safely to dispense with their medieval fortifications and walls and to transform them instead into wide, tree-lined avenues or boulevards (from the same root as the military term, bulwark) which might in turn be given over to the use of leisure (Bernard, 1970, pp.12–14; Parker, 1988, p.43). Similar developments can be seen in the British Isles after 1650 where, with the exception of Ireland, conditions of peace enabled many cities either to dispense with their walls altogether, or to use them for more leisurely activity, such as, for example, promenading in Bath (Jack, 1996, p.14). A major impetus to such changes in the case of Britain was the desire of the restored Stuart monarchy after 1660 to ensure that the towns and cities of the realm could never again be used by the crown's opponents as safe havens from which to wage civil war. As a result, there can be little doubt that the destruction of many city walls by order of the monarchy helped pave the way for that impressive transformation of so many English towns and cities in the late seventeenth and eighteenth centuries, a process described by one historian as constituting nothing less than an urban renaissance (Borsay, 1989).

Transport

Though the age of industrialization is often identified with a revolution in systems of transport, the Early Modern era also witnessed some important developments in this field, some of which undoubtedly helped to shape the European city in this period. Developments in shipbuilding and nautical technology, for example, reduced impediments to long-distance maritime travel, enabling Europeans to trade with, and colonize, far-flung reaches of the globe. In the process, the wealth generated acted as an important spur to urban growth in the Early Modern period. New maritime technology also stimulated the growth of port cities which developed special functions to service this rapidly expanding commerce. The growth of cities such as Rotterdam, for example, which specialized in shipbuilding, is testament to this phenomenon. Technological innovation, moreover, was a feature of the growing Dutch port cities of this period. It had to be if cities such as Amsterdam, inherently unsuitable as a port because of the problems of silting, were to survive and expand. One such innovation of the early seventeenth century was the 'camel' – sealed empty tubs attached to a ship's bottom to provide sufficient buoyancy to traverse the shifting sandbanks which ringed the entrance to Amsterdam's inner harbour (Lambert, 1971, p.190). Another was the mud-mill which was developed by the port authorities at the expense of the city to remove silt from the entrance to Amsterdam's notoriously shallow harbour (Reinders, 1981, pp.229–38; see Figures 6.8 and 6.9). In addition, naval competition between the great powers further stimulated port development: Portsea, just outside Portsmouth in England, was a new town created to meet the demands of England's naval defences in the late seventeenth century (Chalklin, 1974, pp.229–52; Konvitz, 1978).

A more direct relationship between transport technology and urban morphology, however, can be seen in developments in two other areas in the late sixteenth and early seventeenth centuries: the horse-drawn carriage and canalization. The first 'modern' coaches or carriages date from the mid-sixteenth century, but it was not until the second or third decade of the seventeenth century that they began to make a significant impact on the urban landscape, fostering what one urban historian has described as the 'cult of the street' (Gutkind, 1971, p.261). In 1633, a Captain Bailey of the Strand in

Figure 6.8   Dredger (mud-mill) on a canal; drawing by Roelant Savery, c.1610 in the Devonshire Collection at Chatsworth (reproduced by permission of the Duke of Devonshire and the Chatsworth Settlement Trustees; photograph: Courtauld Institute of Art)

Figure 6.9 Mud-mill; illustration from R. van Natrus, *Groot volkoomen moelenbock* (Amsterdam, 1734). The mill was operated by men who walked in two large treadmills, thus moving a chain with wooden boards. The chain turned round the axle of the treadmill and an axle placed at the end of a long, narrow gutter in which the lower part of the chain moved. This structure was placed on an open wooden case with an elongated opening in the centre. The wooden gutter could be lowered or raised through this opening with a windlass on the front of the boat, in order to adjust the mill to the proper depth. Then the boat was slowly pulled through the water, keeping the wooden gutter at its correct depth. The mud that got in front of the gutter was pushed up by the boards on the rotating chain, and discharged into a barge. During the second half of the seventeenth century, human labour was replaced by that of horses. The motion of the circling horses was transmitted by gearwheels to a spindle that by means of further gearwheels was transmitted to the chain (Reinders, 1981, p.234)

London was the first to create an organized service of coaches for hire, and a similar service was soon implemented in Paris based on the French carriage, the *fiacre*.[3] The effect of this development on the urban environment was soon apparent. The roads and streets of most European capitals at this time were ill-equipped to cope with iron-wheeled carriages, and complaints about traffic congestion in the narrow and cramped streets of most major cities were common. The initial response of the authorities was either to ban their use altogether or alternatively to circumscribe their appearance on the streets. However, such was the craze for their use among the wealthier classes that all such measures proved futile, and alternative solutions were sought. One technological response was to seek ways of creating smaller, more comfortable carriages, the second half of the seventeenth century witnessing a number of important changes in vehicular design (see Figure 6.10). Competition, in the form of the sedan chair, another development of the first half of the seventeenth century, also hinted at a more simple and effective form of transport for the wealthy in Europe's congested cities. Ultimately, however, cities rose to the challenge of the increasing use of horse-drawn two- and four-wheeled carriages by adapting their thoroughfares wherever possible to their demands. In particular, horse-drawn carriages reinforced the desire of urban authorities to create, whenever the opportunity arose, straighter and broader streets with more open spaces for public use (Oldenburg, 1965–86, p.32; Straus, 1912, pp.88ff.).

However, where it was not possible to make such adjustments to existing street patterns, the introduction of this new form of conveyance might still prove, in the long run, to be beneficial. The early demographer John Graunt was clearly of the view that the failure of the cramped, pre-fire streets of the

[3]   For this and similar innovations in transport in seventeenth-century Paris, see Bernard's account in the Reader associated with this volume (Chant, 1999).

Coach in the time of Charles I
(From " Coach and Sedan Pleasantly Disputing")

Coach in the time of Charles II
(From Thrupp's " History of Coaches")

Figure 6.10 Developments in carriage and coach design in seventeenth-century England. The middle decades of the seventeenth century witnessed rapid progress in the design of the horse-drawn carriage. Notice in particular the more compact and streamlined nature of the later model (Straus, 1912, p.113)

city of London to accommodate the new volumes of carriages was one of the key factors behind the flight of the wealthy to the new suburbs of the West End. There the newly laid-out streets of Covent Garden and other developments were far more suited to the demands of the coach (see Figure 6.11). (For the failure of the old city to cope with traffic congestion occasioned by coaches, see the comments of John Graunt in Extract 6.1 at the end of this chapter.) The response of the Dutch to the potential problems of traffic congestion was unique. Ever prepared to experiment in difficult situations, the Amsterdam municipal authorities decided in 1615 to implement a scheme of one-way streets for commercial traffic. When this failed to stem the problem which had been considerably exacerbated by the horse-drawn carriage, the

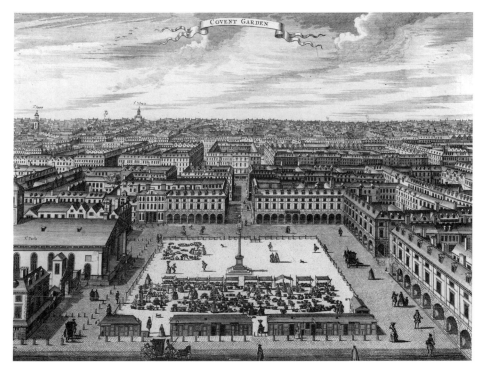

Figure 6.11 Covent Garden, c.1720, from an engraving by Sutton Nicholls after a design by Inigo Jones, 1631 (copyright © The British Museum, Department of Prints and Drawings, Crowle Pennant, vol.VI, no.7, p.200)

ultimate solution was imposed. In 1634, the municipal corporation forbade the use of all private carriages within the city (Zumthor, 1962, pp.15–16). Though in the last resort this proved unworkable, and some minor exemptions were introduced, it does testify to the peculiarly powerful, and even authoritarian, tone of so much Dutch municipal legislation in this period – a factor which was ultimately to drive the evolution of the technologically sophisticated Dutch cities of the second half of the seventeenth century.

Another by-product of the increasing numbers of carriages and coaches was a desire to improve road communication between cities, and not just to tinker with the major arteries of communications within individual cities. However, with a few notable exceptions, very little was achieved, with the result that long-distance travel for people and freight remained slow and uncommon and, in the case of trade, a major obstacle to further economic expansion. A far more effective system than roads for the transportation of bulk loads was river or canal transport, the latter largely undeveloped in this period except in particular cities (such as Venice) or regions (such as the United Provinces or modern-day Netherlands). Access to rivers thus remained a crucial factor in the establishment and growth of Europe's major cities at this time, a point underlined by their additional use as sources of water for both domestic and industrial purposes. The significance of rivers as arteries of transport was not lost on contemporaries. From the mid-sixteenth century onwards, individuals and urban authorities had sought ways to improve the existing waterways with some notable successes. In particular, the introduction of improved barges and pound locks (that is, locks with two gates) had vastly expanded the effective capacity of inland waterways – in the case of England it more than doubled between 1550 and 1700 according to one estimate (Chartres, 1977, pp.42–3). An early example of the effect of this process on an urban community is provided by the city of Exeter where, in the sixteenth century, the city's authorities struggled for over thirty years to apply the new techniques of canal-building and lock construction to create an effective channel linking the new city haven with the sea (MacCaffrey, 1975, pp.127–31).

The clear leaders in water transport technology in Early Modern Europe were the Dutch. From the sixteenth century onwards, Dutch engineers were at the forefront of canal and drainage technology – a necessity in a land which was itself largely created from the sea, and whose wealth and well-being were intimately tied to its ability to control the vagaries of river and maritime flooding (van der Venn, 1993). From the 1570s until the end of the seventeenth century, the very survival of the fledgeling Dutch nation, the United Provinces, was largely dependent on the ability of its people to harness new technology in the management of water. Faced with the threat of Spanish and then French domination from the south, the superior skill of the Dutch in the construction of canals, harbours, sluices and locks was frequently their most potent weapon in the wars of liberation and survival which dominated their history in this period. Water and its control thus played a crucial role in the life of the United Provinces, which by the middle decades of the seventeenth century contained the largest concentration of urban development in Early Modern Europe.

Not surprisingly, water also played a key role in transport, the canalization of its cities and major rivers allowing for unprecedented developments in inter-urban communication. The most startling development was the creation of the passenger transport network known as the *trekschuiten* which dates from the 1630s (see Figure 6.12). The *trekschuit* was a passenger-carrying barge pulled by one or more horses along either pre-existing or new stretches of canal with accompanying towpaths (see Figure 6.13). The novelty of the system lay not so much in its technological components, but rather in the way the canal system, the *trekvaart,* was administered and developed. By the 1660s, most of the major (and many minor) urban centres of the United Provinces were linked to the *trekvaart* which afforded unprecedented standards of dependability and comfort for its customers. In the words of its historian, Jan de Vries:

> the network endowed Holland's numerous cities with a measure of resistance to the powerful forces making them dependent upon and subordinate to Amsterdam. With direct accessibility to the whole region each city could hope to compete with the others and carve out an important role for itself. In short, the *trekvaart* system contributed to the organization of Holland's cities into an urban system.
>
> (de Vries, 1981, p.75)

As always, however, there was a price to pay for such success, and there can be little doubt that the Dutch dependence on water transport produced its own share of problems, particularly within the centres of the rapidly expanding Dutch cities. One particular problem related to the intersection of roads and canals in built-up areas where the traditionally steep-sided bridges of Dutch towns – designed to facilitate the movement of barge traffic through cities to their commercial centres – posed major problems to road traffic. As we have seen, one solution was to limit the movement of people and freight by road. However, where this was impractical – as in the carriage of much small freight – a solution was finally found by the introduction in Amsterdam in 1664 of special sleds or slide carriages. With the structure of a coach on the chassis of a sled, the slide carriage, with its iron slides, greased for extra speed and slowed down by the application of straw, was able to overcome the obstacles posed

Figure 6.12 The Dutch *trekvaart* network after 1665 (de Vries, 1981, p.35, map 2.2)

Figure 6.13 The *trekschuit* (de Vries, 1981, frontispiece)

by the typical hog-backed bridges of Dutch cities – yet another example of Dutch ingenuity in the face of problems associated with rapid urbanization (Zumthor, 1962, pp.16–17).

Hazards of the urban environment

At the beginning of our period, Europe's cities and towns were characteristically dirty, ill-paved and unhealthy places which, in whole or part, frequently burned down because they were largely built in timber and plaster. By 1700, however, a major process of improvement and alteration was under way which would transform the urban environment of most of Europe's cities. Though problems of poor paving and sanitation persisted in the suburbs, the vast majority of urban centres were caught up in an impressive programme of regeneration and reform. Improved systems of water supply enabled many of the larger cities to continue to grow to astonishing proportions – a considerable achievement in the age before steam. But most important of all, a revolution in building materials, combined with the growth of fire-fighting technology and fire insurance, saw an end to the massive conflagrations which had so frequently threatened the life and property of Europe's city-dwellers.

To what extent was the process of urban amelioration indebted to the introduction of new technology? Or would it be more accurate to ascribe the vast majority of the improvements to other factors? As we consider the changes in more detail, we will need to bear in mind debates about the role of political and cultural conditions (as opposed to purely technological ones) in this process. By the mid-eighteenth century, a change in the attitude of Europe's rulers, involving notions of human progress and the feasibility of rational schemes of improvement, was under way, a process which clearly impinged upon contemporary developments in urban planning and administration. However, the changes which I wish to discuss with respect to the environment of the Early Modern city clearly *pre-date* what is known to historians as the age of the Enlightenment (roughly the period from 1720 to 1790). More pertinent is the issue of the role of government (both national and urban) in schemes of urban improvement in this period. Where change did occur, was this largely the product of the action of enlightened rulers supported by powerful bureaucracies? And conversely, were technological initiatives redundant in the face of either official indifference or administrative incompetence? The relative success of the Dutch in promoting urban regeneration provides helpful clues here and contrasts with the relative failure of the French and the British in the same period.[4] Keeping these points in mind, we now look in more detail at developments in the environment of Europe's Early Modern cities to see how the rulers of Europe's cities differed in their response to the problems inherited from their medieval predecessors.

In most major cities at the beginning of our period, responsibility for activities such as street cleaning and paving lay with the individual householder who was responsible for his or her own patch of street. The records of pre-industrial cities, however, are littered with references to the failure of individuals to carry out these duties effectively. As a result, streets constituted a health hazard to pedestrians and residents alike; household waste and refuse were frequently to be found choking the open drains or culverts which typically ran down the middle of city thoroughfares. (A particularly vivid account is provided by Sir Kenelm Digby in his account of Paris in the mid-seventeenth century; see Extract 6.2 at the end of this chapter.) In the absence of proper sewers, water sources frequently became contaminated, with the result that diseases such as dysentery became an everyday hazard of urban

---

[4]   Israel's in-depth discussion of Dutch success in schemes of urban regeneration and its dependence upon new technology is given in the Reader associated with this volume (Chant, 1999).

living. In cities such as London, health problems were compounded by the growing use of coal as a domestic fuel. By the middle decades of the seventeenth century, its widespread use was creating appalling problems of pollution, so much so that, according to some commentators, the air of parts of the city was unbreathable (see Extract 6.2 at the end of this chapter for a graphic description of such problems). Add to these factors the appallingly cramped and overcrowded conditions of so many urban dwellers – themselves the product of the unprecedented growth of many of northern Europe's larger cities in the sixteenth and seventeenth centuries – and one can begin to appreciate the loathing which many felt for urban living in the period.

In the face of these problems, governments, both national and local, were not wholly inactive. In England, for example, the crown repeatedly issued proclamations from the late sixteenth century onwards forbidding the practice of dividing tenements within the city of London, and the city authorities likewise issued numerous ordinances relating to building practice (Knowles and Pitt, 1972, pp.18–36). Similar regulations were promulgated by the government of Henri IV in France in the first decade of the seventeenth century (Sutcliffe, 1993, p.19). The failure of government implied by the frequency with which these edicts were re-issued was not so much its reluctance to act in the face of such problems, but rather its inability to enforce observation of its laws and ordinances. To underline this point, one need look no further than the example of Dutch cities where the strict and successful enforcement of a whole variety of measures designed to make the streets safe and clean was frequently commented on by admiring foreign visitors. Indeed, in seventeenth-century Europe, the cleanliness of Dutch cities was proverbial. How then did the Dutch manage to succeed where everyone else failed? With some reservations, their success owes little to Dutch technical prowess, despite their well-deserved reputation for practicality and innovation. Rather, the solution lies much deeper, embedded in the broad mentality of the wider population (and not just the ruling class) who generally accepted schemes of urban improvement as part of a wider outlook which placed the communal good above the welfare of the individual citizen.

The origins of this approach are almost certainly traceable to the fact that from very early on (certainly before the onset of Dutch urbanization), the Dutch people were conditioned to accept the necessity of pooling individual knowledge and resources in the constant struggle for survival against the sea.[5] It would be a great mistake to assume from this, however, that legislation on behalf of the Dutch people – whether about schemes of land reclamation or urban improvement – was performed in a democratic manner, even by the standards of the seventeenth century. There is of course a delicious irony here in that the country which is seen, quite rightly, as one of the pioneers of modern capitalism was dominated by urban cliques whose authority was vastly more authoritarian than that of their regal counterparts in 'absolutist' France or the England of the 'divine right' Stuarts. The ruling elites of Dutch cities controlled every aspect of city life at this time and were therefore in an extremely powerful position to influence the shape and fabric of their communities. All the evidence suggests that they did so in no uncertain terms. Dutch municipal governors were among the first in Europe to develop compulsory purchase legislation. They were also ingenious in their use of existing laws to enforce their way. Land for building was of course in short supply in the Netherlands. Dutch cities, facing rapid population growth, were faced with a delicate problem: expand or rebuild. Expansion was often out of

---

[5]   The tracing of this 'democratic imperative' in Dutch urban administration to the Dutch people's historic struggle to overcome the threat of inundation from the sea is proposed, for example, by A.E.J. Morris (1985). The significance of the sea, and fear of inundation, in the consciousness of seventeenth-century Dutch men and women is fully explored in Schama (1987, Chapter 1).

the question given that many Dutch cities were so heavily fortified in the new style. Rebuilding thus became the only viable alternative for many. Such, for example, was the route chosen by the burghers of Utrecht, whose skill in overcoming some of the obstacles to demolition and reconstruction posed by recalcitrant citizens exemplified the Dutch approach to urban regeneration:

> New streets were laid out, and the interior of the building blocks was opened up through narrow lanes, passages, blind alleys and closes. These efforts culminated in a comprehensive redevelopment programme, worked out by Burgomaster Moreelse, which occupied the administration from 1640 to 1670. Streets were straightened or widened, individual buildings and rows of houses were demolished, and structures projecting beyond the straight building line were removed. The width of new streets was fixed at 20 to 14 feet [about 6–4 metres], and the width of the Zuijlensteeg, with its 'beautiful large houses', at 35 to 43 feet [about 10–13 metres]. An unusual procedure was followed in 1664: a number of citizens whose houses the Council wanted to pull down to widen the access to the town hall demanded exorbitant prices for their property. As a solution it was decided that the owners would not be allowed to carry out any repairs of their buildings. This produced the desired result: as values decreased, houses were gradually purchased and, finally, in 1700, the last house could be demolished.
>
> (Gutkind, 1971, p.101)

The destruction and renovation of cities on this scale in any other part of northern Europe at this time was largely unthinkable. In England and elsewhere, the only real opportunity to rebuild was provided by major fires such as that which famously destroyed nine-tenths of London in 1666. And even then, as we shall see, it often proved impossible to make anything more than minor modifications to the original street plan.

In those countries where the ability of governments to deliver major schemes of urban transformation was constrained by lack of political will or resources, improvements to the urban environment were none the less a notable feature of the late seventeenth century. Everywhere, brick, stone and tile were replacing timber, plaster and thatch as the dominant building materials. In the process, the adoption of such non-flammable materials substantially reduced the risk of fire. In addition, at every conceivable opportunity civic authorities were attempting to create city panoramas with wider thoroughfares and more open, public spaces and squares for the convenience of their citizens. Concomitant with these developments, urban governors became increasingly interested in schemes to improve the construction, cleansing and lighting of streets. New proposals to pave and cleanse the streets proliferate from the mid-seventeenth century onwards, though in most cases these did not involve technological initiatives so much as administrative reforms (in particular, the transfer of responsibility from the individual householder to the ward or parish). In London, for example, it was proposed in the 1650s to privatize the overseeing of street-cleaning in the capital – one measure which attracted a great deal of interest among followers of the influential German *émigré* Samuel Hartlib (*c*.1600–62). Hartlib's circle were eager to promote the benefits of technology to a largely sceptical audience (Jenner, 1994, pp.343–56).

One aspect of urban improvement did involve technological innovation: the various experiments carried out from the middle decades of the seventeenth century into urban street lighting. Early attempts to provide some form of lighting for the streets of Europe's cities were largely inadequate.[6] However, by the 1650s there was a great deal of interest in various solutions to the problem, much of it stimulated by growing scientific interest in the properties of lenses and mirrors, as well as the increasing accomplishment of those technicians and

---

[6]   Bernard's account of the various experiments carried out in Paris in the second half of the seventeenth century is given in the Reader associated with this volume (Chant, 1999).

craft specialists who made such instruments. Here, undoubtedly, was an area of Early Modern life where the technological and the scientific came together in fruitful partnership (Hartlib, 1634–60, 30/4/67A, 28/2/81B–82A; Oldenburg, 1966, pp.201, 207–8). During the course of the 1660s and 1670s, experimentation finally erupted into practical application when, first in Paris (Bernard, 1970, pp.162–6) and then in Amsterdam (Multhauf, 1985, pp.236–52), the city authorities implemented radical new plans for the lighting of their cities.[7] In Amsterdam, the scheme of the artist and technical whiz-kid Jan van der Heyden (1637–1712) was particularly novel since it was based on the use of oil lamps (see Figures 6.14–6.18 overleaf). None the less, it is possible to take issue with those who, like Jonathan Israel, have seen this as 'one of the outstanding examples of the successful application of technology to daily life' (Israel, 1995a, p.681). Arguably, the greater contribution to the success of the scheme – soon imitated elsewhere – lay in the development of efficient administrative procedures. This may help to explain why the same system appeared to function far more efficiently in Amsterdam in 1670 than it did in London a decade later (Israel, 1995a, p.681). Moreover, not everyone has accepted the view that van der Heyden's scheme marked a technological breakthrough. Malcolm Falkus, for example, in his study of London's street lighting in this period claims that no new technology was involved in the construction of the lamps (Falkus, 1976, p.254).

Another area of urban life in which conspicuous improvement took place in the same period was water supply. As the larger cities of Early Modern Europe continued to expand, the need for access to new and larger supplies of water, both for domestic and industrial use, became urgent. One solution was to pipe water into the cities from the surrounding countryside, as at Worcester and Coventry in England in the sixteenth century. On occasions, it was possible to divert and channel small rivers through narrow culverts and aqueducts to reservoirs located on the outskirts of towns and cities. This was the case in the successful implementation in the reign of King James I (1603–25) of the New River project which conveyed water from Hertfordshire to London (Gough, 1964; Howell, 1657, p.11; see Figure 6.19, p.206). Of course, all such measures ultimately proved incapable of supplying the constantly increasing needs of a city such as London, and ever more grandiose proposals were put forward to alleviate the problem. Part of the difficulty lay in the technological limitations of the means to carry water over long distances. Though lead was used for water-piping in cities, it was far too expensive to be employed over long distances; wooden pipes (usually made of elm in England) were used instead with the result that leakages were high (see Figure 6.20, p.206). Most important of all, however, the inability to pump water effectively from rivers or artificial channels severely limited the amount of water that could be raised and distributed to householders. By the middle decades of the seventeenth century, this problem was arousing widespread interest among engineers and 'projectors' (entrepreneurs), and numerous attempts were made to make more efficient and capacious pumps (culminating in the first experiments with steam around 1700) (Hartlib, 1634–60, 29/5/59B, 28/2/81B, 29/4/2A–B; Jenkins, 1928–9, pp.43–51). By the early decades of the eighteenth century, real progress had been made, with the result that not only cities such as London and Paris were able to keep pace with the growing demand for water, but increasingly provincial towns and cities were able to benefit from improved systems of water supply (Williamson, 1936, pp.43–93).

Advances in hydraulic pumping and water supply during the seventeenth century paved the way for important developments in the technology of fire-fighting. Before the middle decades of the seventeenth century, fire-fighting

[7] These are discussed in more detail by Bernard and Israel in the Reader associated with this volume (Chant, 1999).

Figures 6.14–6.18 Van der Heyden's street-lighting system (Historisch Topographische Atlas, Amsterdam; by permission of the Gemeentearchief, Amsterdam)

Figure 6.14 Van der Heyden's description of the construction and maintenance of street lamps. The inscription on the banner reads 'The light of lamp-lanterns enkindled by Jan van der Heyden, their inventor and supervisor general of the city lanterns of Amsterdam.'

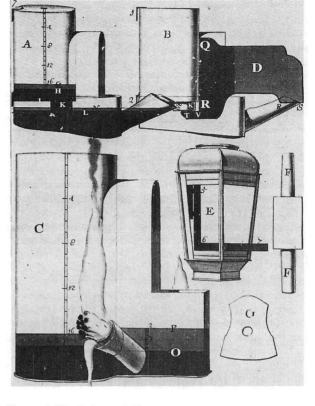

Figure 6.15 C (lower left) represents the unsatisfactory earlier model of the lamp. The oil would rise from O to P and spill out the wick holder. A is the improved design: a closed container is connected with the open area in front by a short pipe (K) and separated from the lower basin by an interior divider (I). When the oil was filled up to G, the heated expanding air would push it back to H, and so the level in the open basin would rise only from L to N, which was below the level of the wick holder. By allowing a broader area in which the oil could expand and a narrow connection between the air container and the oil reservoir below, van der Heyden limited the rise of the oil level at the wick. F and G were used to make the wick holder; the front had to be pinched together to prevent the end of the wick from dipping into the oil. This had happened with some of the first lanterns, and the dripping burning oil set the greasy lamppost alight

equipment was rudimentary and consisted largely of buckets and firehooks; the latter were used to demolish buildings which were either alight or in danger of catching fire (see Figure 6.21, p.206).[8] Primitive fire-engines are first recorded in mid-sixteenth-century German cities such as Nuremberg, but they are not reported in universal use in Europe's cities until the 1620s and 1630s. From this point forward, experimentation with various forms of pump action, allied to innovations in the administration of such appliances and the men who were to use them, became a common feature of urban life (Karslake, 1929, pp.229–38; Jenkins, 1930–31, pp.15–25; Porter, 1996, pp.1–28; see Figure 6.22, p.207). It is difficult to assess precisely the success of these measures, but there can be little doubt that, together with the other improvements in street design and the use of non-flammable building materials, such technological changes assisted in creating safer cities (Porter, 1996, p.153).

---

[8]   More information on this is given by Porter in the Reader associated with this volume (Chant, 1999).

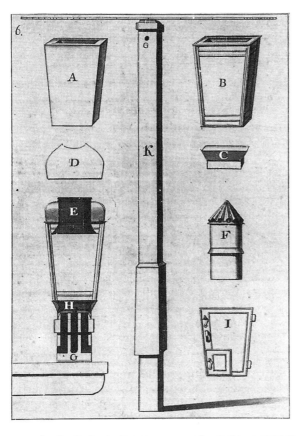

Figure 6.16 Construction of the lanterns with various patterns indicate that van der Heyden was aware of the advantages of interchangeable parts. A is a wooden box twenty-four centimetres high, used to obtain the exact shape. Tin strips for the sides and corners were soldered around it (B). C is the wooden pattern for the foot, and D the pattern for the top. E shows the attached chimney, into which the removable chimney (F) was lowered. When F was removed, all the soot came along and the chimney could easily be cleaned outside the lantern. The air holes (G) were drilled in from the side of the post and then angled up to come out at the top. H shows the base, made with four angled feet to keep it away from the bottom and sides of the lantern so that air and wind, entering through the airholes, were directed sideways along the glass and away from the light. I shows the door at the back of the lantern, which was completely lined with baize; the lamp was lighted through the small lower door

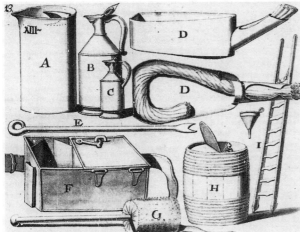

Figure 6.17 The tools of the caretakers. Oil cans came in twenty-two sizes, the smallest holding enough to fuel a lamp for three hours, and increasing by half an hour for each size, up to the largest, at thirteen and a half hours. D designates a copper model that showed exactly how the wick should be trimmed and placed in a lamp; van der Heyden wrote that this could be seen 'much better with those models than could be understood from a description'

Figure 6.18 This medal, designed by von Wermuth for Leipzig in 1702, shows a typical van der Heyden lamppost; the lamp has just been lit by the lamplighter, and a student is reading a book in its light

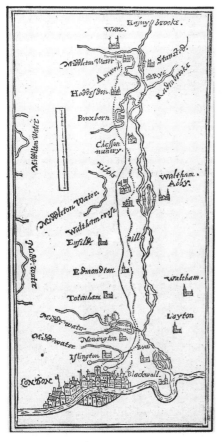

Figure 6.20 Pipe-boring works (Evelyn, 1664; reproduced by permission of The Natural History Museum, London)

Figure 6.19 Myddleton's New River project, from a woodcut of 1641 (reproduced by permission of the Pepys Library, Magdalene College, Cambridge)

Figure 6.21 Woodcut of early fire-fighting techniques at Tiverton, Devon, c.1608 (reproduced by permission of The Guildhall Library, London)

Figure 6.22 Van der Heyden's new fire-fighting system. This beautiful engraving is one of nineteen which the artist van der Heyden published in Amsterdam in 1690 (Rijksmuseum, Amsterdam)

It is probably fair to conclude that by 1700, Europe's cities were cleaner, healthier and safer environments in which to live than those inhabited by earlier generations. Moreover, even where certain problems persisted, unaffected by any new technology, it was assuredly the case that understanding of the root causes of such problems was growing. This is evident in the case of urban pollution which was attracting a great deal of interest in scientific circles after 1660 (Brimblecombe, 1978, pp.123–9). Among devices demonstrated to the Royal Society in 1686 was an 'engine that consumes smoke'. Moreover, interest in such ecologically sensitive issues clearly pre-dated the Restoration of Charles II. In 1638, a Mr Wolfen claimed to have invented a smokeless fuel (Sharpe, 1992, p.258). And twelve years later, Hartlib reported that Sir Hugh Platt had invented a device which reduced pollution from coal smoke in London. He also advertised the invention of a Berkshire man who had 'invented a new kind of chimney wher[e] by a Million pounds might bee saved in fuel to the Nation' (Hartlib, 1634–60, 28/1/42A, 29/6/21B). At the same time, there was greater awareness of the connection between health and urban pollution (see Extract 6.2 at the end of this chapter)[9] as more concerted attempts were made to initiate a public debate on the subject (Jenner, 1995, pp.535–51).

[9]   See also Bernard in the Reader associated with this volume (Chant, 1999).

6.3   Three case-studies: Amsterdam, Paris and London

Introduction

The various processes of urban change, and their relationship to technological innovation, are perhaps best approached from the point of view of individual case-studies which allow the historian to make comparisons across national, political, religious and cultural boundaries. In looking at three of the largest cities in Early Modern Europe, we should thus be in a position to examine whether technological innovation was specific to certain cities and regions with, for example, their own distinctive systems of government or religious belief. Or, alternatively, we might look to see whether the problems facing these cities, and the solutions devised to deal with them, were the same regardless of political, religious and other differences. Either way, in order to assess the impact of such variables on the role of technology in urban development in this period, we shall first need to consider briefly the historical context of our three chosen cities.

Amsterdam, the chief city of Holland (which was itself the most powerful province in the loosely-knit federal republic of the United Provinces), was a largely autonomous city ruled over by a powerful clique of merchant-adventurers – so independent and powerful, in fact, that one historian of Dutch culture has referred to it as 'a virtual city state' (Schama, 1987, p.117). It was, in addition, a haven of religious freedom and home to all manner of religious and political refugees, its espousal of religious tolerance a product of the long Dutch struggle for independence in the War of Liberation against the Spanish (1567–1648).

Paris, on the other hand, was a predominantly Catholic city with little urban autonomy. It was dominated by a succession of powerful monarchs, or their deputies, who increasingly aspired in the seventeenth century to the creation of an 'absolutist' state. While recent historians have pointed to a marked divergence between the reality and theory of Early Modern French absolutism, stressing the limited capacity of the *ancien régime* state to implement the wishes of an authoritarian monarchy, it is none the less the case that France under Louis XIV, the Sun King, was subject to unprecedented levels of centralized royal control. In the case of Paris, the effect was slightly mitigated by Louis's desertion of the capital for a new administrative centre for the nation at Versailles. To what extent this affected Louis's approach to the problems of Paris is a moot point, and one which we will need to consider when studying developments in the capital during his long reign.

London, finally, lies somewhere between the two extremes represented by Amsterdam and Paris. Frequently caught up during the seventeenth century in the religious and political struggles between the monarchy and supporters of Parliament, the city of London was semi-autonomous and governed by a merchant oligarchy that paid lip-service to the monarchy whose power base lay to the west in the city of Westminster. The Stuarts themselves had long harboured plans to exert greater control over the city, none more so than Charles I (1625–49), whose aspirations to a revitalized 'divine right' monarchy ultimately ended in failure and execution. However, strains of such 'absolutist' thinking returned in 1660 with the restoration of Charles's son, Charles II (1660–85), and we shall need to keep such ideas in mind when considering the plans for the rebuilding of London after the Great Fire of 1666.

Another consideration which we shall need to take into account is the extent to which the religious and political conditions prevailing in these three cities may have been more or less conducive to technological innovation. It has long been held, though never satisfactorily proved, that Protestantism – in particular, the more militant, Calvinistic version of the faith – promoted awareness and understanding of scientific and educational reform. If this was the case, then we might reasonably expect London under puritan rule (1642–60), and seventeenth-century Amsterdam, to be far more receptive to new technology. Catholic Paris, on the other hand, might be expected to lag well behind. Recent research into interest in technology in both London and Amsterdam has found some evidence to support the view that both harboured a growing number of intellectuals and artisans who were actively forwarding the cause of the new Baconian science. As informal circles of scholars began to coalesce around influential figures such as Hartlib in London during the 1640s and 1650s, so it became increasingly acceptable and popular to promote and publicize the benefits of new technology which prophets of the new science such as Francis Bacon had predicted would emancipate Early Modern men and women. Similarly in Amsterdam and other Dutch cities, there is a great deal of evidence to support the idea that the religious and political environment was conducive to new technological and scientific initiatives (Davids, 1993, p.94; Davids, 1995, pp.346–7). In the case of England, this growing acceptance of the merits of scientific and technological thinking received official sanction in 1660 with the foundation of the Royal Society – a body specifically set up to promote such activity. Today, its critics tend to be dismissive about its contribution to practical science and the improvement of everyday life in the second half of the seventeenth century. It is none the less the case that many members active in its early proceedings were acutely conscious of the part which science, and its rational application, might play in improving the condition of human society. And many, moreover, as we shall see, were fully alert to the potential of such developments to meet the growing concern of the ruling classes for the environment of England's cities, particularly London. (For contemporary discussions of some of these problems, see Extracts 6.1 and 6.3 at the end of this chapter.[10])

The extent to which these developments – especially those in London – can be attributed to the predominantly Protestant culture of England and the United Provinces has to be measured against the experience of Catholic France. Paris, for example, in the middle decades of the seventeenth century, could also boast a growing interest in the new, pragmatic Baconian science – one which, as in London, saw the formal institutionalization of its worth in the creation of the Académie Royale des Sciences by Louis XIV in 1666. Interestingly, it suffered the same fate as London's Royal Society, condemned as it was for its failure to produce genuine practical benefits for the people of France (Briggs, 1991, pp.38–88). Why these fledgling societies should have succumbed to the same fate is not clear. One circumstance in common, however, was royal patronage. Were the aspirations of two monarchies to absolute rule over their citizens inimical to the pursuit of practical scientific knowledge? Some have argued for this view, contrasting the relative failure of the Stuarts and Bourbons to instigate practical programmes of technological research with the apparent success of the Dutch where no such control over scientific education was exerted by the mercantile ruling elites of the United Provinces. These are important dimensions to our story which we will need to consider in the following case-studies, beginning with the 'economic miracle' of the seventeenth century, Amsterdam.

Amsterdam

Amsterdam's rapid rise to economic prosperity rested on two fundamental considerations. First, it owed its pre-eminence as one of the major trading centres in northern Europe in the seventeenth century to a combination of

---

[10] See also Evelyn in the Reader associated with this volume (Chant, 1999).

Figure 6.23 Amsterdam,
c.1544, by Cornelis Anthonisz
(Burke, 1956, p.142, plate 94)

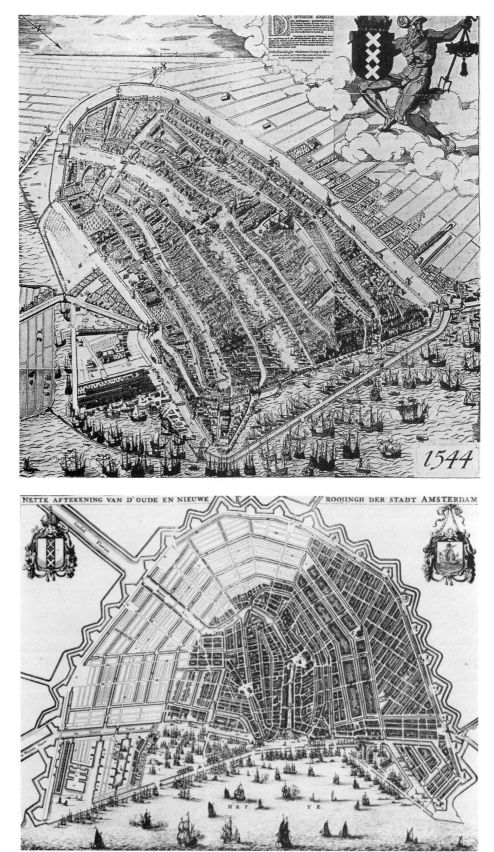

Figure 6.24 Amsterdam, with plan of proposed extensions by city architect, Daniel Stolpaert
(1664). Note the extensive development which had already taken place to the east of the old
city in the new working-class districts of the Joordaan (Kostof, 1991, p.137, plate 135;
photograph: copyright Thames and Hudson)

conditions – economic, religious and political – which had seen it supplant Antwerp as the chief commercial centre of the Low Countries. And second, and more prosaically, it owed its entire existence to the unrivalled skill of the Dutch in controlling and manipulating their watery natural environment. In Amsterdam, as in so many other Dutch towns and cities, water was both an asset and a hindrance to urban success and survival. Land, both rural and urban, was laboriously reclaimed from the sea and marshy river estuaries, and once reclaimed had to be defended with a complex network of dikes and sluices. Furthermore, in the case of land set aside for urban use, it had to be made suitable for habitation through the construction of solid timber foundations – an enormous undertaking exemplified by the city of Amsterdam which was built on a vast number of timber piles. (For a contemporary description of this process see Extract 6.4 at the end of this chapter.) Water was thus both the life-blood and the potential destroyer of cities such as Amsterdam, and it would be foolish to minimize the importance of Dutch technical skill in hydraulic engineering for the urbanization of the United Provinces (see Figures 6.23 and 6.24).

As all accounts of Amsterdam show, the damming of the river Amstel and the subsequent canalization of the city lay at the heart of the city's well-being (see Extract 6.5 at the end of this chapter). Restrictive as this was, it seems to have generated in the citizens of Amsterdam, or at least its ruling elite, a precocious desire for strict control of building within the walled boundaries of the city. Largely rebuilt after two disastrous fires in the late fifteenth century, in 1521 the city passed a building ordinance which required the demolition of all wooden and thatched structures and their replacement with brick and tile constructions. Further ordinances followed in the sixteenth and early seventeenth centuries, underlining the commitment of the city's ruling elite to public health and sanitation. In 1533, multiple occupation of houses, itself a product of Amsterdam's expanding economy and population, was beginning to create a health risk, particularly as upper floors possessed no sanitary provision. Household waste and refuse that was emptied out of windows posed an obvious threat to public health and private safety. The city authorities responded by obliging all house owners to install sinks which emptied, by way of lead soil-pipes, to external drains. It further forbade the building of covered culverts or sewers unless they were fitted at suitable intervals with detachable inspection covers (Burke, 1956, pp.144–5). However, preventing people from emptying waste into the rivers and canals which flowed through the city proved altogether more difficult. Quite frequently, especially at the height of summer, the water in the canals became stagnant and highly polluted, with epidemic disease an inevitable consequence. At Leyden in 1624 over 14,000 people were said to have died from an epidemic caused by the drying-up of polluted waterways (Gutkind, 1971, p.78). One solution was to fill in disused or impractical *grachten* (canals), though some cities devised more sophisticated mechanisms; these involved occasional flushing of the whole urban network of canals by the manipulation of the large sluices which controlled the flow of water through Dutch cities. The process, with its attendant problems, is well described by the English traveller and Fellow of the Royal Society, William Aglionby, in his account of Leyden in the 1660s:

> In summer during the hot weather, these Channels do send forth a noysome smell, particularly when the weather inclines to rain; the reason of it is, the drying up of the Lake of Soetermeer, which did use to cleanse the Town by flowing into it. To prevent this, the Magistrates have caus'd two large Channels to be made, and two Mills to be set upon them, to drive the Water into the Town at one end, and two other that drive or carry it out at the other end; so that by this invention the City is free'd from that noysom and infectious smell, though offen it fail too in the great heat of summer, when there is a great calm, and no winde stirring.
>
> (A(glionby), 1669, pp.250–51)

At Amsterdam, a modified version of this system was introduced which made use of the tidal currents of the IJ. Twice every twenty-four hours at high tide, water was let into the town by means of a number of sluices. The water was then routed around the city in a systematic fashion by the opening and closing of sluices, the new water so introduced driving the polluted water forwards until it was emptied back into the IJ by the use of four 'dirty-water mills'. Though the system was never perfect – a full solution to the problem awaited the introduction of purification plants and sewers in the nineteenth and twentieth centuries – it went some way to reducing the high levels of pollution which threatened the health of the citizens of the city (van der Venn, 1993, p.120).

Amsterdam's early lead in civic legislation concerning the health and environment of the sixteenth-century city is particularly impressive and may have facilitated the unprecedented growth of the city's population at this time. This is certainly the view of Israel who cites the decision of the municipal authorities to introduce a civic garbage and waste-disposal service in the 1590s as one example of the commitment of Amsterdam's governing class to improving the health and sanitation of the citizens. He also refers to recent figures which suggest that Amsterdam's population increased from 30,000 in 1570 to 105,000 in 1622 and 140,000 in 1647. Similar rates of growth are recorded for numerous other Dutch cities in this period, though not on the same scale as that of Amsterdam, which was twice as large as its nearest competitor, Leyden (Israel, 1995a, pp.328–9). Such growth, however, posed other problems and it was soon clear by the early decades of the seventeenth century that an ambitious plan of urban expansion was called for if the city was to accommodate its rapidly growing numbers. Once again, the burgomasters of Amsterdam defied conventional practice by envisaging large-scale growth outside the old city walls in the so-called *drie grachten plan* (plan of the three canals) (see Figure 6.24,p.211), the most novel feature of which was the power it bestowed on the architects and planners to control and regulate the expansion of the city (Burke, 1956, pp.147–53). Most startling of all, perhaps, is the fact that with the exception of the western industrial district known as the Joordaan, the private developers and speculators who carried out this massive programme of urban development abided, almost without fail, by the various covenants, leases and regulations laid down by the civic authorities. It is difficult to envisage a more fitting testament to the ability of the Dutch to marry entrepreneurial individualism with a sense of communal well-being.

Another conspicuous feature of the new Amsterdam, like so many other cities of Holland and Zeeland, was the extent of public-building projects. Unlike many other large and prosperous European cities of the time, such building was largely confined, with the exception of van Campen's magnificent, classical town hall (see Figure 6.25), to projects of social engineering. More than any other people in Early Modern Europe, the Dutch had responded to the problems of poverty, unemployment, homelessness and vagrancy through the foundation of new civic institutions, and the burghers of Amsterdam were no exception. Foreign visitors marvelled at the absence of begging and crime on the streets of the city (a commonplace feature of urban life in the rest of Europe) and they routinely paid homage to Dutch success in dealing with this problem by visiting the various public institutions which had been created in the sixteenth and seventeenth centuries. Thus among the sights of Amsterdam was the *Tugthuis* or *Rasphuis,* a house of correction for young, indigent men which was said to contain a sunken chamber in which those who refused to work for their board and accommodation were

Figure 6.25 Van Campen's town hall, Amsterdam; painting by Jan van der Heyden. As a young man, van der Heyden (1637–1712) was more celebrated as a painter, in particular specializing in cityscapes. Increasingly, however, he turned his attention to the practical problems facing the city of Amsterdam, where he lived from 1661, culminating in his radical proposals for lighting the city and providing it with a fire service (Musée du Louvre, Paris; photograph: Bridgeman Art Library, Giraudon)

deposited. According to the English traveller Edward Browne who visited Amsterdam in 1668, the overseers:

> told us that some that were committed to their charge, and not to be brought to work by blows, they placed in a large Cistern, and let the water in upon them, placing only a Pump by them for their relief, whereby they are forced to labour for their lives, and to free themselves from drowning.
>
> (Browne, 1685, p.97)

A more appropriate metaphor for Dutch civic virtue would be difficult to imagine, the symbolism of the struggle between water and the pump, nature and human artifice, being *the* recurrent image of everyday life in the United Provinces. Other municipal institutions often visited and commented on by the curious traveller included the *Spinhuis* (female house of correction), *Weeshuis* (hospital for sick children), *Dolhuis* (madhouse), *Gasthuis* (general hospital) and *Mannenhuis* (hostel for the old and the poor). There was also a wide variety of places of entertainment and leisure – an indication of the growing demands of a consumer culture – including by one account 'a Musick-house or Entertaining house, where any one is admitted for a Stiver, hears most sorts of Musick, sees many good Water-works and divers Motions by Clock-work, Pictures, and other Divertissements [diversions]' (Browne, 1685, p.100). Seemingly, even in their leisure time, ordinary Dutch men and women were fascinated by technical ingenuity – evidence yet again, perhaps, of the greater integration of the technical arts into Dutch everyday life.

Though new technology may not have been a central feature of this process of urban regeneration, there is little doubt that the existing technology was fully in harmony with contemporary needs and demands. Like so many other Dutch towns and cities of the period, Amsterdam was evidently a paragon of civic virtue. Its houses and public buildings were clearly well cared for and its streets and public places clean and secure. By the 1670s, it had one of the most effective and advanced systems of street lighting in Europe, and its fire regulations were second to none. Even in the one instance where the city lagged behind its counterparts in other parts of Europe, namely in an adequate supply of fresh water, simple solutions prevailed. Drinking-water, as some foreign visitors were only too keen to point out, was in desperately short supply in low-lying Amsterdam (see Extract 6.4 at the end of this chapter). Given Amsterdam's geographical situation, wells, conduits and fountains were all out of the question. As in so much else, the city was reliant on nature and artifice combined to supply the deficiency. Accordingly, most needs were met by the collection of rainwater in roof-top cisterns, found on both public and private buildings. In addition, special boats were used to import fresh drinking-water into the city on a daily basis, and when freezing interrupted supplies in the winter a group of entrepreneurial brewers in 1651 invested in an ice-breaker, fitted with a device first patented in Holland twenty years earlier. The system of water-boats continued until the early nineteenth century when gradually more advanced technological systems were introduced (Davids, 1993, pp.88–9; van der Venn, 1993, p.120).

The city authorities of Amsterdam were practical, made full use of current and developing technologies, and placed pragmatism above idealism in overseeing the management and reconstruction of the city's built environment. They included burgomasters of the stamp of Johannes Hudde (1628–1704) whose many achievements included the commitment of public money to advanced schemes of scientific research, research into methods of improving water supply in the province of Gelderland and the establishment of an accurate method for establishing sea-level throughout the United Provinces (Israel, 1995a, p.907; Lambert, 1971, pp.219–20; Struik, 1981, p.86). The genius of men such as Hudde and his colleagues in government lay not so much in their support for technological innovation *per se,* but rather in their ability to create visionary urban bureaucracies and to pass progressive legislation, while at the same time securing the support and approval of those whom they governed – a goal which proved highly elusive elsewhere in Europe at this time. The 'democratic' image of the Dutch in their 'golden age' is a deceptive one which cannot sufficiently explain the success of cities such as Amsterdam. On the contrary, in many respects it was the 'absolutist' qualities of Dutch urban government which provided the vital spark. Unburdened by a centralized tier of national government, the authority vested in the city authorities was typically wide-ranging and unconditional. It is therefore no coincidence that such vital tools of the modern urban planner as compulsory purchase orders were widely used here in the sixteenth century, to be invoked, alongside a whole stream of municipal legislation, whenever the need arose. And yet, the dictatorial powers of an urban elite would not by themselves have guaranteed success; it was also necessary that the vast majority of their fellow citizens subscribed to, and underwrote, the aims of the governors. Once again, we return to that sense of corporate pride and partnership which a number of historians have earmarked as the fundamental source of Dutch success – a fierce commitment to the public good upon which, the Dutch intuitively understood, their past success and present survival depended. In the last resort, the success of Dutch cities such as Amsterdam was based not simply on the well-founded pragmatism of its people as expressed in the hydraulic skills of its engineers; rather it is to be located in the cultural make-up of the Dutch – a facet of the Dutch character most evident, perhaps, in their attitude to civic cleanliness and public hygiene. This trait of the Dutch people is admirably captured by the historian Simon Schama:

> The threshold of shame about tainting the neighbourhood was very low indeed in Holland. In some towns it was the neighbourhood authority, the *buurt*, that was formally responsible for keeping streets, and sometimes canals, clean. Because their jurisdiction was so small and their watch so omniscient, it was relatively easy to identify offenders against the communal canon. And there were legal sanctions to back up neighbourhood odium. Any person injured or soiled by refuse and filth thrown from a window to the common street might claim double reparations before the magistrate … [By] the eighteenth century, moralists accused those who polluted canals and avenues of, in effect, a kind of social treason: of being in league with the civic enemy, contagion that stood at the gates. To be filthy was to expose the population to the illicit entry of disease and the vagrant vermin that were said to be its carriers. To throw a dead cat in the canal, to harbour an illegal immigrant, or to neglect one's duty of washing the pavement were all tantamount to delinquency – as if one had opened the gates to an enemy of infected marauders.
>
> (Schama, 1987, p.378)

Where cleanliness was equated not simply with godliness but also with civic patriotism, technological solutions to the everyday problems of town life were never likely to be central to the concerns either of citizens or of urban governors. While never neglecting such solutions where practicable, the burgomasters of Amsterdam do not appear to have been excessively active in the propagation of technological initiatives. No formal societies for the promotion and pursuit of scientific research were established in any of the towns and cities of the United Provinces. On the other hand, it should be emphasized that none of the prejudices against a technical education prevalent in other European countries, particularly among the wealthy, are evident in Dutch elite culture. In this sense, it might be said that technology was better assimilated by the Dutch, its benefits always likely to receive a warmer reception in the Netherlands than in other parts of Europe where any form of manual labour was sneered at. And to understand this fact, we need look no further than the seas which consistently threatened to overwhelm them since in large part they were held in check by the Dutch genius for techniques in water management.

Paris

Unlike Amsterdam, the political authority wielded by the city of Paris had long been established by the end of the sixteenth century. Its place in the economic, social, religious, political and cultural life of the people was also very different from that of Amsterdam. Whereas Amsterdam existed largely as a minor 'city state', formally part of the United Provinces, but in practical terms semi-detached from its highly federalized structure, Paris was first and foremost a capital city and the home of the French monarchy. Though only loosely governed by the indifferent Valois monarchs in the first half of the sixteenth century, Paris remained, in theory at least, a royal city in which the authority of the king was absolute. The day-to-day running of the city was delegated to an urban patriciate, but in real terms the power of the leading citizens of Paris to control their own affairs and to innovate in the life of the city was highly circumscribed and could only be invoked when weak or uninterested monarchs allowed the Parisian ruling class a free rein. Such conditions had prevailed for much of the sixteenth century, but in 1589 this state of affairs came to an end with the accession of the first Bourbon monarch, Henri IV.

Under Henri IV (1589–1610), the transformation of Paris was begun, a process not completed until the long reign of Louis XIV (1643–1715). For both men, the renewal of Paris, devastated during the sixteenth-century Wars of Religion, was envisaged as part of a much wider programme of reform, intended to reflect the growing power and authority of the French monarchy. It was the product of an increasingly centralized system of royal administration which historians have characterized as 'absolutism'. The debate currently raging in historical circles with reference to this term 'absolutist' need not concern us unduly here. Suffice it to say that many believe it to be simplistic and anachronistic, particularly in ascribing the language of totalitarianism to individuals and governments of the Early Modern era. It is none the less true that by the end of the seventeenth century the French monarchy had evolved into one of the most powerful regimes yet seen in Europe, its ability to act without recourse to constitutional checks and balances making it a formidable institution. The case of seventeenth-century Paris, therefore, is an intriguing one, for it provides us with another model for the politics of urban regeneration and its relationship to the role of technological innovation in this period.

The first phase in the renewal of Paris was largely the responsibility of Henri IV (see Figure 6.26). When Henri finally entered Paris in 1594, he discovered a ravaged city, the product of many years of neglect and siege warfare. Despite these problems, Paris had continued to grow rapidly in the last half of the sixteenth century (c.350,000 by 1600) and, in common with many other European towns and cities, much of the growth consisted of poverty-stricken country people in search of work. Henri was determined to improve the condition of the city, not simply to give his new and insecure regime a veneer of respectability, but also to provide a boost to the depressed economy of Paris. Consequently, under the able stewardship of his trusty minister, Maximilien de Béthune, Duke of Sully (1559–1641), the government of Paris was firmly subordinated to the rule and will of the new king and an unprecedented programme of public works and urban renewal was launched. Under Henri IV, for example, several royal palaces, including the old Louvre, were rebuilt and expanded and new town squares were constructed according to the newly fashionable principles of Renaissance town-planning. A new bridge, the Pont Neuf, was completed in 1606 (see Figure 6.27) and a year later construction began of the Hôpital Saint-Louis (situated outside the city walls to the north of the city), the 'first monumental hospital in Europe for exclusive treatment of the plague' (Ballon, 1991, p.166; see Figure 6.28). In addition, new and wider roads were constructed, initiatives in street paving and cleansing were undertaken, and a new building code was introduced in 1607 which prohibited construction in wood and directed owners of empty property lots to build along the street edge in order to create a more uniform street frontage. Typically, the responsibility for the introduction and overseeing of these measures was transferred from the municipal authorities to Henri's chief minister, Sully (Sutcliffe, 1993, p.19).

There can be little doubt that Henri's prime motivation in all this activity was to add lustre to the new regime, a political ploy to strengthen his fragile claim to the French throne (indeed, he was assassinated in 1610 before many of his projects came to fruition). Nevertheless, Henri was also genuinely concerned with the urban fabric and environment of his capital city and was acutely aware of the economic significance of his programme of public works, as is evident in the construction of the Place Royale (now the Place des Vosges; see Figure 6.29, p.218). This much is apparent from the research of Hilary Ballon who has concluded that the Place Royale was not originally conceived as an aristocratic residential square (its ultimate destiny), but rather as a commercial square which represented 'the centre piece of a royal campaign to stimulate

Figure 6.26 South-facing view of Paris, 1615, by Mathieu Merian (Collection Bibliothèque Nationale de France, Paris; B.N. Est. Va 419j)

Figure 6.27 The Pont Neuf and Place Dauphine, situated at the northern end of the Ile de la Cité; engraving by Israël Sylvestre, *c.*1660 (Collection Bibliothèque Nationale de France, Paris; B.N. Est. Ed 45)

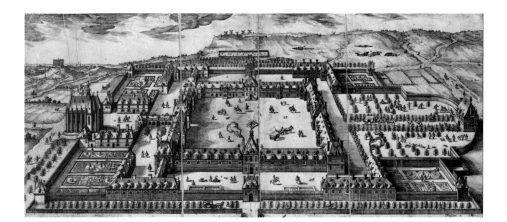

Figure 6.28 Hôpital Saint-Louis (built 1607–12); engraving by Claude Vellefaux, after Claude de Chastillon, *c.*1615. The plague hospital was situated outside the city walls and its construction was part of a much wider scheme of social, medical and urban reform initiated by Henri IV (Collection Bibliothèque Nationale de France, Paris; B.N. Est. rés. Ve9)

Figure 6.29 The Place Royale (Place des Vosges). Built by Henri IV as an innovative project to attract new commercial ventures to Paris, it soon became an aristocratic residential square used for spectacular royal ceremonies, as in this painting which depicts celebrations in honour of the impending marriage between Louis XIII and Anne of Austria (Collection Musée de la Ville de Paris/Musée Carnavalet; photograph: Giraudon)

French manufacturing' (Ballon, 1991, p.58). From the perspective of the present series, then, Henri IV is an interesting figure since his vision of the urban regeneration of Paris was informed by far more than simple political showmanship or aesthetic taste. He was essentially a practical man who both understood, and revelled in, the practical arts which he encouraged whenever possible, partly with the city of Paris in mind. Accordingly, in 1601 he created a new Commission on Commerce which was designed to promote domestic industry and to encourage skilled craft specialists from other parts of Europe to settle in Paris and other French cities. Henri's famous declaration of religious freedom for the minority of French Protestants can be seen as part of this general plan, in much the same way that guarantees of religious liberty in the Netherlands encouraged the settlement of skilled artisans and religious refugees in Dutch cities. The attempt to establish a silk industry in Paris, with one side of the Place Royale set aside for a large complex of modern workshops, was a product of this thinking.

How successful was Henri in achieving his ambition of a commercially prosperous and technologically advanced Paris? As was so often the case in Early Modern Europe, ambition tended to outstrip achievement. Henri's new silkworks survived just a few years before being moved to the outskirts of the city. Other well-meaning schemes also foundered on the rocks of civic conservatism or Renaissance utopianism. The Commission on Commerce, for instance, vetted a range of interesting and potentially lucrative schemes for the amelioration of the Parisian and French economy. Among other things, improved water pumps, windmills, looms for disabled people to use, and a canal linking the Atlantic to the Mediterranean were brought before the Commission for its inspection and approval. But even where they did get off the drawing board, their effectiveness was frequently undermined by external criteria involving royal protocol and privilege. A case in point was Henri's attempt to improve the supply of water to the citizens of Paris. In 1604, the king commissioned the Flemish engineer Jean Lintlaer to build a pump to siphon off water from the Seine. Housed in a three-storey building called the Samaritaine, these new waterworks were loudly trumpeted as yet another example of the king's munificence, the name Samaritaine carrying clear overtones of royal charity. In the event, precious little benefit was to accrue to the people of Paris. The pump's prime function was to deliver water to the fountains and gardens of the Tuileries and only a privileged few who were granted a royal dispensation were allowed to use the water for its original purpose (Ballon, 1991, pp.122–4; Carré, 1932, pp.179–81; Mundy, 1907, p.125; see Figure 6.30).

Over the long term, it is possible to take a more sympathetic view of Henri's contribution to the regeneration of Paris. The general philosophy of improving the urban fabric of the city which lay at the heart of the king's political vision was influential in paving the way for further and faster change in the second half of the century. The construction of the great plague hospital, the Hôpital Saint-Louis, for example, undermined the medieval and Renaissance image of the walled city whose fortifications and gates had partly functioned as a barrier against disease and infection. At the same time, the important programme of fortification undertaken by the king on France's vulnerable north-eastern border simultaneously undercut the military and strategic function of the old walls of the city. These had now become, in the words of Paris's historian:

> vestigial, losing both their functional and symbolic value as the constituent element holding Paris together. Henri IV's urbanism had moved Paris towards becoming an open city, a city whose boundaries no longer needed to be marked.
>
> (Ballon, 1991, p.197)

A related and important facet of this programme of urban renewal was the unleashing of a rash of private speculation which continued from the reign of Henri IV to that of Louis XIV. In this period, private citizens and courtiers, as

Figure 6.30 The Pont Neuf and Château-d'Eau de la Samaritaine, oil painting by Nicolas Ragenet (1715–93) (Collection Musée Carnavalet, Paris; photograph: Photothèque des Musées de la ville de Paris)

well as the crown, became increasingly interested in developing the built
environment of Paris, and in the process demonstrated a willingness to employ
both existing and new technology in their speculative schemes. Examples
include a proposition to build a navigable canal around Paris which would
follow, and largely replace, the redundant walls. A refinement to this scheme
envisaged using the water of the river Marne rather than that of the Seine in
order to capitalize on its faster flow. This would make it easier to keep the
channel clear of debris and in-fill; it was also proposed to use it as a water
supply to fountains, a source of power, and a channel for disposing of sewage
(Ballon, 1991, pp.161, 206; Bernard, 1970, p.8). Despite the ultimate failure of
these schemes – often due to engineering problems which would not be
resolved until the nineteenth century – the shape and fabric of Paris was being
transformed through the pressure of royally sponsored and privately financed
projects of urban regeneration. In 1612, a private consortium built a new
bridge over the Seine, the Pont Marie, and began the development of the Ile
Saint-Louis, utilizing a gridiron pattern of regular streets with fine aristocratic
houses. And in the 1620s, the crown sold off land next to the Tuileries Palace
which was subsequently developed into a prestigious residential area, the
Faubourg Saint-Germain. However, these developments were destined to pale
into insignificance when compared with the renewal of the city under the rule
of Louis XIV.

The transformation of Paris under Louis XIV is at first sight rather puzzling.
Louis was largely uninterested in the day-to-day affairs of the city, and his
priority for much of his long reign (he assumed control when he came of age
in 1661) was to focus royal time, energy and resources on Versailles, situated
on the outskirts of Paris. Here he succeeded spectacularly in projecting an
image of royal authority fully in line with his own absolutist aspirations.
Playing second fiddle to Versailles, the citizens of Paris might well have
expected the city to succumb once more to royal apathy. However, this was
not to be, largely through the foresight of the king in devolving all authority for
the government of Paris to the care of his chief minister, Jean-Baptist Colbert
(1619–83). Under the pragmatic and far-sighted administration of Colbert, Paris
was transformed. In some respects, it is possible to see the reforms initiated by
Colbert and his willing deputies as a continuation of those begun by Henri IV
sixty years earlier. Further new areas were laid out for modernization and

Figure 6.31 The Porte St
Martin (1674). The Porte St
Martin was one of the many
monumental arches erected in
honour of Louis XIV. This
particular engraving gives a
good impression of the new
openness of Paris with greater
freedom of movement for the
horse-drawn coach (by
permission of Cabinet des
Estampes, Bibliothèque
Nationale de France, Paris)

development, more squares and public places were built (often with statues and memorials glorifying the absent monarch), and new and wider roads and promenades were constructed (see Figure 6.31). Stricter building regulations were introduced, and in the 1670s, the ramparts and walls on the right bank were demolished and filled in, to be replaced by a tree-lined boulevard thirty metres wide. Not completed until the end of the century, it provided the wealthier citizens of Paris with a vast public space in which to congregate, promenade and ride their coaches. And even though the bulk of royal revenue was spent on Versailles, monumental public buildings continued to pay lip-service to the charity of the crown, as in the construction of the Invalides, a home for retired soldiers of whom there were many in Louis's war-torn reign.

In one crucial respect, however, Louis's development of Paris differed from that undertaken by his illustrious predecessor. According to Leon Bernard, the foremost historian of Louis XIV's Paris, the change which was to come over the city:

> was more basic than a physical change. One does not have to await the famous Baron Haussmann in the nineteenth century to encounter a systematic, rational campaign of modernization in the affairs of Paris … One of the most respected historians of European cities, Pierre Lavedan, dates the start of modern urbanism to the seventeenth century. By 'urbanism' he means the overall view of the city and its problems, something much more than the old notion that the embellishment of a community could be achieved by a few fine buildings and houses. The new urbanism demanded that city magistrates plan the total needs of the community in terms of water and sanitation, circulation, recreation, provisioning, open spaces, and the rest.
>
> (Bernard, 1970, p.26)

The various improvements to the physical infrastructure of Paris described by Bernard are indeed impressive for the time. But to what extent, we might ask, were they driven by a technological impulse? As hinted earlier in this chapter, two developments might lead us to propose a far greater technological drive to urban change at this time compared with earlier periods. First, the Baconian-inspired scientific revolution which placed such emphasis on the need for scientists to apply their knowledge to the common good was by the 1660s receiving royal support in the shape of new institutions set up specifically to advance the cause of science. In France this took the form of the Académie Royale des Sciences established by Louis XIV in 1666. And second, now that the pursuit of practical science was officially recognized as a worthy occupation, it was increasingly likely that the rapid advances in many fields of scientific endeavour would meet with far greater encouragement from royal appointees who controlled the government of Paris. For one thing is clear from the work of Bernard and others: reform of the French capital on the scale envisaged by Colbert could only be implemented by rational administrators who were willing to envisage a coherent plan for the built environment of the whole city.

Evidence for a technological imperative in the work of the Parisian authorities of this period is particularly apparent in the contributions of two men, François Blondel and Pierre Bullet. Blondel, an engineer and mathematician by training, was appointed *directeur* of the newly established Royal Academy of Architecture in the 1670s. In 1675, he published his *Cours d'Architecture,* a work which has been described as 'one of the first theoretical considerations of modern urbanism' (Bernard, 1970, p.27). Largely devoted to technical aspects of the art of building, the *Cours* broke new ground in those chapters which focused on the programme of public works then in full swing in Paris. Street improvements, new quays, pumps, sewers and water-lines were all considered an essential aspect of the work of the architect. In the course of this section, Blondel also referred to a new city plan for Paris which, when

Figure 6.32 Paris (c.1676) by François Blondel and Pierre Bullet de Chamblain (Archives Nationales, Paris; photograph: Lauros-Giraudon)

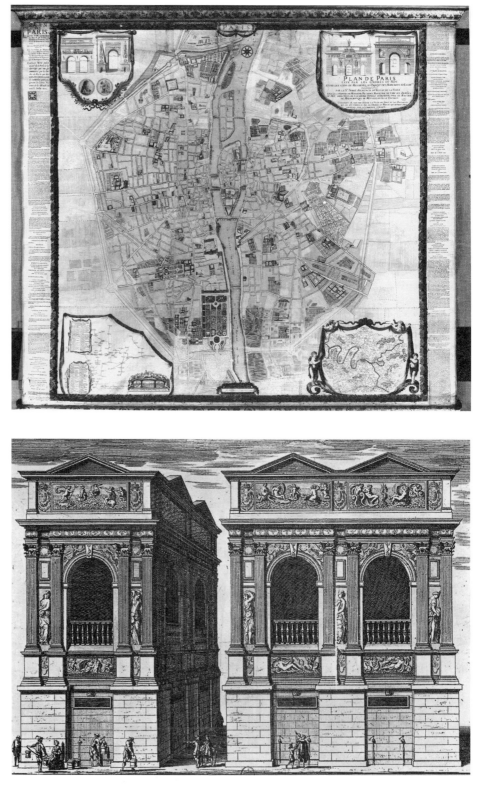

Figure 6.33 Holy Innocents Fountain. One of sixteen new fountains constructed in the reign of Louis XIV. The limitations of the system are well illustrated by the fact that so many of Paris's citizens had to fetch and carry water from public fountains (by permission of Cabinet des Estampes, Bibliothèque Nationale de France, Paris)

published a year later, broke new ground by depicting not only existing features of the Parisian landscape but also projected schemes of improvement (see Figure 6.32). It even included a map-within-a-map showing the system of water-pipes for the city with the location of forty fountains, sixteen of which were new (see Figure 6.33). Whether there was any substance to Blondel's claim that all future improvements to the city's fabric must conform to this masterplan is debatable. There can be little doubt, however, that under his authority the Royal

Academy of Architecture was transformed into an important forum charged with discussing technical issues as complex as computing the precise water requirements of the population of the city. In this work he was ably assisted by his junior partner, Pierre Bullet, who was probably responsible for actually drawing up the plan of 1676. Bullet, who combined an interest in aesthetics with a lively concern for public welfare, was well known for his technological wizardry. He even claimed to have invented a machine that was guaranteed to dissipate bad odours from cesspools and privies, described at length in a pamphlet entitled *Observations sur la nature et sur les effets de la mauvaise odeur* ('Observations on the nature and consequences of bad smells') (1695).

Despite the efforts of men such as Blondel and Bullet, many of the more grandiose schemes of the would-be urban reformers of Louis XIV's Paris failed to get beyond the drawing board. One explanation for this failure has highlighted the lack of practical endeavour on the part of the newly founded Académie Royale des Sciences. Despite its avowed Baconianism, the new college promoted few schemes of technological merit – behaviour which the historian Robin Briggs has partly attributed to the interference of Colbert who, he suggests, was more interested in exercising political control over the new society than in promoting innovative schemes within its ranks. Moreover, on those rare occasions when the Académie did concern itself with projects of wide practical concern and application, the end result is revealing. One example cited by Briggs was the urgent need to increase supplies of fresh water to the growing population of Paris. A solution was at hand in the form of the 'machine de Marly' which cost upwards of 4,000,000 livres and took over seven years to construct (1679–86). Requiring a permanent maintenance staff of sixty men, it consisted of fourteen water-wheels and three pumps with the capacity to lift about 3,000 cubic metres of water a day more than 150 metres above the level of the Seine. Though never as efficient as originally planned, the 'machine de Marly' none the less represented a potentially useful addition to the water supply of Paris. Typically, however, the machine was never used for this purpose. Instead, it was diverted to the exclusive use of the court at Versailles where it supplied water for drinking and display. And moreover, despite the government's sponsorship of research into hydraulic engineering within the Académie, there is, according to Briggs, no clear evidence that the 'machine de Marly' was the product of the work of academicians (Briggs, 1991, p.47). The ultimate beneficiaries of these schemes, as with the Samaritaine engine of Jean Lintlaer (see pp.218–9 above), were not the people of Paris but rather the king and his intimate circle of friends and courtiers. In the circumstances, bodies such as the Académie Royale and the Royal Academy of Architecture seem little more than utopian think-tanks in the service of monarchical glorification.

Other examples cited by Briggs highlight the continuing conflict in Louis XIV's France between technological innovation on the one hand, and vested interest and aristocratic privilege on the other. Lawsuits effectively destroyed one of the first timetabled public transport systems in Europe, the *carrosses à cinq sous* (five-penny coaches), the brainchild of the mathematician Blaise Pascal.[11] A similar fate awaited the scheme of Pascal's partner, the duc de Roannez, when he attempted to build a canal, with waterbuses, connecting Paris to Troyes – a scheme which envisaged the use of cost-saving inventions such as new sluice-gates and water-wheels (Briggs, 1991, pp.57–60). As in the past, private enterprise was frequently defenceless in the face of such concerted opposition. But above all, such examples accentuate the critical role played by government in the Early Modern era in sheltering and promoting new initiatives – a facet of Colbert's administration which was clearly more important in the regeneration of Paris than any specific technological advances. Consortia of

---

[11] Details of this transport innovation are given by Bernard in the Reader associated with this volume (Chant, 1999).

private speculators were changing the appearance of small sectors of Paris, but much more important for the 'modernization' of Paris after 1660 were the reforms carried through by Colbert aimed at the city's confused and multi-layered system of government. In many respects, the creation of the Conseil de Police in 1666, an advisory group set up to explore all aspects of the city's organization, marks the beginning of a new era in the history of Paris. Individual committee members were allotted specific tasks (including investigation of street cleaning and street lighting) and ultimately the recommendations of the whole body resulted in the creation of a new executive authority for Paris under the control of a Lieutenant of Police. Alongside these changes, financial reforms were instituted and a new tax introduced, the *taxe des boues et lanternes* (mud and lantern tax), the income from which was set aside to pay for street cleaning and the erection and maintenance of 5,000 new street lanterns.[12] It was far from perfect as a system of urban government. The Lieutenant of Police and his forty-eight commissaries were given far too much business to deal with, including responsibility for sanitation, control of disease, regulation of the water supply and new building. One is tempted to conclude that the marked improvement in the urban environment of Louis XIV's Paris was therefore more indebted to these administrative reforms than it was to the piecemeal adoption of technologically-driven change. But it might also be the case that, as with the 'machine de Marly', genuine technological innovation tended to be reserved for grand schemes sponsored by the crown. Another example is provided by the construction of the grand colonnades on the east front of the Louvre, begun in 1667. Not only were special machines constructed to lift the huge weight of the stone columns, but a revolutionary system of hidden iron tie-rods and cramps was integrated into the building works in order to overcome the inherent instability of the design (see Figures 6.34 and 6.35). Technological innovations on this scale are rarely evident in the many schemes of public works which so transformed Paris in the second half of the seventeenth century (Berger, 1994, p.35).

Figure 6.34 Construction of the east front of the Louvre (1667–77); engraving (by permission of Cabinet des Estampes, Bibliothèque Nationale de France, Paris)

---

[12] See Bernard in the Reader associated with this volume (Chant, 1999).

Figure 6.35 Galleries inside the entablatures on the east front of the Louvre. Louis XIV, on inspecting the progress of the work on the east front, is said to have remarked: 'If only Versailles could have been built like that' (Berger, 1994, p.36, figure 24; photograph: Rowland J. Mainstone)

London

During the course of the sixteenth century, London, like Amsterdam and Paris, experienced extraordinary levels of population growth. Between 1550 and 1650, the population tripled from 120,000 to over 375,000 and continued to grow thereafter, touching half a million by the end of the seventeenth century. Population growth in London, unlike that in Amsterdam and Paris, was exceptional in a number of respects. First, it continued well into the eighteenth century when the population of most Dutch cities, including Amsterdam, began to decline. Second, the proportion of the national population in London was far greater than that in Paris. And third, the bulk of this growth was not in the old walled city itself, but rather in the suburbs where all attempts to check the flood of illegal incomers proved futile. Some of the problems that resulted were common to London's continental rivals. An inadequate supply of water, traffic congestion, fire risk, and issues relating to health and sanitation were regularly the cause of concern for both the monarchy and the municipal authorities in the seventeenth century (Weinstein, 1991, pp.29–40). Some, however, were peculiar to the English capital. Pollution, caused by the burning of coal for domestic and industrial purposes, was a growing health hazard during this period, and one which frequently attracted comment from contemporaries (see Extracts 6.1(b) and 6.2 at the end of this chapter). How did the authorities respond to these problems, and to what extent was the urban fabric of London altered in the process? Before we can answer these questions, we shall need to consider briefly the government of London in the seventeenth century and to focus in particular on the curious relationship between the central and municipal administration which lay at its heart.

The city of London, strictly defined, was confined to the square mile established by the Romans. Here during the course of the Middle Ages, the powerful guilds of merchants and craft specialists had established an ascendancy in the government of the capital through the instrument of the Common Council (based on the rule of a mayor and aldermen at the Guildhall). This body was responsible for protecting and overseeing the monopolistic powers wielded by the guilds (for example, in the regulation of the number of apprentices and businesses in specific fields of the urban economy). It also held legislative authority in respect of issues relating to the condition of the streets, water supply and public health which on a day-to-day basis was delegated to the individual wards and parishes that constituted the old city. Central government, on the other hand, in the shape of the monarch, the court, Parliament and the lawcourts, was situated outside the city in its own enclave, the city of Westminster. The history of the relationship between the two cities is complex; central government interference in the affairs of the city tended to wax and wane depending on the strength and character of individual monarchs as well as the pattern of national events. It is probably fair to conclude that by the beginning of the sixteenth century the ruling merchant oligarchy wielded a fair degree of autonomy in the daily administration of the city – a situation which was slowly to change under the reformist influence of the Tudors in the sixteenth century.

One reason for this growth of interest on the part of central government in the internal affairs of the city of London was the rapid and illegal growth of the city outside its walls – a concern shared by the city authorities because it threatened the economic well-being of the guilds. Central government, however, was far more anxious at the prospect of the creation of large urban ghettos on its doorstep where the rule of law was either rudimentary or non-existent. The Tudor obsession with law and order, and the threat to it represented by the unchecked spread of London beyond its traditional boundaries, was manifest in the issuing of various royal proclamations prohibiting any further incursions into the countryside surrounding London. Not surprisingly, all failed (see Figures 6.36 and 6.37).

During the course of the seventeenth century, further attempts were made to regulate this expansion, most notably during the reign of Charles I, whose loathing for disorder was matched by his knowledge of, and enthusiasm for, classical architecture. Once again, various royal proclamations were issued on a variety of urban matters, including illegal building, though opinions differ as to their effectiveness. Some have concluded that the fines that were imposed on offenders were little more than expedients devised by Charles's government to shore up its weak financial condition. More recently, however, it has been claimed that the growing concern of Charles's government to legislate more effectively on the urban environment of London was part of a wider programme of royal reform which was genuinely attempting to wrestle with such problems as overcrowding and poor hygiene (Sharpe, 1992, pp.403–12). In the event, little of lasting value was achieved by Charles's government. But this should not detract from the fact that central government in England was manifesting a growing and commendable interest in the beautification and improvement of London – a development evident in Charles's support for such varied schemes as a new and more efficient water supply (including the installation of lead piping), a radical redesign of major thoroughfares, and the construction of a new London Bridge (Sharpe, 1992, pp.404–6). Moreover, in the very first year of Charles's reign, the king created a commission for the regulation of all building, which attempted to establish strict controls over all new building projects in the capital. This eventually bore fruit in the form of the new development at Covent Garden, outside the old city, where not only were the new Italianate styles of architecture introduced, but stone and brick provided a practical, as well as an aesthetic, model for future development in the city (see Figure 6.11, p.197). The rash of private speculation that followed testifies to the success of this scheme. It pioneered the idea of building on a large scale according to an overall plan, and was imitated in the new squares and straight, broad streets which subsequently sprang up in the prosperous enclaves of the West End (Stone, 1980, pp.167–212; Summerson, 1969, pp.29–34).

The outbreak of the Civil War in 1642 temporarily halted many of these developments. The success of the armed forces of Parliament, backed by the

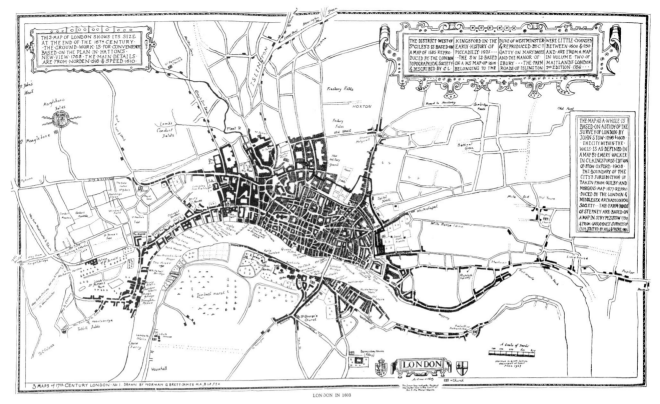

Figure 6.36 London, *c.*1600 (Brett-James, 1935, p.78)

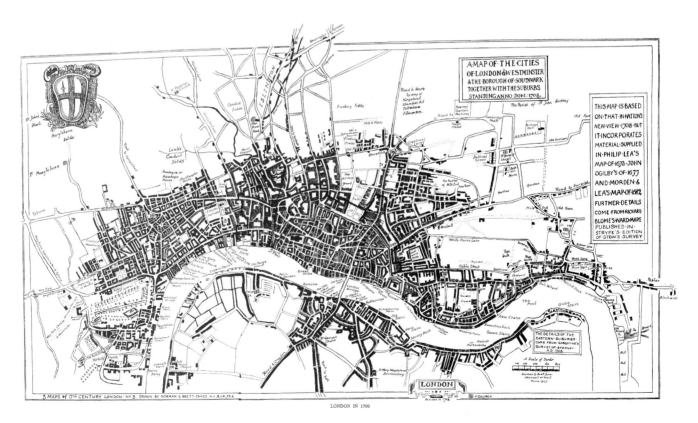

Figure 6.37 London in 1700. Most notable in terms of the expansion of the city is its growth to the east, in the mainly working-class and industrial enclaves of the East End; and to the west, where the city now merged, almost seamlessly, with the city of Westminster. Between the two now lay the many aristocratic developments around Covent Garden, Lincoln's Inn and Bedford Square, forming the nucleus of the new West End (Brett-James, 1935, p.494)

religious fervour of the puritans, paved the way for a new form of political authority in England, Cromwell's republic (1649–60). Some historians of science and technology have persuasively argued that this period also witnessed a remarkable revival of interest among intellectuals and educated artisans in the implementation of a wide-ranging programme of reform based on the utilitarian principles of Francis Bacon (1561–1626). It is now generally accepted that the middle decades of the seventeenth century were noteworthy for schemes of social improvement reflecting the contemporary emphasis on the improvability of the quality of life and the place of science and technology in such a vision. In particular, like-minded thinkers began to form organized pressure groups devoted to popularizing the merits of Baconian science, and many insinuated themselves into positions of importance either on the fringes of government or in the reconstituted universities. The result was impressive, not so much in terms of real, practical achievement, but rather in promoting the potential role of useful knowledge in all facets of Early Modern life, including urban living.

Countless examples of this process can be found in the papers of Hartlib who during this period performed a crucial role in co-ordinating the efforts of fellow Baconians from his base in London. Among the schemes which attracted Hartlib's attention, there were proposals for improving the water supply of London, new and more sophisticated forms of fire-fighting equipment, lanterns and lamps for public street lighting and a host of other suggestions for the improvement of the urban environment. One such scheme has recently received detailed attention. In the 1650s John Lanyon, a Londoner with a long-standing reputation for technological innovation, proposed to reform London's street cleaning and maintenance practice. This had long been a cause of concern to the authorities, particularly the failure of the locally appointed parish officers, known as scavengers, to oversee the work of those paid to clear the muck and detritus from the streets. It was also an issue which attracted the attention of Hartlib and his technophile associates, Lanyon himself being well known to Hartlib's circle (in the 1650s, for example, he was reported by Hartlib to have invented a new water-mill and augur to improve the water supply of the city; Hartlib, 1634–60, 28/2/14B, 15B, 18A). In 1654, Lanyon proposed to the city authorities that he should take over sole responsibility for the cleansing of the capital's streets. This reform – radical within the context of a highly conservative city administration which was normally loath to delegate such powers to private individuals – was primarily administrative in kind and involved no new technology. It was, however, a major break with tradition, its subsequent failure a timely reminder of the enormous problems faced by anyone wishing to transform the environment of Early Modern London (Jenner, 1994, pp.343–56).

The restoration of the monarchy in 1660 did not bring an end to the growing fascination with technological and scientific experimentation in England. Within months of the new king's return, science received the royal seal of approval with the foundation of the Royal Society, which had Baconian utilitarianism ostensibly high on its official agenda. Through its meetings and official publications, the potential benefits of the new science received wide publicity, while individual members canvassed technological solutions to all manner of problems including the ills of urban society. In 1664, the Mechanical Committee of the nascent Society was investigating current research on hydraulics with a view to improving London's water supply (Hunter, 1989, pp.115–18). Two years later, when nine-tenths of the city was consumed by fire, the Royal Society was offered a unique opportunity to implement its programme of practical reform in the rebuilding of the city (see Figure 6.38). Urban destruction wrought by fire always provided Early Modern cities with opportunities to incorporate new thinking into the rebuilding process. The Great Fire of London of September 1666 gives the historian a chance to judge the extent to which technology might flourish in an urban context in the brave new world of post-Restoration England.

Figure 6.38 London after the Great Fire of 1666; engraving by Wenceslas Hollar, 1667. Easterly winds ensured that most of the old city was destroyed, including enclaves to the west of the city walls. The only area to escape within the city itself was a small area in the north-eastern quarter of the city between Aldgate and Bishopsgate (copyright: The British Museum)

Of the five plans drawn up for the rebuilding of London in the immediate wake of the Great Fire, three were the products of members of the Royal Society. Besides John Evelyn, both Robert Hooke and Christopher Wren submitted plans (see Figures 6.39–6.43 overleaf). The proposal of the obscure Valentine Knight (Figure 6.43) is interesting since it resurrected the idea of a canal running in an arc from the Thames at Billingsgate to the River Fleet at Holborn Bridge. Those of Richard Newcourt and Hooke (Figures 6.41 and 6.42) most closely adhere to the gridiron system favoured by Renaissance town-planners. In the process, they also pay the least attention to the problem of topography. Squares, piazzas and long, straight lines feature in all five plans. Hooke was one of the foremost scientists of his day, whose *Micrographia,* published in 1665, stands as one of the great achievements of scientific publishing in this period. As a professional mathematician, moreover, he was to become actively involved in the actual work of rebuilding after 1666, work which prompted, among other things, his invention of numerous surveying devices (Bennett, 1980, p.38). Likewise Wren, more famous today as an architect, was chiefly recognized in the mid-1660s for his precocious endeavours in the field of astronomy. It seems beyond dispute, therefore, that when Charles II sat down to consider the shape which his new capital city might take, he was kept fully abreast of the latest technological thinking with regard to urban design. Individually, and as a group, members of the Royal Society were in a powerful position to affect the new layout of the city. What is less certain, however, is the extent to which their voice was heard and acted upon in the actual reconstruction after 1666.

Although there is little evidence for the incorporation of new technological thinking in the rebuilding of the old city, there can be little doubt that the

Figure 6.39 Wren's plan for the new London (National Monuments Library, London)

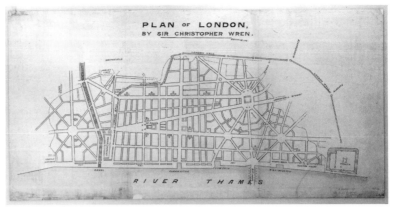

Figure 6.40 Evelyn's third and final plan for the new London (National Monuments Library, London)

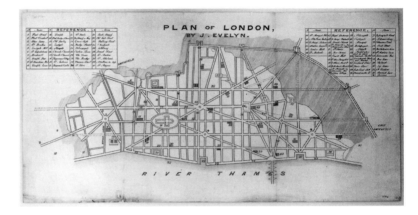

Figure 6.41 Newcourt's plan for the new London (reproduced by permission of The Guildhall Library, London)

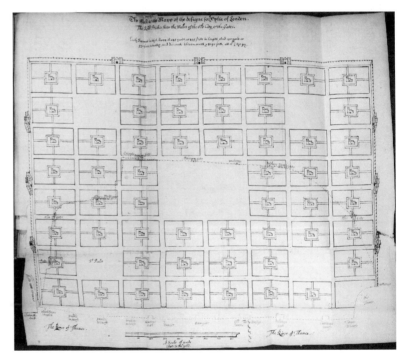

appearance of the city, as it took shape in the decade or so after 1666, was very different from that of its predecessor. The most obvious difference is discernible in the new building regulations which were now enforced both by royal edict *and* parliamentary legislation, thus ensuring a more fireproof city constructed out of brick, stone and tile. Uniformity of street design and building façade was also guaranteed by the new, stricter regulations which in turn allowed greater scope for housing speculators to use methods of mass production in the building

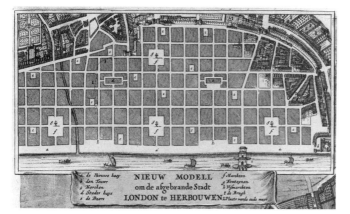

Figure 6.42 Hooke's plan for the new London (reproduced by permission of The Guildhall Library, London)

Figure 6.43 Knight's plan for the new London (reproduced by permission of The Guildhall Library, London)

trade. Foremost among such people was the ingenious financier and builder Nicholas Barbon. The son of a fiery Cromwellian lay preacher and MP, he was clearly a man for the new age. An advocate of free-market economics, he rebutted all charges that the rebuilding of new houses in London would lower property values in the old city. He was also a pioneer of fire insurance. From the perspective of this study, however, perhaps his main contribution was the way in which he used and encouraged the adoption in the building trade of large-scale methods of production, particularly in the construction of internal timber-frames and carpentry work (Summerson, 1969, p.45; see Figure 6.44 overleaf). Here again, however, we would appear to have another good example of the application of an administrative, rather than a purely technological, initiative. As McKellar points out, though London's new houses looked very different on the outside from their timber-framed predecessors, internally little had changed. In her view, the adoption of a brick veneer represented little more than a 'transitional technological stage' in the building process. The growing demand for brick and other building materials, on the other hand, did stimulate some technological innovation in the manufacturing process; the introduction for example of the pug mill – a horse-powered mixing machine which brought the clay to the right consistency before it was baked into bricks (McKellar, 1992, pp.192, 217–18). By far the biggest changes, however, affected the way in which the building industry operated. McKellar thus concludes her study of the late seventeenth-century London house by affirming that:

> there were no fundamental technological changes despite a transformation in the form of the house. There was a shift in production methods towards a seemingly widespread use of standardized parts. However this was not a change in the processes of production but rather in the scale of manufacture … If the technological and constructional processes remained similar, the structure within which the building industry operated did not.
>
> (McKellar, 1992, p.265)

Other developments which transformed the appearance of the new city included wider and more airy streets with fewer impediments and bottlenecks, both for pedestrians and horse-drawn carriages. Indeed, a prominent feature of the rebuilt city was the way in which foot and vehicular traffic were separated by the creation of raised pavements, which in turn facilitated the cleaning of streets and thoroughfares. In addition, the streets were made cleaner and more healthy by the abolition of projecting windows, jetties and spouting gutters, the last of which were to be replaced by pipes running down the sides of houses (see Figure 6.45, p.233). Further developments of a similar nature soon followed. Two Acts in 1707 and 1709 reflected the trend towards greater fire safety in the new city. One outlawed the construction of wooden eaves and ordered their replacement with stone and brick parapets. The other stipulated the recession of

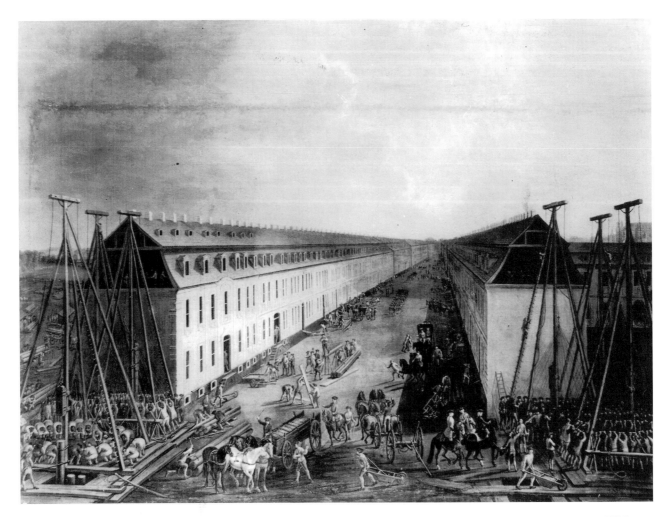

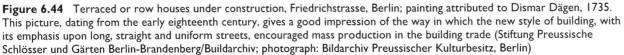

Figure 6.44 Terraced or row houses under construction, Friedrichstrasse, Berlin; painting attributed to Dismar Dägen, 1735. This picture, dating from the early eighteenth century, gives a good impression of the way in which the new style of building, with its emphasis upon long, straight and uniform streets, encouraged mass production in the building trade (Stiftung Preussische Schlösser und Gärten Berlin-Brandenberg/Buildarchiv; photograph: Bildarchiv Preussischer Kulturbesitz, Berlin)

all timber-framed windows, a development which was given added impetus by the introduction in the 1660s of the one genuine architectural innovation of the seventeenth century, the sash window (Louw, 1983, pp.49–71; Summerson, 1969, pp.68–9). At the same time, more and more open-air water conduits were placed underground, and most important of all, authority for drains, sewers, street cleaning and various other urban services was placed in the hands of a single body of commissioners for the city of London.

However, there were also some notable failures. Foremost among them were the proposals for a Fleet Canal, as well as a project for the creation of a lavish new quay (see Extract 6.6 at the end of this chapter), though here again it should be stressed that the principal cause of failure was not so much technological as financial and administrative. Indeed, in many respects, finance and administration were truly the key to both success and failure with regard to the rebuilding of the city. On the one hand, given the much lower levels of bureaucratic development and resources in seventeenth-century England, much was achieved in a relatively short space of time. The need to balance the interests of individual householders and tenants who had lost their original properties in the blaze with the desires and requirements of the capital as a whole was always likely to tax the authorities to the full. Moreover, conditions were unfavourable for such a large-scale change: the government was already reeling in 1666 from the disastrous consequences of the last great outbreak of

Figure 6.45 Cheapside, London, c.1750. Fire-proofed brick or stone houses form a continuous line, with vehicular traffic separated from pedestrian walkways by way-posts. Note also the lead down-pipes which had replaced spouting gutters (reproduced by permission of The Guildhall Library, London)

bubonic plague in London in the previous year, and was simultaneously attempting to finance a catastrophic naval war against the Dutch (1665–7). In the circumstances, one can only applaud the speed with which the authorities responded to the Great Fire. On the other hand, an unprecedented opportunity for the radical redesign of the city was missed in the aftermath of 1666. The street pattern of the old city was, by and large, repeated in the new London and, with the exception of St Paul's Cathedral and other Wren churches, public monumental building was low-key and generally uninspired. To sum up, practicality and economy won out over the competing demands of the court and its architects for a new Baroque city on the scale of Rome; the city of London, in the words of its historian Reddaway, bore more 'the imprint of Dutch burgherdom ... than the grandeur of absolutism' (Reddaway, 1951, p.298).

6.4 Conclusions

Urbanization in Early Modern Europe was constrained by the state of existing technology. But as our three case-studies have shown, we should not infer from this that the built environment of the city in this period remained unaltered – quite the reverse, in fact. The physical appearance of large cities such as Amsterdam, Paris and London was undergoing fundamental change. The increasing use of brick, stone, tile and slate and the wide acceptance of an aesthetic revolution which emphasized uniformity and geometrical precision in street layout helped to create urban communities which were very different in appearance from their medieval predecessors. Such changes also produced healthier environments in which to live and work. The threat of bubonic plague, for example, was almost certainly removed from Europe's cities in the second half of the seventeenth century as a result of these changes to the fabric of the urban environment (Slack, 1985, pp.152, 322–3).

The failure to detect new technology at work in this process of urban amelioration should not lead us to conclude that such change took place in a technological vacuum. More often than not, the means to meet the challenges posed by rapid urban growth were provided by existing technology, for instance in Amsterdam a combination of factors – socio-economic, religious, political and cultural –created an atmosphere fully conducive to the development and application of technical solutions to the city's diverse problems.

Amsterdam was not unique, however, in this respect. Indeed, the changes to the built environment of the Early Modern city were widespread in Europe at this time, the vast majority dating from the middle decades of the seventeenth century. Why should this be? One explanation lies in the existence of a much more complex web of international communication between technologically oriented scholars, writers and artisans than has hitherto been recognized by historians. This process of technology transfer is exemplified in the European network of correspondence surrounding Hartlib. But it can also be found throughout northern Europe at this time where, as in the Netherlands, it was often a by-product of economic and industrial advance (Davids, 1995). Moreover, where differences in physical form and fabric can be discerned between cities, these can rarely be attributed to specific technological approaches unique to those cities. Instead, the explanation is generally to be found in the contrasting nature of the administrative and political structures of the respective cities. The case of Amsterdam is particularly salient here, for in many respects it too suffered from the failure of technology to provide solutions to some of the city's most pressing problems, notably an inadequate water supply. Its success – as the work of Israel makes clear – lay in the ability of its ruling oligarchy to wield, unchecked, enormous authority in the field of municipal government. Much the same might have been achieved in Louis XIV's Paris if not for the fact that so much of the nation's energy and resources (including technological) had been diverted into the construction of grand royal projects such as that at Versailles.

If the role of new technology was minimal, and that of religion and politics often immaterial, in the transformation of the built environment of the Early Modern city, where can we locate the dynamo of change? The answer probably lies in that barely perceptible shift in attitude among urban dwellers and their rulers in favour of schemes of improvement and reform – principles, that is, most frequently associated with the era of the Enlightenment, but which clearly had their roots firmly in the preceding century. Indirectly, science and technology played a role in this important cultural shift. The idea that humankind might apply scientific knowledge to the improvement of the individual and his or her environment was a radical and liberating notion in the seventeenth century, even if the actual practical benefits which resulted from the activity of groups such as the Royal Society and the Académie Royale des Sciences were less than earth-shattering. The critical function of these groups was to act as a stimulus to innovative thinking about the way people lived, and how changes might be effected for the sake of both the individual and the common good. Clearly, the benefits of such thinking bore greatest fruit in the period after 1750 with the onset of the Industrial Revolution when, arguably, technology did acquire a momentum of its own and the pace of urban development quickened as never before. Before this, however, this study underlines the idea that technology was fully subservient to broader cultural and political imperatives which, though they may have differed from country to country, were largely responsible for setting the agenda for urban change in the Early Modern period.

Extracts

6.1   Graunt, J. (1662) *Natural and Political Observations ... upon the Bills of Mortality ... with reference to the Government, Religion, Trade, Growth, Ayre, Diseases, and the Several Changes of the said City*, London, Thomas Roycroft, pp.55–6, 70

John Graunt was a prominent London citizen and one of the first accurately to record statistical information about the city in the mid-seventeenth century and provide a sophisticated analysis of the shifting demographic patterns. He was also acutely interested in recent developments in the general field of natural philosophy, which knowledge he brought to bear on his studies of London's population.

(a)   On the westward expansion, outside the walls, of the city of London

The general observation which arises from hence is, That the City of London gradually removes Westward, and did not the Royal Exchange, and London Bridg stay the Trade, it would remove much faster …

The reasons whereof are, that the King's Court (in old times frequently kept in the City) is now always at Westminster. Secondly, the use of Coaches, whereunto the narrow streets of the old City are unfit, hath caused the buildings of those broader streets in Covent-Garden &c.

Thirdly, where the Consumption of Commodity is, viz among the Gentry, the vendors of the same must seat themselves.

Fourthly, the cramming up of the voyd spaces, and gardens within the Walls, with houses, to the prejudice of Light and Air, have made men Build new ones, where they less feer those inconveniences.

Conformity in Building to other civil Nations hath disposed us to let our old Wooden dark houses fall to decay, and to build new ones, whereby to answer all the ends above-mentioned.

Where note, that when Lud-gate was the onely Western gate of the City, little building was Westward thereof. But when Holborn began to encrease New-gate was made. But now both these Gates are not sufficient for the Communication between the Walled City, and its enlarged western Suburbs, as dayly appears by the intolerable stops and embarasses of Coaches near both these Gates, especially Lud-gate.

(b)   On the worsening health of the population of London

When I consider, That in the Country seventy are Born for fifty-eight Buried, and that, before the year 1600 the like happened in London, I considered, whether a City, as it becomes more populous, doth not, for that very cause, become more unhealthfull, I inclined to believe, that London now is more unhealthfull then heretofore, partly for that it is more populous, but chiefly, because, I have heard, that 60 years ago few Sea-Coals were burnt in London, which now are universally used, for I have heard, that Newcastle is more unhealthfull then other places, and that many People cannot at all endure the smoak of London, not onely for its unpleasantness, but for the suffocations which it causes.

6.2   Digby, Sir K. (1658) *A Late Discourse Made in a Solemne Assembly ... Touching the Cure of Wounds by the Powder of Sympathy*, London, R. Lownes and T. Davies, pp.38–42

Sir Kenelm Digby was a wealthy courtier and amateur scientist who was later to play a prominent role in the Royal Society, and wrote a number of pamphlets concerned with scientific subjects before his death in 1665. His views were extremely eclectic, as is evident in this small tract which attempted to validate certain occult beliefs – in this case, the ability to heal wounds at a distance by anointing the weapon concerned with a special unguent – while at the same time espousing support for the revived doctrine of atomism, closely associated with the philosopher René Descartes.

We have in London an unlucky, and troublesome confirmation of this doctrine [of atomism], for the air useth to be full of such atomes. The material then whereof they make fire in that great City, is commonly of pit coal, which is brought from Newcastle, or Scotland. This cole hath in it a great quantity of volatil salt very sharp, which being carried on by the smoak useth to dissipate it self, and fill the air, wherewith it doth so incorporate, that although we do not see it, yet we find the effects, for it spoils beds, Tapistries and other houshold stuffs, that are of any beautiful fair colour for the fuliginous [sooty] air doth tarnish it by degrees: and although one should lock up his Chamber, and come not thither a good while, and keep it never so clean, yet at his returne, he will find a black kind of thin soot cover all his houshold-stuff … The said coal-soot also gets abroad, and fouls clothes upon hedges, as they are drying, as also in the Spring time, the very leaves of the trees are besooted therewith. Now, in regard that it is this air which the lungs draw for respiration among the inhabitans, therefore the flegme and spittle which comes from them, is commonly blackish and fuliginous. Moreover, the acrimony of the soot produceth another funestous [fatal] effect, for it makes the people subject to inflammations, and by degrees to ulcerations in the lungs. It is so corrosive, and biting, that if one put gammons of bacon, or beef, or any other flesh within the chimney, it so dries it up, that it spoiles it. Wherefore they who have weak lungs, quickly feel it, whence it comes to passe that almost the one half of them who dye in London, dye of ptisical, and pulmonicall distempers, spitting commonly bloud from their ulcerated lungs. But at the beginng [sic] of this malady, the remedy is very easie; It is but to send them to a place where the air is good: many do usually come to Paris, who have means to pay the charge of such a journy, and they commonly use to recover their healths in perfection. The same inconveniences are also, though the operations be not so strong in the City of Liege … Paris her self also, although the circumambient [living environment] be passing good, yet is the subject to incommodities of that nature. The excessively stinking dirt and channels of that vast City, mingleth a great deal of ill allay with the purity of the air, stuffing it every where with corrupted atoms, which yet are not so pernicious as those of London. We find that the most neat and polished silver, exposed to the air, becomes in a short time livid, and foul, which proceeds from no other cause, then from those black atoms, the true colour of putrefaction [decomposition] which stick unto them. I know a person of quality … who is lodged in a place, where on the one side a great many poor people do inhabit, where few Carts use to passe, and fewer Coaches: his neighbours behind his house empty their filth and ordures in the middle of the street, which useth hereby to be full of mounts of filth, which is used to be carried away by Tombrells; when they remove these ordures you cannot imagine what a stench, and a kind of infectious air is smelt thereabout every where. The servants of my said friend, when this happens, use to cover their plate, and andirons [supports for burning wood] of polished brasse, with other of their fairest houshold-stuff, with cotton or course-bayes, otherwise they would be all tarnished; yet nothing hereof is seen in the air: yet these experiences do manifestly convince, that the air is stuffed with such atoms.

6.3   A.R. Hall and M.B. Hall (eds) (1966–7) *The correspondence of Henry Oldenburg*, Madison, University of Wisconsin Press, vol.3, pp.230–31

Letter to Robert Boyle, London, 18 September 1666

Henry Oldenburg was the first secretary of the newly founded Royal Society whose copious letters testify to his tremendous energy and enthusiasm in promoting scientific and technological innovation. In this letter to his friend and colleague, the eminent scientist Robert Boyle, he refers to a recent meeting with another member of the society, the precocious Christopher Wren, whose new design for the city of London in the wake of the Great Fire was the subject of much interest.

Dr Wren has since my last [10 September], drawn a model for a new city, and presented it to the King, who produced it himself before his council, and manifested much approbation of it.

I was yesterday morning with the doctor, and saw the model: which, methinks, does so well provide security, conveniency, and beauty, that I can see nothing wanting to those three main articles; but whether it has consulted with the populousness of a great city, and whether reason of state would have that consulted with, is a query to me. I then told the doctor, that if I had had an opportunity to speak with him sooner, I should have suggested to him, that such a model, contrived by him, and received and approved by the Royal Society, or a committee thereof, before it had come to the view of his majesty, would have given the Society a name, and made it popular, and availed not a little to silence those who ask continually what they have done? He answered, that he had been so pressed to hasten it, before other designs came in, that he could not possibly consult the Society about it. However, since it is done without taking in the Society, it must suffice, that it is a member thereof that hath done it, and, by what I see, hath done it so, that other models will not equal it; and I hope, when it comes to be presented to the Parliament, the author will be named, so his relation to the Society will not be omitted.

6.4 Brereton, Sir W. (1844) *Travels in Holland, the United Provinces, England, Scotland and Ireland, 1634–5*, E. Hawkins (ed.), Chetham Society Publications, vol.1, pp.65–6

And although this [i.e. Amsterdam] be a most flourishing city, which maintains as great a trade as any city in Christendom, yet most inconveniently seated in many respects, the air so corrupt and unwholesome, especially in winter-time, when most part of the country round about over-flowed. Here is no fresh-water, no water to brew withal, but what is fetched from Weesoppe (i.e. Weesp) six English miles distant. Hence they have much beer … no water to wash withal but rain-water preserved in rain-bags [i.e. large tubs or vats]; little fire to be afforded in this country except turf … The most of the wood burnt here brought out of Denmark, Norway, which is here used: the coals come from Newcastle … And although here they now build most glorious and spacious houses, yet it must needs be at a most excessive charge, as not only wanting all materials within themselves, no timber not stone in the land, no brick burnt near them, but they are constrained to undergo a great addition to these charges in making firm and secure foundations in this boggy, mareish [i.e. miry] soil. They are enforced to drive trees to the head which are fourteen or sixteen yards long, a yard distant round about the ground whereupon is intended the house to be built; upon the heads of these trees placed planks, whereupon the walls and foundations are erected; this are forced into the ground by a great heavy piece of wood, which is drawn up in a frame for this end contrived, and the weight thereof falling upon the tree drives into the ground.

6.5 Burke, G.L. (1956) *The Making of Dutch Towns: a study in urban development from the tenth to the seventeenth centuries*, London, Cleaver-Hume Press, pp.41–2

[At Amsterdam] the flow of the river was directed along alternative channels to the sea, the original downstream portion serving as an outer harbour and the upstream portion between the dam and the alternative channels as an inner harbour. The dikes constructed to define and contain the alternative channels provided the main circulation routes for land traffic, and on their berms [narrow ledges] were sited buildings directly concerned with external trade such as warehouses and offices. The dam itself constituted a valuable central space and was treated as such; it provided the setting for public buildings like the town hall and weighing hall and, occasionally, a church. The 'Dam' of Amsterdam has always been the civic centre and principal place of public assembly for the City. The lower ground lying between the streams was used for purposes less directly connected with external trade as, for example, shops, workshops, social buildings and houses; its layout form was in long, narrow blocks typical of the 'water town'.

6.6   Reddaway, T.F. (1951) *The Rebuilding of London after the Great Fire*, London, Edward Arnold, pp.284–300

Ten years after the Fire the secular work was complete, and the citizens could take stock of the changes. These were striking. The new city, if not unrecognizable, was very different from the old. It had been restored rather than replanned, but restoration had been accompanied by a purge in which every effort had been made to reduce shortcomings and abolish evils. Most of the reforms now seem commonplace but to contemporaries they were almost revolutionary. First and foremost was the wholesale, compulsory adoption of a superior building material … The new houses were all constructed of brick or stone … Linked with this was the logical but drastic enforcement of better housing standards … The provisions of the Act [of 1667] covered everything that could, with advantage, be regulated, be it the position of cellar flaps or the construction of sewers. Scantlings were fixed for the woodwork, beams and girders, laths and rafters alike. The thickness of the walls, the height of the rooms, and the levels of basement and street were all prescribed. Finally, the appointment of surveyors, sworn to discover infringements, securely removed the whole work from the usual realms of ineffectual paternalism … Ordered lines of regulation but seemly houses replaced the picturesque inconveniences of the pre-fire days. Jetties, bulks, projecting shop-fronts, and water-pipes gouting on to the passers-by almost entirely disappeared … At heavy cost, in time and money, the city was recreated in brick and stone, without hovels, without ill-planned, ill-executed temporary buildings, and without slums.

The change, though almost unbelievable, passed without comment … For the London area it was only an extension of the standards of the new estates in Westminster and its outskirts. Significantly it was accompanied by an evolutionary administrative reform of the highest importance. Authority over the vital but mismanaged concerns of drains and sewers, street paving, street levels, and street cleaning was handed over to a single body of commissioners. Appointed by the City, with exclusive control, they had power to impose taxes and to distrain if they were not paid. A central body, though left to work through the old machinery of ward and parish, they were in a position to overcome unneighbourly jealousies and local conservatism, transcending parochial divisions in the interests of a wider project or of general uniformity. Naturally the innovation was not wholly successful, yet it was a real improvement, with results to justify its creation. Paving still disintegrated under the pressure of iron tyres, and the authorities had still to consider 'the universal Complaint made through the whole City of the great and continuall neglect of cleansing the streets'. But the conditions of the period were largely responsible for these failings, and there was much to counter-balance them. Foot-passengers owed the security of the post-protected pedestrian tracks along the sides of the 'High Streets' to the Commissioners' work. Their authority, to take one example, lay behind the construction of the great new drain which carried away sewage from the houses in Fleet Street and so added to the amenities of twelve hundred people in an important area …

Better housed, better drained, better administered, freed from Fire and Plague, brought into line with the needs of its own growth, the new city was in all things an improvement on its predecessor. It was not the dream creation of any one architect, but to its citizens it was … a better place in which to live. There can be few planners who would not accept with pleasure such a verdict upon their work.

References

A(GLIONBY), W. (1669) *The Present State of the United Provinces of the Low-Countries*, London, John Starkey.

BALLON, H. (1991) *The Paris of Henri IV: architecture and urbanism*, New York, MIT Press.

BARBOUR, V. (1966) *Capitalism in Amsterdam in the Seventeenth Century*, Ann Arbor, University of Michigan Press.

BENNETT, J.A. (1980) 'Robert Hooke as mechanic and natural philosopher', *Notes and Records of the Royal Society of London*, vol.35, pp.33–48.

BERGER, R.W. (1994) *A Royal Passion: Louis XIV as patron of architecture*, Cambridge, Cambridge University Press.

BERNARD, L. (1970) *The Emerging City: Paris in the age of Louis XIV*, Durham, NC, Duke University Press.

BORSAY, P. (1989) *The English Urban Renaissance: culture and society in the provincial town, 1660–1770*, Oxford, Clarendon Press.

BRAUDEL, F. (1981) *The Structures of Everyday Life: civilization and capitalism, 15th–18th century*, London, Fontana Press, vol.1.

BRERETON, SIR W. (1884) *Travels in Holland, the United Provinces, England, Scotland and Ireland, 1634–5*, E. Hawkins (ed.), Chetham Society Publications, vol.1, pp.65–6.

BRIGGS, R. (1991) 'The Académie Royale des Sciences and the pursuit of utility', *Past and Present*, vol.131, pp.38–88.

BRIMBLECOMBE, P. (1978) 'Interest in air pollution among early Fellows of the Royal Society', *Notes and Records of the Royal Society of London*, vol.32, pp.123–9.

BROWNE, E. (1685, 2nd edn) *A Brief Account of Some Travels in Divers Parts of Europe*, London, Benjamin Tooke.

BUISSERET, D. (1968) *Sully and the Growth of Centralized Government in France, 1598–1610*, London, Eyre and Spottiswoode.

BUISSERET, D. (1984) *Henri IV*, London, George Allen and Unwin.

BURKE, G.L. (1956) *The Making of Dutch Towns: a study in urban development from the tenth to the seventeenth centuries*, London, Cleaver-Hume Press.

CARRÉ, H. (1932) *Sully, sa vie et son œuvre 1559–1641*, Paris, Payot.

CHALKLIN, C.W. (1974) 'The making of some new towns' in C.W. Chalklin and M.A. Havinden (eds) *Rural Change and Urban Growth 1500–1800: essays in English regional history in honour of W.G. Hoskins*, London, Longman, pp.00–00.

CHANT, C. (ed.) (1999) *The Pre-industrial Cities and Technology Reader*, London, Routledge, in association with The Open University.

CHARTRES, J.A. (1977) *Internal Trade in England 1500–1700*, London and Basingstoke, Macmillan.

DAVIDS, K. (1993) 'Technological change and the economic expansion of the Dutch Republic, 1580–1680' in K. Davids and L. Noordegraaf (eds) *The Dutch Economy in the Golden Age*, Amsterdam, Het Nederlandsch Economisch-Historisch Archief, pp.79–104.

DAVIDS, K. (1995) 'Shifts of technological leadership in Early Modern Europe' in K. Davids and J. Lucassen (eds) *A Miracle Mirrored: the Dutch republic in European perspective*, Cambridge, Cambridge University Press, pp.338–61.

DE LA BÉDOYÈRE, G. (ed.) (1995) *The Writings of John Evelyn*, Woodbridge, Boydell Press.

DE VRIES, J. (1981) *Barges and Capitalism: passenger transportation in the Dutch economy, 1632–1839*, Utrecht, H. and S.

DIGBY, SIR K. (1658) *A Late Discourse Made in a Solemne Assembly of Nobles and Learned Men at Montpellier in France … Touching the Cure of Wounds by the Powder of Sympathy*, London, R. Lownes and T. Davies.

DIJKSTERHUIS, E. (1970) *Simon Stevin: science in the Netherlands around 1600*, The Hague, Martinus Nijhoff.

DUFFY, C. (1996) *Siege Warfare: the fortress in the Early Modern world 1494–1660*, London, Routledge (first published 1979).

EVELYN, J. (1664) *Sylva Sylvarum*, London, Jo. Martyn & Ja. Allestry.

FALKUS, M. (1976), 'Lighting in the Dark Ages of English economic history: town streets before the Industrial Revolution' in D.C. Coleman and A.H. John (eds) *Trade, Government and Economy in Pre-Industrial England: essays presented to F.J. Fisher*, London, Weidenfeld and Nicolson, pp.248–73.

FRIEDRICHS, C.R. (1995) *The Early Modern City, 1450–1750*, London, Longman.

GOUGH, J.W. (1964) *Sir Hugh Myddelton: entrepreneur and engineer*, Oxford, Clarendon Press.

GRAUNT, J. (1662) *Natural and Philosophical Observations Mentioned in a Following Index, and Made Upon the Bills of Mortality ... With Reference to the Government, Religion, Trade, Growth, Ayre, Diseases, and the Several Changes of the Said City*, London, Thomas Roycroft.

GUTKIND, E.A. (1971) *Urban Development in Western Europe: the Netherlands and Great Britain*, London, Collier-Macmillan, vol.6.

HALL, A.R. and HALL, M.B. (eds.) (1966) *The Correspondence of Henry Oldenburg*, Madison, Wisconsin, University of Wisconsin Press, vol.1.

HALL, A.R. and HALL, M.B. (eds.) (1966–7) *The Correspondence of Henry Oldenburg*, Madison, Wisconsin, University of Wisconsin Press, vol.3.

HARTLIB, S. (1634–60) 'Ephemerides of Samuel Hartlib', *Hartlib Papers*, Sheffield University Library.

HOWELL, J. (1657) *Londinopolis: an historical discourse or perlustration of the City of London*, London, J. Streater.

HUNTER, M. (1989) *Establishing the New Science: the experience of the Royal Society*, Woodbridge, Boydell Press.

ISRAEL, J. (1995a) *The Dutch Republic: its rise, greatness, and fall 1477–1806*, Oxford, Clarendon Press.

ISRAEL, J. (1995b) 'A Golden Age: innovation in Dutch cities, 1648–1720', *History Today*, vol.45, pp.14–20.

JACK, S.M. (1996) *Towns in Tudor and Stuart Britain*, Basingstoke, Macmillan.

JENKINS, R. (1928–9) 'A chapter in the history of the water supply of London: a Thames-side pumping installation and Sir Edward Ford's patent from Cromwell', *Transactions of the Newcomen Society*, vol.9, pp.43–51.

JENKINS, R. (1930–31) 'Fire-extinguishing engines in England, 1625–1725', *Transactions of the Newcomen Society*, vol.11, pp.15–25.

JENNER, M. (1994) '"Another Epocha"? Hartlib, John Lanyon and the improvement of London in the 1650s' in M. Greengrass, M. Leslie and T. Raylor (eds) *Samuel Hartlib and Universal Reformation: studies in intellectual communication*, Cambridge, Cambridge University Press, pp.343–56.

JONES, E.L. and FALKUS, M.E. (1979) 'Urban improvement and the English economy in the seventeenth and eighteenth centuries', *Research in Economic History*, vol.4, pp.193–233.

KARSLAKE, J.B.P. (1929) 'Early London fire-appliances', *The Antiquaries Journal*, vol.9, pp.229–38.

KNOWLES, C.C. and PITT, P.H. (1972) *The History of Building Regulation in London 1189–1972*, London, Architectural Press.

KONVITZ, J.W. (1978) *Cities and the Sea: port city planning in Early Modern Europe*, Baltimore, Johns Hopkins University Press.

KOSTOF, S. (1991) *The City Shaped: urban patterns and means through history*, London, Thames and Hudson.

KOSTOF, S. (1992) *The City Assembled: the elements of urban form through history*, London, Thames and Hudson.

LAMBERT, A. M. (1971) *The Making of the Dutch Landscape: an historical geography of the Netherlands*, London and New York, Seminar Press.

LOUW, H.J. (1983) 'The origin of the sash window', *Architectural History*, vol.26, pp.49–71.

MacCAFFREY, W. (1975, 2nd edn) *Exeter, 1540–1640: the growth of an English county town*, Cambridge, Mass., and London, Harvard University Press.

McKELLAR, E. (1992) 'Architectural practice for speculative building in late seventeenth-century London', unpublished Ph.D. thesis, Royal College of Art, London.

MONTAGU, LADY M.W. (1887) *The Letters and Works*, London, George Bell and Sons, vol.1.

MORRIS, A.E.J. (1985) 'Historical roots of Dutch city planning and urban form' in A.K. Dutt and F.J. Costa (eds) *Public Planning in the Netherlands: perspectives and change since the Second World War*, Oxford, Oxford University Press, pp.50–71.

MULTHAUF, L.S. (1985) 'The light of lamp-lanterns: street lighting in 17th-century Amsterdam', *Technology and Culture*, vol.26, pp.236–52.

MUNDY, P. (1907) *The Travels of Peter Munday in Europe and Asia, 1608–1667* (ed. R. Temple), Hakluyt Society, 2nd series, 17.

PALLISER, D. (1974) 'Dearth and Disease in Staffordshire, 1540–1670' in C.W. Chalklin and M.A. Havinden (eds) *Rural Change and Urban Growth 1500–1800: essays in English regional history in honour of W.G. Hoskins*, London, Longman, pp.00–00.

PARKER, G. (1988) *The Military Revolution: military innovation and the rise of the West, 1500–1800*, Cambridge, Cambridge University Press.

PORTER, S. (1996) *The Great Fire of London*, Stroud, Sutton Publishing.

REDDAWAY, T.F. (1951) *The Rebuilding of London After the Great Fire*, London, Edward Arnold.

REINDERS, R. (1981) 'Mud-works: dredging the port of Amsterdam in the seventeenth century', *International Journal of Nautical Archaeology and Underwater Exploration*, vol.10, pp.229–38.

SCHAMA, S. (1987) *The Embarrassment of Riches: an interpretation of Dutch culture in the Golden Age*, London, Collins.

SHARPE, K. (1992) *The Personal Rule of Charles I*, New Haven and London, Yale University Press.

SLACK, P. (1985) *The Impact of Plague in Tudor and Stuart England*, Oxford, Clarendon Press.

STONE, L. (1980) 'The residential development of the West End of London in the seventeenth century' in B.C. Malament (ed.) *After the Reformation: essays in honor of J.H. Hexter*, Manchester, Manchester University Press, pp.167–212.

STONE, N. (ed.) (1989, 3rd edn) *The Times Atlas of World History*, London, Times Books.

STRAUS, R. (1912) *Carriages and Coaches: their history and their evolution*, London, Martin Secker.

STRUIK, D.J. (1981) *The Land of Stevin and Huygens: a sketch of science and technology in the Dutch republic during the Golden Century*, Dordrecht and London, D. Reidel.

SUMMERSON, J. (1969, rev. edn) *Georgian London*, London, Pelican Books.

SUTCLIFFE, A. (1993) *Paris: an architectural history*, New Haven and London, Yale University Press.

TEMPLE, SIR W. (1673) *Observations Upon the United Provinces of the Netherlands*, London, Samuel Gellibrand.

VAN DER VENN, G.P. (1993) *Man-Made Lowlands: history of water management and land reclamation in the Netherlands*, Utrecht, Ustgeverij Matrijs.

WEBSTER, C. (1975) *The Great Instauration: science, medicine and reform, 1626–1660*, London, Duckworth.

WEINSTEIN, R. (1991) 'New urban demands in Early Modern London', *Medical History*, supplement no.11, pp.29–40.

WILLIAMSON, F. (1936) 'George Sorocold of Derby: a pioneer of water supply', *Journal of the Derbyshire Archaeological and Natural History Society*, vol.10, pp.43–93.

ZUMTHOR, P. (1962) *Daily Life in Rembrandt's Holland*, London, Weidenfeld and Nicolson.

Chapter 7: CITIES OF THE NEW WORLD

by David Goodman

7.1 Cities of the New World

Columbus's four Atlantic voyages of discovery (1492–1502) revealed a new continent, populated by inhabitants whose very existence and utterly unfamiliar ways threatened to undermine the image of humanity long established in the Old World. The Amerindians, it is now widely assumed, are a Mongoloid people who, in remote pre-history, migrated from Asia to America, crossing at the narrowest point of separation of the two continents: the Bering Straits. By the Early Modern period, when they were first encountered by Europeans, they had long spread throughout the continent to form societies of varying degrees of development (see Figure 7.1). The people Columbus described were sedentary Caribbean islanders who lived in villages in simple log huts thatched with straw and practised rudimentary agriculture. The Portuguese navigators soon discovered the food-gatherers of Brazil. And later, in North America, French explorers would be confronted by hunters, and the English colonizers by semi-nomadic tribes who lived in camps. But the most astonishing of all these early European contacts with Amerindians came in the period 1519–32 when the Spanish extended their explorations from the islands of the Antilles to the American mainland. It was then that the *conquistadores* (conquerors) encountered new civilizations, densely populated empires based on large cities. The Spanish soldiers who marched with Hernán Cortés to the Valley of Mexico simply could not believe what they were seeing; some asked 'whether it was not all a dream' (Díaz del Castillo, 1963, p.214).

7.2 Pre-Columbian cities

Mesoamerica, the now-established term for the high urban culture of central America, has ancient origins. Archaeological investigations at La Venta, a site close to the coast of the Mexican Gulf, show that perhaps as early as 800 BCE, the Olmec people were already erecting monumental religious buildings in a planned central space of large squares. Their influence spread to the peoples of the central plateau, to the Valley of Mexico, where, much later, it culminated in the building of the great city of Teotihuacán. This city had steadily grown over the centuries to reach its peak around 500 CE, when it covered an area greater than imperial Rome and had an estimated population of between 50,000 and 200,000. By causes unknown, the city was destroyed around 750 CE. Excavations, from the 1960s, have revealed the full extent of urban development here (see Figure 7.2). From the ceremonial centre of giant pyramids and broad avenues, thousands of buildings, including one-storey houses and some schools, stretched out for miles to the distant hills, all arranged on a precise grid aligned with the city centre. The foundations of the houses were made of a type of moisture-resistant concrete, prepared by mixing the local volcanic rock with lime and earth. The houses were arranged around patios, each of which was provided with its own drainage system. There were numerous workshops, in concentrations of particular crafts.

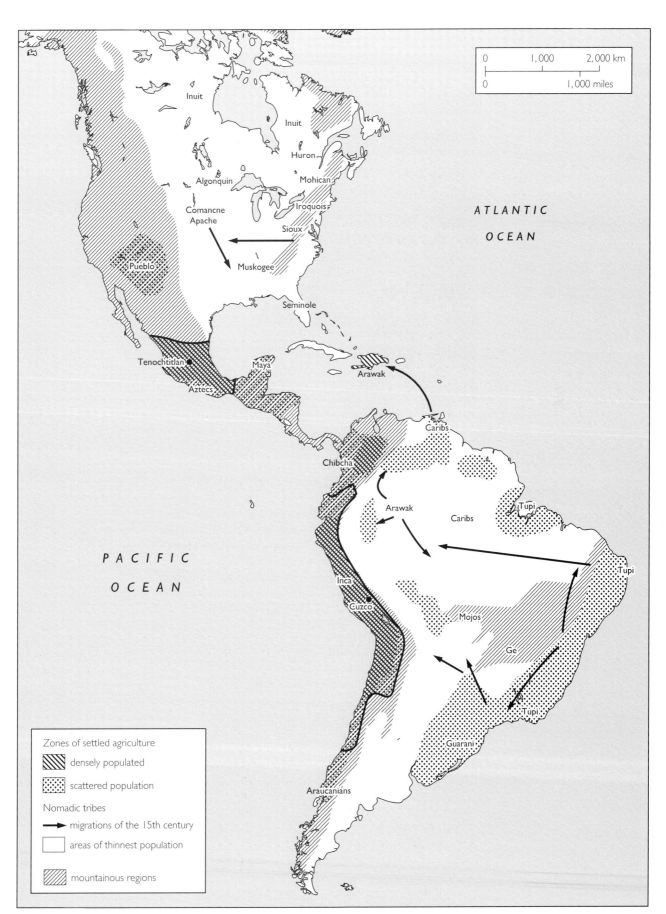

Figure 7.1 Amerindian peoples at the time of Columbus (Morales Padrón, 1988, p.56)

Figure 7.2 Ceremonial heart of the city of Teotihuacán: broad avenues, enclosures, two large pyramids and the remains of many other religious structures (photograph: Professor René Millon)

Particularly conspicuous here were craftsmen working the locally mined obsidian, a volcanic glass used for cutting-tools and the points of weapons (Millon, 1967, 1970). This led to speculations that the flourishing obsidian industry was the original spark that fired the growth of what has been described as America's first metropolis. But others point to a different technological stimulus: irrigation works in the agriculturally rich surroundings (Millon, 1974).

Teotihuacán was never rebuilt. But just forty kilometres from its deserted ruins, in a terrain of lakes, another city with similar features grew up from around 1325.

The semi-nomadic Aztecs had invaded the Valley of Mexico from the desert to the north, subjugating peoples to form an extensive empire. The Aztec capital, Tenochtitlán, was built on rocks in the middle of a lake, partly for reasons of defence (see Figure 7.3). By 1519, when Cortés and his 600 Spanish soldiers reached the city after their long march, it had grown into one of the world's most populous cities – some estimates put it at 300,000.

The arriving Spaniards, the first intruders from the Old World to witness the city, were utterly amazed. The scale of construction was breathtaking. Massive temple towers and enormous urban spaces far exceeded in magnitude any such structures in the largest cities of Spain. The quality of the stonework was unmatched by anything Cortés had seen. And he was astonished by the width of the stone causeways, raised embankments which continued as streets leading to the city centre. They took pedestrians across the lake to the mainland in a dead-straight line. The numerous city canals were spanned by well-constructed wide timber bridges. And the bridges were removable. Here Cortés's admiration mingled with apprehension – he was in the city as an invited guest, but was secretly planning its conquest. Recognizing the strength of this provision for defence in the city's design, he feared that his invading force might be starved by being cut off from the mainland if the bridges were removed. Cortés was particularly impressed by the city's water-supply system. Two aqueducts (built

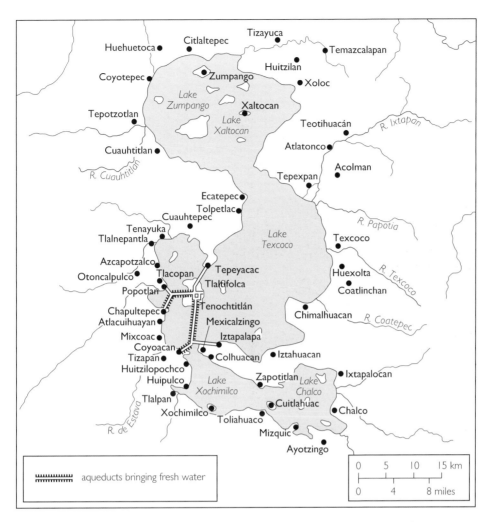

Figure 7.3 Tenochtitlán and its surroundings. Chapultepec and Coyoacan were the sources of the capital's drinking-water (Grosser, 1997, p.100)

of terracotta, and extending more than four kilometres to the spring source) brought fresh drinking-water, which was collected and sold throughout the city using the efficient, urban canoe transport (Cortés, 1969, pp.101–12).

The striking technology of the Aztec capital is vividly portrayed in the map allegedly produced on Cortés's orders and subsequently printed along with his letter to Charles V describing the city (see Figure 7.4). It shows the causeways which connected the island city to the mainland and which continued as streets intersecting at right angles at the central square where the main pyramidal temple stood. Clearly visible along the causeways are the bridges which could be removed when enemy attack threatened. And at one extremity is depicted the city's defence from floods: a barrage of wooden piles and containers filled with rubble and clay. But the map could not display two other remarkable technological achievements. The drinking-water brought by earthen aqueducts from a distant spring was distributed not only by ubiquitous canoes but also by a network of pipes to supply the temples, Montezuma's palace and chieftains' houses. And an ingenious system of intensive aqueous horticulture provided the basis for urban subsistence, as well as a beautiful green appearance. The aim was to create farmland over the lake and anchored to the bottom. Fertile soil from the mainland was brought by canoe to the lake, where it was deposited in shallow water on a trellis of reeds, secured by wooden posts to the lake bed. The earth was dropped until it formed a mound rising nearly a metre above the surface of the water. Willow trees, planted at the edge of the sunken trellis, took root and consolidated the terrain. Even during drought, the lagoon kept the earth mounds moist, a highly fertile medium for maize, fruit and vegetables. These artificial gardens, called *chinampas*, covered an estimated area of 128 square kilometres around the city and, on recent calculations, could have produced enough food for a population of 100,000 (Armillas, 1971; Musset, 1991a, p.236). Other food came in the form of the abundant fowl attracted to the lake.

Still today, the *chinampas* provide valuable market produce, as well as a popular leisure amenity of a great city (see Figure 7.5). In fact these *chinampas* are a rare material survival of the Aztec capital. The city was razed to the ground by Cortés's forces, swelled by an alliance with other Amerindian tribes intent on

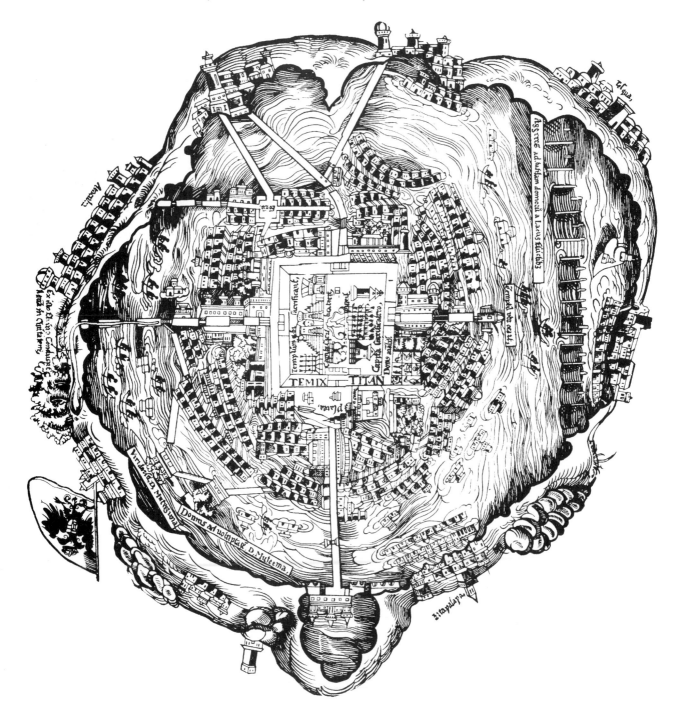

Figure 7.4 The Aztecs' capital: plan of Tenochtitlán, which first appeared in the *Atlas* of Benedetto Bordone (Venice, 1528) and is a simplification of that attributed to Cortés (Second and Third Reports to the King, 1524, Nuremberg; Bordone, 1528, p.x)

Figure 7.5 *Chinampas*, artificial lake-gardens of the Aztecs, have survived into the present day (photograph: South American Pictures)

removing the yoke of the hated Aztecs. After a siege of almost 100 days, the city was conquered. European technology had been decisive. Dominance of the lake had been secured by the building of large vessels armed with gunpowder weapons. That cut off food supplies to the densely populated island-city. Starvation forced the surrender of Tenochtitlán. On the site of the destroyed capital, Cortés built Mexico City, a Spanish city, which proclaimed the total subjugation of the Aztecs. In the great central square an enormous Christian cathedral rose on the site once occupied by the temple of the Aztec war-god, the foundations of the triumphant Christian edifice consisting of broken stones of Aztec idols. On the opposite side of the great square, Cortés built his palace, close to where the palace of Montezuma had once stood. But the layout of the Aztec city remained: the causeways and – so one expert alleges – the *chinampas* imposed 'an urban scheme of regular lines' on Mexico City (Hardoy, 1973, pp.173, 181).

The monumental works of Aztec construction had been achieved in spite of severe deficiencies in their technological resources. The Aztecs had no knowledge of the wheel (therefore no pulleys for lifting heavy objects, as well as no vehicular road transport), no horse, no other draught-animals or beasts of burden, no iron tools (only bronze axes and knives, and stone hammers and knives) and no acquaintance with the arch. Exactly the same limitations (except for the availability of the llama as a pack-animal) affected the technology of the Incas of Peru. And the same alternative means, the massing of human muscle power, was again used to great effect, bringing more gasps of amazement from the invading Spaniards.

The power of the Incas steadily grew from the early thirteenth century until, by the fifteenth century, they controlled an empire which stretched almost the

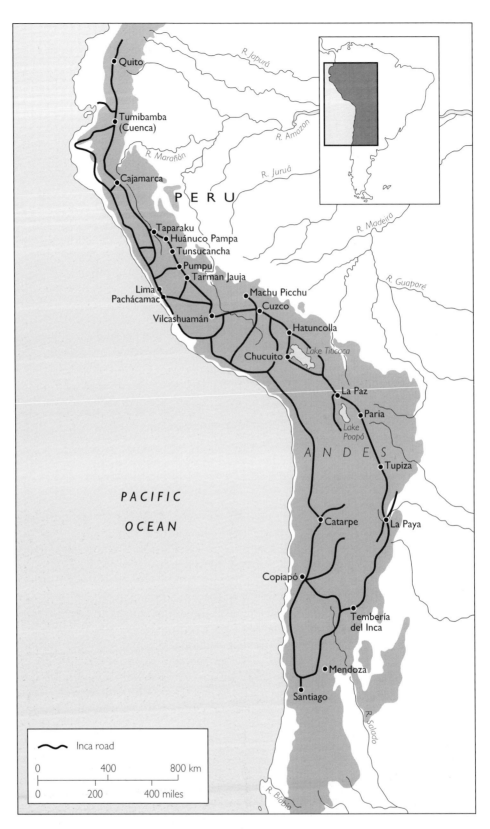

Figure 7.6 Extent of the Inca road network (Fernández-Armesto, 1991, p.35)

entire length of South America, from modern Ecuador to Chile. This extensive empire was held together by a communications system, a vast network of some 22,400 kilometres of roads, generally seven and a half metres wide (and many still in use), built in the world's most variable and challenging terrain, passing from coastal plain to jungle and soaring Andean heights (see Figure 7.6). The hub of the empire and of the road network was the Inca capital, Cuzco, which in the fifteenth century had been transformed from a village of straw and clay to a city of exquisite stone, befitting its role as the ceremonial and administrative centre. Reports from the outlying empire reached Cuzco by relay-runners covering around 240 kilometres a day. The necessary provisions or troops were then sent from the capital.

By the 1520s news had reached the Spanish in Panama of the existence of a rich state to the south. Spanish adventurers began to prepare for a voyage of conquest. In 1531 Francisco Pizarro set sail with 180 men and some thirty horses. By 1533 this minute Spanish force had conquered an empire gravely weakened by an internal crisis, a dispute over the succession to the vacant Inca throne. The Spanish had to fight fierce battles, but victory in the end was assured by the possession of steel swords and horses.

The fabric of Cuzco presents such precision in the shaping and fitting of stones that all who have seen it, from the *conquistadores* to present-day archaeologists, wonder how it could have been achieved. The skills are evident in the stone walls erected to protect the city from flooding (see Figure 7.7). But the finest work was reserved for the most prestigious buildings (see Figures 7.8 and 7.9). The most impressive of all buildings in Cuzco is the monumental structure of Sacsahuaman, situated on a hilltop near the city (see Figure 7.10). Its function, probably a fortress and perhaps a temple as well, has not been established. But the first Europeans to see it, the *conquistadores*, could scarcely believe that giant boulders, more than six metres long, could have been positioned by humans, and set (without cement) so delicately that it was impossible to fit a coin between them (see Figure 7.11). In 1970 an archaeologist doubted if there was anything in ancient Egypt or Renaissance Rome to rival the 'monstrous magnificence' of the stonework (Gasparini and Margolies, 1980, p.282).

Figure 7.7 Wall defending the Inca capital from floods (Gasparini and Margolies, 1980, p.57; photograph: Abraham Guillen)

Figure 7.8 Fine cutting, precise fitting and enhanced surface-treatment to distinguish a prestigious building in Cuzco (photograph: Hans Mann)

Figure 7.9 Precision fitting: stone wall of the Temple of the Sun, Cuzco (Gasparini and Margolies, 1980, p.230; by courtesy of Professor G. Gasparini and Professor L. Margolies, Ciudad Universitaria, Caracas, Venezuela)

Figure 7.10 The hill 'fortress' of Sacsahuaman (by permission of Professor Craig Morris of the American Museum of Natural History and Emeritus Professor Donald E. Thompson of the University of Wisconsin; photograph: American Museum of Natural History)

There are strong similarities to the Egyptian pyramids. Like the Pharaohs, the Incas imposed a system of compulsory labour on tens of thousands of captives. They hauled huge boulders from quarries and, as in ancient Egypt, used log rollers, inclined planes and bronze crowbars to move them. The precise shaping and fitting is thought to have been achieved by the constant pounding of the boulders with harder stones, a continuous action maintained by a force of thousands of labourers working in shifts (Gasparini and Margolies, 1980, p.324; Hardoy, 1973, p.465).

Other notable Inca urban technology is evident at Huánuco Pampa, the best preserved of the numerous administrative centres established at intervals of four or five days' travel on the road network. Here, situated between Cuzco and Quito, was a town with a vast rectangular space measuring 540 metres by 370, surrounded by one-storey buildings. It is thought that the size of the central square may have been designed as a display of Inca power. One side of the square has a series of distinctive buildings, perhaps reserved for officials, with perfectly aligned doorways, producing a powerful perspective effect which was not missed by a Spanish settler (see Figure 7.12 and Gasparini and Margolies, 1980, p.103). The site also has remains of the *quollqua*, the typical Inca storehouse of stone walls and thatched roof, used to stock provisions. The storehouse here has remains of potatoes, a staple part of Inca diet. Unlike grain, which can be kept for long periods without deterioration, tubers soon sprout and become inedible, unless preserved by controlling the temperature so that it does not rise above 3°C. At Huánuco Pampa, situated at an altitude of 1,830 metres, the temperature frequently falls to 3°C, so providing a natural refrigerator for potatoes. The problem for the Incas was to prevent the temperature in the storehouse rising above that critical level. They did it by insulation (thick walls and a thatched roof) and ventilation (windows oriented to prevailing winds and set on opposite sides to provide draughts). The stone floor had ducts to the exterior, opened and closed by stone plugs. These, and the windows, were opened at night to bring in cold air. It has all been described as ingenious 'warehouse engineering' (Morris and Thompson, 1985, pp.104–6).

Figure 7.12 Perfect alignment of doorways on the same axis at the Inca administrative centre of Huánuco Pampa (Gasparini and Margolies, 1980, p.106; by courtesy of Professor G. Gasparini and Professor L. Margolies, Ciudad Universitaria, Caracas, Venezuela)

7.3 Hispano-American cities

Unlike the later English colonizers, who sailed from a country characterized by rural centres of aristocratic influence, the Spanish arrived in America with a strong sense of urban identity. In sixteenth-century Spain the nobility lived in cities, along with prelates, merchants and the organizers of agricultural activity in the surrounding countryside. The upper nobility and clergy were the largest landowners. Huge landed estates had existed in Spain since the Roman period, but many more were created during the Christian reconquest of Spain, when vast areas of territory regained from the Muslims were donated by royal grants to the upper nobility. Later, in the fifteenth and sixteenth centuries, nobles acquired other large estates more gradually by amalgamating small parcels of land or simply by usurping neighbouring property. Whatever the means of acquisition, vast areas came into the possession of a few aristocratic families. And since their landed estates were frequently on the edge of Spain's largest cities, such as Seville and Salamanca, the privileged nobility were able to combine the pleasures of the countryside, indulging in the sport of hunting, with enjoyment of the amenities of urban life. Such was the dominance of the city in Early Modern Spain that Spaniards, whether they lived in the city itself or in its subject countryside, thought of themselves as belonging not to a nation, but to the province of a particular city. When they settled in America they brought this urban mentality with them, permanently shaping the face of the continent.

From the start, the city became the prime instrument of Spanish conquest. The city was the unit of Spanish colonial settlement, the base for exploiting the economic resources of discovered territories, the strategic weapon for converting the idolatrous Amerindians, and a highly effective means of controlling the vast spaces of the New World. There is an unmistakable urban pattern in the sequence of Spanish expansion. The first European cities in the New World had been created by the Spanish in the 1490s on Hispaniola, the Caribbean island (today divided between Haiti and the Dominican Republic) which was their first administrative centre. From there the *conquistadores* had sailed west and south to conquer Cuba and the Panamanian isthmus, and new cities were immediately founded. Next, from Havana, Cortés sailed further west to the mainland, where he conquered the Aztecs, creating new cities and renaming existing ones; here the Amerindian urban tradition reinforced Spanish practice. And from the new Spanish city of Panama, Pizarro sailed south to conquer Peru and establish new cities. In turn, the cities of New Spain (Mexico) and Peru became bases for further Spanish exploration, conquest and city-founding. By the end of the sixteenth century the Spanish had created around 250 new cities in America (see Figure 7.13, which shows many of them) with populations ranging from the several hundreds of Buenos Aires to some 80,000 (predominantly Amerindian) in Mexico City.

Figure 7.13 Multiplication of Hispanic cities in the New World, with foundation dates (Morales Padrón, 1988, p.286)

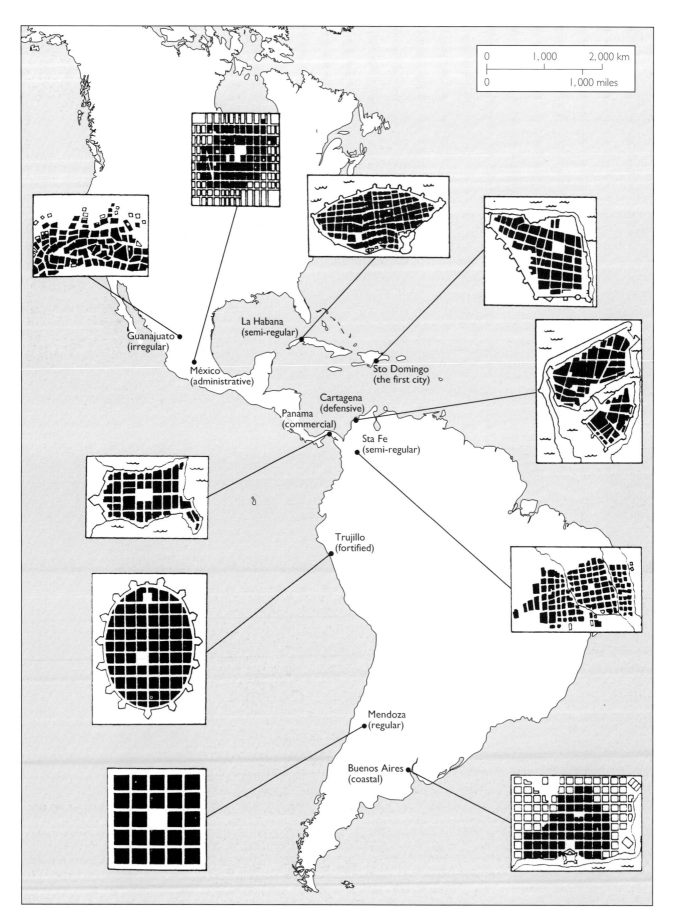

Figure 7.14 Hispano-American city forms: variations of a grid (Morales Padrón, 1988, p.283)

The foundation of cities in Spanish America was a solemn act performed with religious and legal ceremony. The leader of a band of *conquistadores* would initiate the foundation with a standard formula, invoking the power held in the name of the King of Spain, and then proceed to create a municipal authority, the *cabildo*. And from 1502, when it was first implemented at Santo Domingo (Hispaniola), a grid plan became a common physical characteristic of these new colonial cities throughout the continent (see Figures 7.14–16). Various suggestions have been made to explain the source of this Spanish urban chessboard. Some see its origin in the camps of the ancient Roman soldier-farmer colonists; others refer to grid cities in peninsular Spain (there are very few) such as Santa Fé, the royal military camp erected for the conquest of Granada. Still others detect the influence of Italian Renaissance theories of the ideal city or of Amerindian designs. And there are those who insist on the influence of *bastides*, those new towns of grid plan and a central market square that sprouted in south-west France in the late Middle Ages, princely creations intended to function as strongholds (Montauban, near Toulouse, is an outstanding example) or merely as unfortified commercial centres. The debate is inconclusive. What is clear is that frequently the grid form was a convenient way of distributing building lots to settlers. This required exact measurement, and the municipal councils of new cities such as Puebla, New Spain (founded in 1531), had custody of standard measures of length, thought to have been made of cord and leather, to regulate the length of square lots and the width of streets (Kubler, 1948, p.84). Elsewhere, in Chile, where the Spanish settlers were engaged in a fierce and protracted war against the native Araucanians, the grid plan of at least 100 settlements was the product of the spatial deployment of soldiers in a fortress. And in Guatemala, Amerindians living in hamlets scattered over the countryside were forcibly concentrated in grid towns, seen by the Spanish missionaries as the most expeditious way to Europeanize, civilize and Christianize.

Figure 7.15 Grid plan of Santo Domingo, showing a centrally placed cathedral (Morales Padrón, 1988, p.288)

Figure 7.16 Layout of the city of San Cristóbal de las Casas (Mexico) revealed from the air. The original grid of 1528 has expanded to its present-day city limits (Markman, 1984, figure 12, by permission of the American Philosophical Society; photograph: Fotos Kramsky, San Cristóbal de las Casas)

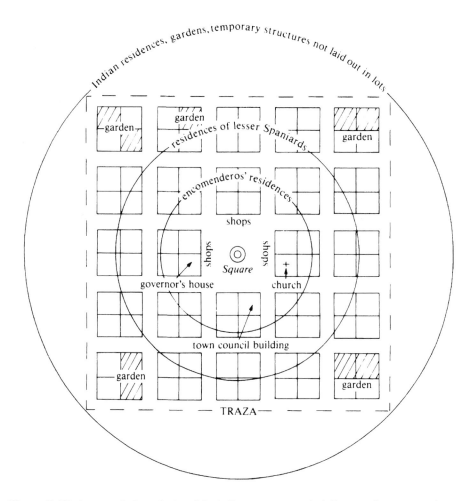

Figure 7.17 Layout of a Spanish city of the Indies, conquest period. *Encomenderos* means those 'entrusted' with Amerindians; in other words, those entitled to exploit native labour (Lockhart and Schwartz, 1983, p.67)

Decades of colonial urban organization culminated in codification, the proclamation by Philip II in 1573 of a set of official rules for the foundation and layout of Spanish cities in the Indies. They contained nothing new, merely sanctioning and standardizing existing practice. In the centre of a new city, a large rectangular space, the *plaza mayor*, was first laid out as the site of the most prestigious buildings: the house of the king's representative, the town hall, the main church. Arcades with shops ringed the square. This was the centre of the grid, the point of intersection of the four main streets. Receding from the centre, further square lots of diminishing importance were distributed until, at the periphery, ground was left for common pasture, or, as in the case of Mexico City, Amerindians were housed in huts, the unplanned district of the city's labour force. The layout therefore represented a hierarchy of status (see Figure 7.17).

A perfect grid could not be imposed everywhere, because of topographical constraints. There is no hint of a grid in Zacatecas, on the northern frontier of New Spain. There could be no geometrical regularity in a city built high on the slopes of a deep, narrow ravine. Instead there are narrow, crooked streets climbing over steep hills. Rich silver-ore deposits had been discovered here in the 1540s and a boom mining-town rose up. In the same decade, a silver mountain was discovered in remote Potosí (viceroyalty of Peru). Here a regular grid was possible on an Andean terrace – at 4,000 metres, one of the world's highest cities. Silver was the greatest prize in the Indies. It was shipped to metropolitan Spain to pay for the imports of manufactures which Madrid's legislation prevented its colonies from producing for fear of competition.

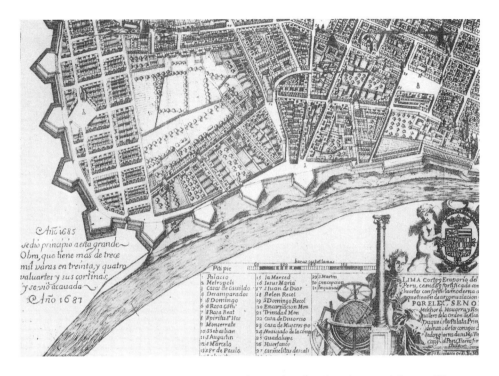

Figure 7.18 Lima in 1687, showing an expanding grid and outlying bastion defences (Chueca Goitia and Torres Balbás, 1981, p.291; by permission of Archivo General de Indias, Seville)

'Do not select sites that are too high up because … access and service to these are difficult' (quoted in Mundigo and Crouch, 1977, p.252). This was one of Philip II's town-planning ordinances, which dwelt on the choice of suitable sites for building new cities. This particular consideration had already caused Pizarro to reject Cuzco as the capital of Spanish Peru. The Incas' capital was in the high sierra, over 3,350 metres above sea-level. Instead, in 1535, Pizarro founded Lima on a site close to the Pacific coast, with road communications to its nearby port of Callao, soon established on a fine natural harbour. Although much smaller than Mexico City – its population, in 1600, of about 40,000 was half that of Mexico City's – Lima had a higher political status because of Peru's richer silver resources. A map of 1687 shows an expansion of Lima's grid plan, reaching to the recently built extensive line of bastion fortifications (see Figure 7.18).

But some new cities had to be built in the mountains to exploit the silver treasure of New Spain and Peru. Mining operations at Zacatecas would have been impossible without wagon traffic bringing in supplies and taking out refined silver ingots to Mexico City, over 650 kilometres to the south. The long road was built in the 1540s under the supervision of a Franciscan monk, Sebastián de Aparicio. He, like other missionaries, had given road-building a high priority because of the need to reach and convert isolated tribes in New Spain. Aparicio built wagons and is thought to have been the first European in America to yoke bullocks to a wheeled vehicle, astounding the Amerindians who had never seen either (Kubler, 1948, p.162). European technology for road- and city-building had to be transmitted to the Amerindians, because they were the servile population required to do the manual labour which the conquering Spanish despised. It took years for them to learn how to handle horses and mules. They had not acquired these skills in the 1520s when 20,000 Amerindians were employed to build Mexico City over the ruins of the Aztec capital. Much of the material – stone, lime and sand – was carried by labourers in hods (Kubler, 1948, p.161). While canoe transport was invaluable, it was reduced by the filling-in of the city's canals already begun by Cortés – the Spanish were unaccustomed to lake-cities and regarded the ubiquitous water as a source of pestilential vapours. Formal training in basic European

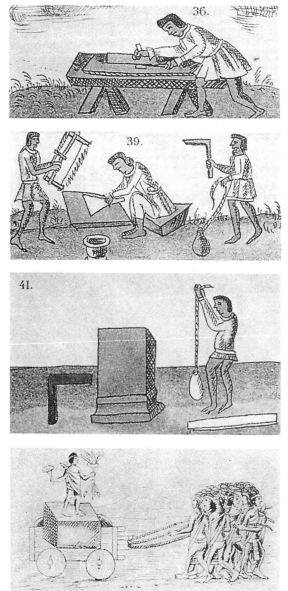

Figure 7.19 Transmission of European tools to Amerindians for the building of cities for the Spanish conquerors (Kubler, 1948, plate 41)

technology was given by Spanish Franciscan monks, and by the 1570s the use of metal picks, saws, chisels and planes had been widely assimilated (see Figure 7.19), as well as the method of building arches (Kubler, 1948, pp.152–8).

But of all urban technology, it was hydraulic engineering which presented the greatest challenge to the Spanish in America. The availability of fresh water, recognized in Philip II's ordinances as a basic requirement for any city site, soon became an acute problem for Mexico City. The introduction of cattle, the expansion of agriculture and the growth of the city's population increased demand for water to a level for which the Aztec aqueducts were no longer adequate. And the regrouping of the city's indigenous population into outlying segregated zones created additional shortages of drinking-water. The city's causeways could not bear the weight of the Aztecs' terracotta aqueducts, and they frequently cracked, causing the loss of drinking-water. From the 1550s the Spanish began to introduce an improved technique for carrying the aqueducts over the lakes in a way the Aztecs had never imagined: the construction of an extensive series of supporting arches. The most successful of these engineering projects, delayed by lack of funds but completed in 1620 and still in service in the nineteenth century, was an aqueduct of 1,000 arches four and a half metres high, bringing water from a source some ten kilometres to the south-west (see Figure 7.20). Here was a powerful new influence on the axes of city growth. The rich moved to the west of the city, towards the aqueduct terminal, paying for a water concession and the installation of piping to bring water to their homes. The east and north of Mexico City became sectors of poverty and water shortage. Although public fountains were built in these areas, they were inadequate. And the carrying of water from the aqueduct by canoe gradually diminished as the canals were filled in; in these deprived sectors, water-carriers were still conspicuous at the end of the nineteenth century, filling their containers at public fountains. Accentuating the lower status of the east side, the leper hospital was moved there in the sixteenth century, from the vicinity of the western aqueduct, to free the rich area from a supposed public-health risk (Musset, 1991a). The technology of aqueducts is also thought to have been responsible for the residential pattern in the city of Santiago de Guatemala. There, too, the elite were concentrated close to the point of entry of a new aqueduct; by the end of the seventeenth century just four per cent of the city's population were receiving in their homes ninety-two per cent of the water supply. The other zones were poor and very short of water (Webre, 1990).

Mexico City was confronted by a yet more urgent problem than water shortage. This recurring problem was so pressing that, in the 1630s, serious consideration was given to moving the entire city to a new location, an idea which had been firmly rejected by Cortés in order to derive political gain from building a Spanish city on the very site of the capital of the conquered Aztecs. Flooding of the great city had been experienced by the Aztecs, but it was nothing compared with the inundations of the post-conquest period. The change was due to the disturbance of the delicate ecological balance in the area of the lake-city, brought about by a new, Spanish way of life. The conquerors filled in many of the Aztec canals which had assisted drainage. On Cortés's orders many thousands of trees on the edges of the lake and in the surrounding foothills were felled to provide timber

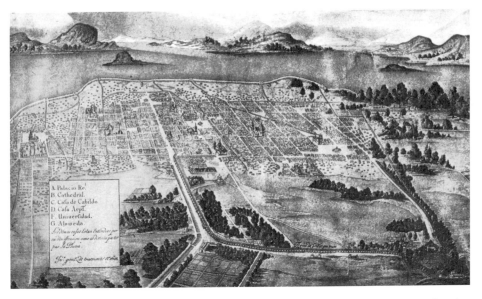

Figure 7.20 Painting of
Mexico City (1628) showing the
surrounding lake, an extended
built-up area and (lower) the
recently completed arched
aqueduct from Santa Fé
(*Historia General de España y
América*, 1985, p.439)

for the rebuilding of the ruined Aztec capital, and to clear forested areas for the
sowing of wheat. This immense deforestation resulted in soil erosion and silting
of the lake, causing the lake bottom and the water-level to rise. And the silting
was increased by the introduction of pastoral farming. From the other side of the
Atlantic, sheep and cattle were brought to the New World and bred in large
numbers. The herds grazing in the foothills around Mexico City killed the surface
vegetation and its roots, which anchored the topsoil. This erosion caused further
silting, as loose earth was carried down to the lake by wind and rain. The city
was now much more vulnerable to flooding, and there was no river system to
carry surplus water to the sea. Unusually heavy rains in 1607 and again in 1629
left much of the city under water. The authorities rushed to the engineers to see
what could be done to prevent any such occurrence in the future.

The subsequent ambitious drainage project entailed the construction of a
tunnel of length unparalleled since ancient Rome. But work on the drainage
scheme, begun in 1607, would not be completed until 1789. Why did it take so
long? The causes, elucidated by Hoberman (1980), were a combination of
technological, economic and political obstacles. The consulted expert engineers
disagreed on the remedy, nor did they yet possess the necessary knowledge to
solve the complicated problem. The project was prohibitively expensive and
unacceptable to the tax-paying elite. Nor were they willing to release their
thousands of labourers for a prolonged public-works project. And so there was
a lack of accord between the municipal council, representing the elite, and the
viceroy, the king's representative, who was instructed from Madrid to view the
problem from a wider perspective.

The acute difficulties of funding a huge drainage project and of building a
tunnel in crumbly terrain were not overcome until 1789, and not perfectly
resolved until 1900. Mexico City had survived and expanded, but at great
environmental cost. Already in the eighteenth century there were regrets for the
loss of greenery, due to building over the *chinampas* and clearance of forests.
The drainage work had serious long-term effects on the subsoil; so, at the end
of the twentieth century, the world's most populous city was sinking. And
another undesirable consequence plagues today's citizens: the drainage created
large areas of alkaline lake bed which fill the city with dust when strong winds
blow (Mathes, 1970).

The Spanish conquerors had seeded the American continent with a host of
new cities, reflecting the pronounced urban tradition characteristic of much of
Europe. In so doing, the settlers soon experienced the same need to resort to
technology to satisfy the basic requirements for life in large cities.

Extract

7.1 Cortés, H. (1969) *Five Letters of Cortés to the Emperor* (trans. and ed. J. Bayard Morris), New York and London, W.W. Norton & Company, Inc. and Routledge, pp.101–110

Most Powerful Lord, in order to give an account to Your Royal Excellency of the magnificence, the strange and marvelous things of this great city of Temixtitan and of the dominion and wealth of this Mutezuma, its ruler, and of the rites and customs of the people, and of the order there is in the government of the capital as well as in the other cities of Mutezuma's dominions, I would need much time and many expert narrators. I cannot describe one hundredth part of all the things which could be mentioned, but, as best I can, I will describe some of those I have seen which, although badly described, will, I well know, be so remarkable as not to be believed, for we who saw them with our own eyes could not grasp them with our understanding …

Before I begin to describe this great city and the others which I mentioned earlier, it seems to me, so that they may be better understood, that I should say something of Mesyco, which is Mutezuma's principal domain and the place where this city and the others which I have mentioned are to be found. This province is circular and encompassed by very high and very steep mountains, and the plain is some seventy leagues in circumference: in this plain there are two lakes which cover almost all of it, for a canoe may travel fifty leagues around the edges. One of these lakes is of fresh water and the other, which is the larger, is of salt water. A small chain of very high hills which cuts across the middle of the plain separates these two lakes. At the end of this chain a narrow channel which is no wider than a bowshot between these hills and the mountains joins the lake. They travel between one lake and the other and between the different settlements which are on the lakes in their canoes without needing to go by land …

This great city of Temixtitan is built on the salt lake, and no matter by what road you travel there are two leagues from the main body of the city to the mainland. There are four artificial causeways leading to it, and each is as wide as two cavalry lances. The city itself is as big as Seville or Córdoba. The main streets are very wide and very straight; some of these are on the land, but the rest and all the smaller ones are half on land, half canals where they paddle their canoes. All the streets have openings in places so that the water may pass from one canal to another. Over all these openings, and some of them are very wide, there are bridges made of long and wide beams joined together very firmly and so well made that on some of them ten horsemen may ride abreast.

Seeing that if the inhabitants of this city wished to betray us they were very well equipped for it by the design of the city, for once the bridges had been removed they could starve us to death without our being able to reach the mainland, as soon as I entered the city I made great haste to build four brigantines, and completed them in a very short time. They were such as could carry three hundred men to the land and transport the horses whenever we might need them.

This city had many squares where trading is done and markets are held continuously. There is also one square twice as big as that of Salamanca, with arcades all around, where more than sixty thousand people come each day to buy and sell, and where every kind of merchandise produced in these lands is found; provisions as well as ornaments of gold and silver, lead, brass, copper, tin, stones, shells, bones, and feathers. They also sell lime, hewn and unhewn stone, adobe bricks, tiles, and cut and uncut woods of various kinds. There is a street where they sell game and birds of every species found in this land …

There are streets of herbalists where all the medicinal herbs and roots found in the land are sold. There are shops like apothecaries', where they sell ready-made medicines as well as liquid ointments and plasters. There are shops like barbers' where they have their hair washed and shaved, and shops where they sell food and drink. There are also men like porters to carry loads. There is much firewood and charcoal, earthenware braziers and mats of various kinds like mattresses for beds, and other, finer ones, for seats and for covering rooms and hallways. There is every sort of vegetable, especially onions, leeks, garlic, common cress and watercress, borage, sorrel, teasels and artichokes; and there are many sorts of fruit, among which are cherries and plums like those in Spain …

There are, in all districts of this great city, many temples or houses for their idols …

Amongst these temples there is one, the principal one, whose great size and magnificence no human tongue could describe, for it is so large that within the precincts, which are surrounded by a very high wall, a town of some five hundred inhabitants could easily be built. All round inside this wall there are very elegant quarters with very large rooms and corridors where their priests live. There are as many as forty towers, all of which are so high that in the case of the largest there are fifty steps leading up to the main part of it; and the most important of these towers is higher than that of the cathedral of Seville. They are so well constructed in both their stone and woodwork that there can be none better in any place …

There are in the city many large and beautiful houses, and the reason for this is that all the chiefs of the land, who are Mutezuma's vassals, have houses in the city and live there for part of the year; and in addition there are many rich citizens who likewise have very good houses. All these houses have very large and very good rooms and also very pleasant gardens of various sorts of flowers both on the upper and lower floors.

Along one of the causeways to this great city run two aqueducts made of mortar. Each one is two paces wide and some six feet deep, and along one of them a stream of very good fresh water, as wide as a man's body, flows into the heart of the city and from this they all drink. The other, which is empty, is used when they wish to clean the first channel. Where the aqueducts cross the bridges, the water passes along some channels which are as wide as an ox; and so they serve the whole city.

Canoes paddle through all the streets selling the water; they take it from the aqueduct by placing the canoes beneath the bridges where those channels are, and on top there are men who fill the canoes and are paid for their work. At all the gateways to the city and at the places where these canoes are unloaded, which is where the greater part of the provisions enter the city, there are guards in huts who receive a *certum quid* of all that enters … Yet so as not to tire Your Highness with the description of the things of this city (although I would not complete it so briefly), I will say only that these people live almost like those in Spain, and in as much harmony and order as there, and considering that they are barbarous and so far from the knowledge of God and cut off from all civilized nations, it is truly remarkable to see what they have achieved in all things.

Touching Mutezuma's service and all that was remarkable in his magnificence and power, there is so much to describe that I do not know how to begin even to recount some part of it; for, as I have already said, can there be anything more magnificent than that this barbarian lord should have all the things to be found under the heavens in his domain, fashioned in gold and silver and jewels and feathers; and so realistic in gold and silver that no smith in the world could have done better, and in jewels so fine that it is impossible to imagine with what instruments they were cut so perfectly; and those in feathers more wonderful than anything in wax or embroidery …

He had, both inside the city and outside, many private residences, each one for a particular pastime, and as well made as I can describe – as is befitting so great a ruler. The palace inside the city in which he lived was so marvelous that it seems to me impossible to describe its excellence and grandeur. Therefore, I shall not attempt to describe it at all, save to say that in Spain there is nothing to compare with it.

He also had another house, only a little less magnificent than this, where there was a very beautiful garden with balconies over it; and the facings and flagstones were all of jasper and very well made. In this house there were rooms enough for two great princes with all their household. There were also ten pools in which were kept all the many and varied kinds of water bird found in these parts, all of them domesticated. For the sea birds there were pools of salt water, and for river fowl of fresh water, which was emptied from time to time for cleaning and filled again from the aqueducts …

References

ARMILLAS, P. (1971) 'Gardens on swamps', *Science*, vol.174, pp.653–61.

BORDONE, B. (1528) *Libro de Benedetto Bordone nel quale si ragione de tutta l'isole del mondo con li loro nomi antichi y moderni,* Venice, Book I.

CHUECA GOITIA, F. and TORRES BALBÁS, L. (eds) (1981, 2nd edn) *Planos de Ciudades Iberoamericanas y filipinas existente en el Archivo de Indias,* Madrid, Instituto de Administración Local.

CORTÉS, H. (1969) *Five Letters of Cortés to the Emperor* (trans. and ed. J. Bayard Morris), New York and London, W.W. Norton & Company, Inc. and Routledge.

DÍAZ DEL CASTILLO, B. (1963) *The Conquest of New Spain* (trans. J.M. Cohen), Harmondsworth, Penguin Books.

FERNÁNDEZ-ARMESTO, F. (ed.) (1991) *The Times Atlas of World Exploration,* London, HarperCollins.

GASPARINI, G. and MARGOLIES, L. (1980) *Inca Architecture* (trans. P. Lyon), Bloomington and London, Indiana University Press.

HAMMOND, N. (ed.) (1974) *Mesoamerican Archaeology: new approaches,* London, Duckworth.

HARDOY, J.E. (1973) *Pre-Columbian Cities,* London, Allen and Unwin.

Historia General de España y América (1985) Madrid, Ediciones Rialp, vol.9, part 1.

HOBERMAN, L.S. (1980) 'Technological change in a traditional society: the case of the *desagüe* in colonial Mexico', *Technology and Culture,* vol.21, pp.386–407.

KUBLER, G. (1948) *Mexican Architecture of the Sixteenth Century,* New Haven and London, Yale University Press, vol.1.

LOCKHART, J. and SCHWARTZ, S.B. (1983) *Early Latin America: a history of colonial Spanish America and Brazil,* Cambridge, Cambridge University Press.

MARKMAN, S. (1984) *Architecture and Urbanization in Colonial Chiapas, Mexico,* Philadelphia, American Philosophical Society.

MATHES, W. (1970) 'To save a city: the *desagüe* of Mexico-Huehuetoca, 1607', *The Americas,* Academy of American Franciscan History, Washington D.C., vol.26, pp.419–38.

MILLON, R. (1967) 'Teotihuacán', *Scientific American,* vol.216, pp.38–48.

MILLON, R. (1970) 'Teotihuacán: completion of map of giant ancient city in the Valley of Mexico', *Science,* vol.170, pp.1077–82.

MILLON, R. (1974) 'The study of urbanism at Teotihuacán, Mexico' in N. Hammond (ed.) *Mesoamerican Archaeology: new approaches,* London, Duckworth, pp.335–62.

MORALES PADRÓN, F. (1988) *Atlas Histórico Cultural de América,* Las Palmas de Gran Canaria, Consejería de Cultura y Deportes, vol.1.

MORRIS, C. and THOMPSON, D.E. (1985) *Huánuco Pampa: an Inca city and its hinterland,* London, Thames and Hudson.

MUNDIGO, A.I. and CROUCH, D.P. (1977) 'The city planning ordinances of the laws of the Indies revisited', *Town Planning Review,* vol.48, pp.247–68.

MUSSET, A. (1991a) *De l'eau vive à l'eau morte: enjeux techniques et culturels dans la vallée de Mexico (XVIe–XIXe siècles),* Paris, Éditions Recherche sur les Civilisations.

MUSSET, A. (1991b) 'De Tlaloc à Hippocrate: l'eau et l'organisation de l'espace dans le bassin de Mexico (XVIe–XVIIIe siècles)', *Annales: économies, sociétés, civilisations,* vol.46, pp.261–98.

STIER, H.-E. *et al.* (1963) *Westermanns Atlas zur Weltgeschichte,* Georg Westermann Verlag, Braunschweig.

WEBRE, S. (1990) 'Water and society in a Spanish American city: Santiago de Guatemala, 1555–1773', *Hispanic American Historical Review,* vol.70, pp.57–84.

Part 3
PRE-INDUSTRIAL CITIES IN CHINA AND AFRICA

Chapter 8: FIVE CHINESE CITIES BEFORE 1840

by Arnold Pacey

8.1 Preliminary note

The five cities discussed in this chapter can be located on the map in Figure 8.1 overleaf, while Table 8.1, on p.265, is a guide to the spelling and pronunciation of their names. The modern (1970s) Pinyin convention for spelling Chinese words is used for names throughout this chapter, with two exceptions. One is that, following practice common in newspapers and some books, the Yangzi (Yangtse) and Yellow Rivers are referred to here by these names, which are more familiar to Westerners, rather than by the Pinyin forms. The other exception is that where quotations, names or book titles are drawn from sources which use older spelling conventions (usually the Wade–Giles convention), the original spelling is retained here to ensure that the quotations are still recognizable.

To help with these other spellings, Table 8.1 gives the traditional (Wade–Giles) versions of the names of cities as well as the Pinyin form, while on p.287, Table 8.3 gives older spellings for the names of dynasties. In the latter table, and in the text when each name first appears, traditional spellings are given in parentheses after the Pinyin. The names of some dynasties, though, and a few other names are spelled the same way in both conventions. In many instances, the older spelling is a better guide to pronunciation for English-speakers, as for example with the Qing (Ch'ing) dynasty.

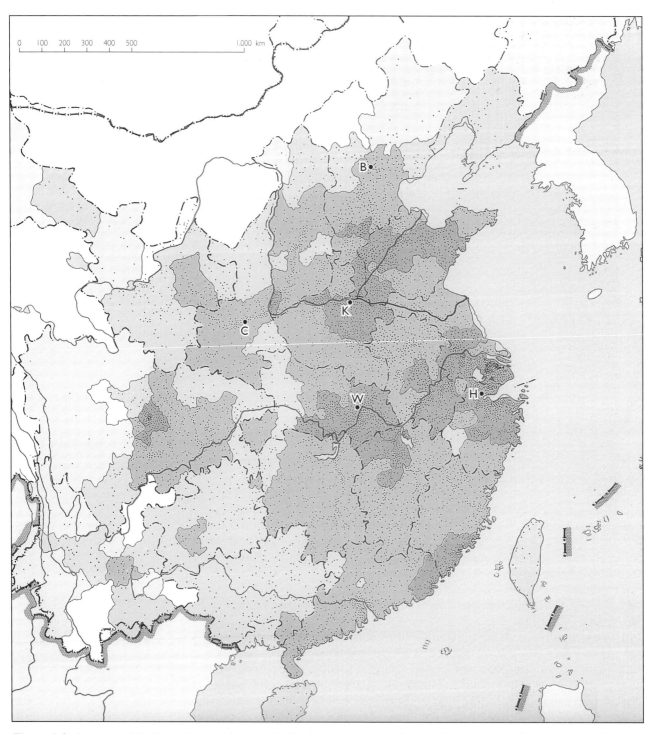

Figure 8.1 Location of the five main cities discussed in this chapter, on a map showing the distribution of population in China in about 1820. Each dot represents 50,000 people, and the darkest shaded areas have population densities above 200 people per square kilometre. The two chief rivers shown are the Yangzi, crossing central China, and the Yellow River (Pinyin form: Huanghe) in the north, with different outlets to the sea indicating changes of course in 1324 and again in the nineteenth century

Key B: Beijing; C: Chang'an; H: Hangzhou; K: Kaifeng; W: Hankou (Wuhan) (adapted from Population Census Office of China, 1987; by kind permission of Professor Liu Yue of the Institute of Geography, Chinese Academy of Sciences, Beijing, People's Republic of China)

Table 8.1 Names of cities

	Chinese name of city	Modern (Pinyin) spelling	Traditional (Wade–Giles) spelling	Approximate guide to pronunciation	Meaning of name or other notes
1	北京	*Beijing*	Peking (or Peiping)	bay-jing	'northern capital'
2	長安	*Chang'an* now rebuilt as the modern city of:	Ch'ang-an	chang-ann (*or* jang-ann)	'eternal peace'
	西安	Xi'an	Sian	she-ann	'western peace'
3	杭州	*Hangzhou*	Hangchow	hang-joe	the ending 'zhou' means 'city'
4	漢口	*Hankou* now often referred to by the name of the conurbation:	Hankow	han-coe (han-cow is a common Anglicization)	'mouth of Han (River)'
	武漢	Wuhan	Wuhan	oo-han	contraction of names of three cities: Wuchang, Hankou, Hanyang
5	開封	*Kaifeng*	K'aifeng	kie-fung	no specific meaning

8.2 Introduction: different kinds of city

China's history was much influenced by the distance of its main cities from other major civilizations on the Eurasian landmass, all of which lay to its west. Travel in that direction was impeded by mountains and deserts, and sea routes were long and circuitous. That did not prevent some trade from developing at an early date, especially with Persia (which is now Iran), but contacts with other cultures were sufficiently limited for a distinctive culture to develop, at first in north China, from where the south of the country was slowly colonized. Meanwhile, Korea and Japan were developing their own cultures, partly under Chinese influence, and at one time using the Chinese script to write their quite different languages.

'China is ... the agricultural state *par excellence*' (Bray, 1984, p.1). The fundamental occupation throughout Chinese history was agriculture, and horticulture was a popular and basic art. Even the emperor was expected to perform ceremonies related to agriculture at altars dedicated to the spirits of land and grain, and in the spring he ploughed the ritual first furrow of the new season. Not surprisingly, most people lived in the countryside. But from early times, a small number of large capital cities (including some provincial capitals) were of great symbolic importance, and during periods of commercial prosperity, smaller towns and cities that lived by trade grew rapidly in population. One of the best examples of the latter is Hankou (Hankow), which is close to the geographical centre of China. It is situated on the west bank of the Changjiang, better known in the West as the Yangzi, which is one of the country's two biggest rivers. Today Hankou is part of the conurbation of Wuhan, which is a major industrial centre, a focal point for rail, road and river transport, and the seat of a famous university. But even in 1800, it was one of China's largest urban centres.

Figure 8.2 Plan of the Wuhan cities in the late nineteenth century

Key 1–3: offices of the provincial government in Wuchang;
4, 5: prefecture offices in Hanyang;
6: Moon Lake; 7: Tortoise Hill;
8: riverside quarter of Hanyang – houses and moorings;
9: customs office, submagistrates' offices in Hankou, under supervision of magistrate at Hanyang; 10: salt-trade superintendent's office;
11–14: gates in small nineteenth-century city wall; 15: temple, and landing-place for ferries;
16: temple; 17: Huizhou (Hui-chou) Guildhall;
18–21: other guildhalls; A: lifeboat headquarters; B: shipbuilding yards; C: docks/piers;
D: navigable canal, made to improve drainage of city
(adapted from Rowe, 1984, p.22;
© 1984 The Board of Trustees of the Leland Stanford Junior University)

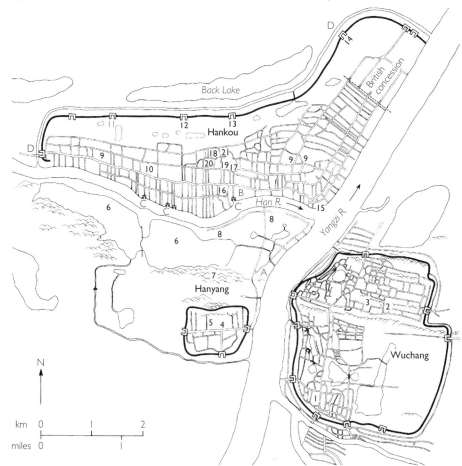

Figure 8.3 Bird's-eye view of the Wuhan cities in 1818. A gate in Wuchang city wall and a pagoda (tower) within that city are at bottom right, and Hankou is in the background. Hanyang is shown in more detail

Key 1: Hanyang walled city;
2: riverside quarter, with boats moored in the Han River;
3: Moon Lake; 4: Tortoise Hill (with temple pavilion on summit); 5: monastery (probably Buddhist);
6: the 'Clear Stream Pavilion', a fifteenth-century monument used by Hankou guilds as headquarters of a lifeboat service
(*Han-yang hsien-chih* (*Gazetteer of Hanyang County*), 1818)

Wuhan formerly comprised three distinct cities, because the Yangzi is about a kilometre wide at this point, and in bad weather conditions the ferry crossing was unpleasant and sometimes dangerous. So until the river was bridged in 1957, the cities on opposite banks tended to lead separate lives. On the east bank, Wuchang was an old administrative centre with a market serving its immediate hinterland; it was the capital city of Hubei Province. Across the river, Hanyang was another administrative centre, but only of a prefecture, not a province. Separated from Hanyang by the Han River was Hankou, an enormous, bustling, commercial city which grew in size very rapidly during the eighteenth and nineteenth centuries. As its plan shows (see Figure 8.2), it was a city of quite different character from the other two. It was a riverport, and it stretched along the banks of the Yangzi and Han Rivers to achieve the maximum length of moorings and wharfage. Its principal streets ran parallel with the riverfront, with narrow lanes at right angles to them giving access to warehouses and landing-stages – and to boats moored three or four abreast.

In 1840, the main street of Hankou was famous for its shops (Huc, 1859; Rowe, 1984), and after 1900, having passed through many vicissitudes, it again presented 'four miles ... of almost continuous shops' (Rattenbury, 1944). The largest and most imposing were those selling silks (for clothing) and herbal medicines. There were also shops selling foods of many kinds, also furniture, stationery, porcelain and crockery – and there were tea-houses, wine-shops and restaurants. The more seamy side of life was also notoriously well developed, and there were gaming-houses, opium dens and a well-known red-light district (Rowe, 1989).

On the plan in Figure 8.2, part of Hankou is marked as 'British concession'. This refers to the semi-colonial phase in Chinese history which began in 1842, after China had lost a naval war against Britain. The war began because opium, grown in Bengal, was being sold to the Chinese by British merchants in exchange for tea. The Chinese government became alarmed by the scale of this illegal drug-trafficking and appointed a commissioner to stamp it out. The British responded to his confiscations of illegal opium by sending a naval task force, which was called off only after the Treaty of Nanjing (1842) was signed, allowing the British to establish small colonies, or 'concessions', in each of five ports. Even before the treaty, in 1841, the British had occupied Hong Kong Island. Other countries insisted on similar concessions, and in 1857, Britain and France staged another war to gain access for trade over a wider area of the country. It was after this, in 1858, that Hankou became a 'treaty port', and soon after began to acquire a quota of European buildings.

In this chapter, we are concerned with Chinese cities before they were touched by any such Western influence, which in general means before the treaty of 1842. Figure 8.3 captures something of the character of the three cities of Wuhan a few years before this. Wuchang and Hanyang both have several monumental buildings and imposing city walls, but Hankou, in the distance, is just a sea of roofs. That indicates numerous dwellings and shops, but none of the ornate buildings found in the more traditional cities, such as gateways in city walls, bell-towers and pagodas. Indeed, it is said that warehouses and guildhalls were the dominant architectural forms in Hankou.

Masts of many boats moored in the Han River can also be seen in Figure 8.3. One visitor in 1850 described Hankou as a 'great port' that was 'literally a forest of masts' (Huc, 1859, p.364). It was 'astonishing to see vessels of such size in the very middle of China'. In fact, Wuhan is more than 750 kilometres from the sea, but ocean-going vessels (today of up to 8,000 tonnes displacement) can come upriver, and in 1850 could come under sail. Historically, the port developed here because this was the point where deep-draught vessels transhipped cargo into smaller boats for onward transport up the Yangzi or its tributaries. The Chinese transport system, both for freight and for passengers, depended on rivers and canals, and this was its central interchange and entrepôt. In the eighteenth and early nineteenth centuries, indeed, Hankou is said to have functioned as a central commodities market for the Chinese economy in much the same way as Chicago did later for the US economy (Rowe, 1984, p.23).

The other point we need to notice, though, is that Wuhan offers a contrast between markedly different types of city. We have already noted this in comparing the traditional architecture of Wuchang and Hanyang with the plain roof-tops of Hankou, which reflects the difference between the administrative functions of the two former cities and the commercial function of Hankou. It would be nice to think that further exploration of this contrast might tell us all we need to know about Chinese cities. However, in a country as large as China, there are also many regional differences, including significant contrasts between metropolitan centres and provincial outposts. It would certainly be desirable to discuss regional variants, such as the great southern city of Guangzhou (better known in Europe as Canton), and a variety of provincial centres, such as Yuxian in Henan Province, where the noted mathematician Li Zhi was magistrate in the thirteenth century (Ho, 1973); or the small town of Yincheng in Shanxi Province where almost everybody worked at iron-smelting (Wagner, 1997); or the remote city of Zhaotong in Yunnan Province, administrative centre of a strikingly multi-ethnic county (and birthplace of the present author). However, there is space here to discuss only a few variants among the many types of Chinese city, and in a study of cities and technology, one obvious choice – because of its technological interest – is Hangzhou, one of the 'canal cities' of the Yangzi delta and its southern fringes. Other cities are discussed here mainly because they provided precedents (Chang'an) or models (Beijing) for Chinese traditions in urban design. If that seems a biased sample, it at least represents some of the more densely populated areas (see Figure 8.1).

Table 8.2 gives more detail about the populations of the sample cities. It uses the 'conservative' figures given by Skinner (1977), except for Kaifeng, where Hartwell (1966) better demonstrates the rapid growth of the city while giving data not greatly different from Skinner's. Some larger, quite incredible figures are sometimes claimed, especially for Beijing. Skinner emphasizes that early *capitals* were sometimes much larger than the biggest cities of the early nineteenth century, but points out that there were many more large urban centres during the latter period.

Table 8.2 Population estimates for the five cities discussed in this chapter

City	Date(s) when the city was at the height of its importance	Population at the date cited	Population in 1843
Chang'an (or later, Xi'an)	eighth century	c.1,000,000	300,000
Kaifeng	1021 1078	510,000 750,000	190,000
Hangzhou	c.1270	1,200,000	500,000
Beijing	fifteenth to eighteenth centuries	never greater than in 1843	850,000
Hankou	see Table 8.4		see Table 8.4

8.3 Chang'an: an administrative city

While commercial centres such as Hankou developed in somewhat informal
and unregulated ways, ancient principles governed the planning of most
administrative cities. Classical Confucian writings set out these principles very
clearly, but much precedent also stemmed from the city of Chang'an, as it was
rebuilt from 618 CE. At this time, China was recovering from a long period of
division and conflict, and Chang'an was chosen as a new seat of government for
the reunified nation because it had once before served as the capital (Twitchett
and Fairbank, 1979). It was near a tributary of the Yellow River, in a part of
northern China that archaeologists regard as the cradle of early Chinese
civilization. The modern city of Xi'an (Sian) now occupies part of the site.

The plan of the rebuilt Chang'an was nearly square, and the south, east and
west walls each had three gates (see Figure 8.4). The walls are described as built
of earth, which means that they were embankments, with a vertical outer face,
probably formed by tamping the earth firmly down behind wooden formwork,
or else by building up the face with mud-brick and stone rubble. When a little
lime is mixed with soil, tamped or rammed earth (sometimes referred to as *terre
pisée*) can be a surprisingly durable form of construction.

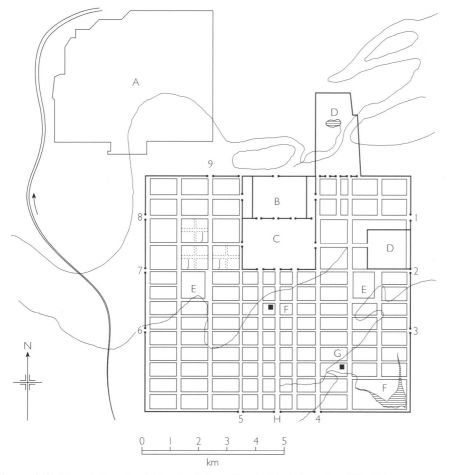

Figure 8.4 Plan of Chang'an during the Sui and Tang (T'ang) dynasties, 589–904 CE

Key A: site of earlier city (capital from about 200 BCE); B: the emperor's palace enclosure;
C: administration area, government ministries and offices; D: the emperor's pleasure palaces
and gardens; E: the two markets; F: parkland and temple precincts (with a surviving pagoda
indicated by a black square, and with a lake to the south-east); G: the Great Wild Goose
Pagoda and its monastery; H: the Great South Gate on the central axial street (other main
gates are numbered 1–9); J: blocks in the main grid were often subdivided into four by narrow
lanes (based on information in Wu, 1963, plate 113, by permission of the author; also Boyd,
1962; Skinner, 1977; Willetts, 1958; local guidebooks)

The regular geometry of the plan was meant to be an improvement on the irregular shape of old Chang'an (also shown on Figure 8.4), and to a large extent it conformed to much earlier concepts of an ideal city, which formed part of Confucian tradition (Wright, 1977; Wu, 1963). The square layout not only marked off a rational and ordered human world from the irregular and unknown, but its four sides represented the four seasons, and also symbolized different aspects of cold and heat – *yin* and *yang*. In these ways, urban design could incorporate cosmological references, and the strict north–south orientation of the city was intended to express a sense of cosmological order also. As building proceeded, whole villages were relocated, land was levelled and trees were planted in orderly rows along the streets. According to legend, 'there was one great old locust tree that was not in line'. It was allowed to survive because the architect-general had often sat under it to watch progress of the work, and a special order from the emperor spared it (Wu, 1963).

The emperor's palace was at the northern end of the main axis, in its own rectangular walled enclosure. To the south of this, in another rectangular enclosure, was the administrative area, where the government offices lay. To the south again, symmetrically placed, were two market-places. The city grew to be a great cosmopolitan centre, famous throughout Asia, and was the model for other Chinese cities, as well as for two in Japan: Nara, founded in 710, and Kyoto in 790 (Chang, 1977; Wu, 1963).

A simplified version of the same concept was the basis for planning smaller administrative cities, where again the main street would run precisely north–south, and the administrative buildings, as at Chang'an, would be in the northern sector. Comparison of this ideal form (see Figure 8.5) with the later plan of Hanyang (on the left of Figure 8.2) demonstrates both the persistence of the basic idea, and the scope for adjustment in adapting to a difficult site. It is also worth noting that, in being sited south of Tortoise Hill, and with a small stream along its south side, Hanyang was well placed according to the theory of *feng shui* (pronounced 'fung shway'), which dealt with the way building should relate to natural energies.

Not only was Chang'an aligned strictly north–south so as to be in harmony with the cosmos, but an important aspect of the emperor's duties involved astronomy, and the Astronomical Bureau was an important ministry in his government. An emperor was expected to promulgate a calendar that would accurately predict equinoxes, the solstices and the phases of the moon. There was also a problem of fitting lunar months into a solar year, which had practical consequences, because orderly administration depended on having a calendar that kept in step with the climatic seasons. Beyond that, though, Chinese emperors were supposed to be able to mediate between earth and heaven (*tian*). If they could not even issue an accurate calendar of events in the heavens (also *tian*), their standing in the eyes of their subjects suffered. They lost credibility and legitimacy. There was a strong incentive, therefore, to get the calendar right, and eventually, nearly all celestial events could be predicted,

Figure 8.5 One of several possible ideal plans for a small administrative city, such as the capital of a prefecture or county. The plan is strictly square and aligned north–south. The yamen, or government office, consists of a series of buildings arranged around courtyards. Compare the less regular plan of Hanyang in Figure 8.2 (based on several ideal plans in Chang, 1977)

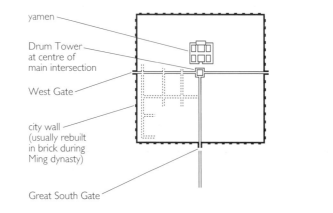

yamen

Drum Tower at centre of main intersection

West Gate

city wall (usually rebuilt in brick during Ming dynasty)

Great South Gate

N

although 'solar eclipses remained just outside the realm of certainty and hence remained ominous' (Cullen, 1990).

Because it was important for the emperor to get the best available calendrical advice, Indian astronomers, known for their mathematical expertise, were invited to work in the Astronomical Bureau in Chang'an. Needham and Wang Ling (1959) trace the possible influence of Indian mathematics in China, notably in the way the zero was used in decimal numbers. Contacts with India also brought knowledge of the chemistry of saltpetre, which contributed to the invention of gunpowder, and there may have been stimulus also for the most important Chinese invention of this period: printing on paper (Pacey, 1990, pp.16–17).

However, Indian astronomers were not the only foreigners in Chang'an – there were merchants and monks as well, representing most parts of Asia. Indeed, the cosmopolitan character of the city demonstrates one of the ways in which its life was stimulating for technology. Cities were focal points for specialized occupations and expertise, in this case including mathematicians. Contacts between people with different skills and different knowledge in a big city were, throughout history, a stimulus to innovation.

In this context, the geographical position of Chang'an needs to be emphasized. It was far to the west of most cities of importance in later Chinese history, and was the natural terminus of what was later called by romantics the 'Silk Road'. This was the ancient trade route linking Baghdad and Persia to China via the Central Asian city of Samarkand and desert tracks well to the north of the Himalayas and Tibet. The distance from the coast of Syria to Chang'an is 8,000 kilometres, and it is doubtful whether many traders went all the way in a single journey (which could have taken six months or more). Even so, there was a chain of trading links through which China exported silk and imported a few luxury goods from Persia, and horses from the Central Asian steppes.

In 635 CE, travellers to China along this route included Nestorian Christians from Syria and Persia, disturbed by the rise of Islam in those countries. The emperor, impressed by their 'luminous religion', as he called it, was moved to issue an edict in 638 allowing them to establish a Christian monastery in Chang'an (which survived for two centuries). Communications with India were far more important, however, and while there was some sea-borne trade between India and south China, most travellers going to India from Chang'an followed the Silk Route for about 3,000 kilometres, and then turned south through parts of what are now Afghanistan and Pakistan. Some trade went this way, too, with goods originating in China being taken by sea from Indian ports to the Persian (Arabian) Gulf.

One result of travel and trade along these routes was that Chang'an became a centre for the translation of Buddhist manuscripts brought from India. Efforts to disseminate the translations may have been among many impulses behind the development of printing, for the oldest surviving book-length work produced by printing is a scroll of Buddhist scripture known as the Diamond Sutra. It dates from 868 CE (Liu and Zheng, 1985; Tsien, 1985).

One of the few surviving buildings of this period in Chang'an is the Great Wild Goose Pagoda. It is part of a Buddhist monastery founded in 646 CE by the monk Xuan Zang, who spent seventeen years travelling and studying in India, and came back with a vast collection of Buddhist manuscripts and an interesting diary of his travels. Pagodas were often built to commemorate and house sacred relics, and Xuan Zang began the Great Wild Goose Pagoda in 652 as a place to store his precious Indian manuscripts. Because he wanted a fireproof place of storage, he built in fired brick. The pagoda was originally a five-storey structure, but in 704 it was enlarged to become a seven-storey tower, sixty-four metres in height. Its outer walls are of yellowish brick, and its internal passageways and the staircase shaft are lined with the same material. But behind the thick skin of brickwork, the mass of the structure is again tamped earth, making for an extremely heavy and durable building (see Figure 8.6 overleaf).

Figure 8.6 The Great Wild Goose Pagoda at Chang'an (now in the city of Xi'an) as it was completed in 704 (top left), compared with two Indian buildings which may have influenced its design. These are the shrine at Sarnath, Varanasi (bottom left, showing a conjectured restoration of the seventh-century building), and a tower at Bhitargaon, near Kanpur (bottom right, dating from close to 400 CE) (Willetts, 1958; © Penguin Books Ltd)

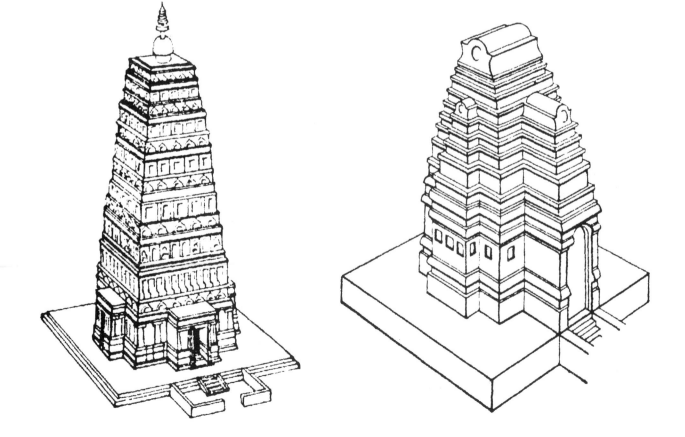

Figure 8.7 Part of a drawing engraved on stone in the Great Wild Goose Pagoda at Chang'an, thought to date from 704. The building depicted stands on a stone or brick platform and has posts or columns, presumably of timber, connected together near their tops by doubled wooden beams. Above them, elaborate decorative brackets support the projecting eaves of the tiled roof. The curving roof-line and the timber-frame construction persist through most succeeding phases in the history of Chinese architecture (Wu, 1963, plate 115)

Willetts (1958) describes several buildings in India that were constructed on similar principles, and there is written evidence that Xuan Zang wished to follow Indian precedent. Indeed, the very name of the pagoda refers to an Indian legend about a wild goose, and there was an Indian temple with the same name. Needham *et al.* (1971) describe a commemorative arch that was a 'gift from Indian to Chinese architecture' and this pagoda can be described in the same way.

Inside the Great Wild Goose Pagoda is a mural drawing depicting another building, which is probably more representative of Chang'an at this time. It is an open-sided hall with a group of figures celebrating a Buddhist ritual. There is a row of columns along the front and there are wide, overhanging eaves. The high roof has the curves familiar from more recent Chinese architecture, and the whole building is recognizable from later structures as a typical Chinese timber-framed hall (see Figure 8.7). This type of building is rectangular in plan, and has been adapted to a great many uses. Such a hall may be the central building in a Buddhist temple (as in this illustration), or the main reception room in a private house, or where audiences are held in the imperial palace.

The persistence over many centuries of this basic style of rectangular building with timber columns is a strong expression of continuity in Chinese culture, but the argument that its structure provides evidence of a static, unchanging technology is mistaken. Craft traditions in technology were always evolving, albeit slowly, and the craft of making timber-framed buildings in China shows ample evidence of regional variation and change over time. For example, Figure 8.11 (on p.278) illustrates several timber-framed buildings of a later period with roof-frames incorporating triangulated bracing. This is otherwise rather unusual in Chinese framed building, and may represent a localized experiment, or an unrecognized regional tradition. Surviving buildings that have been studied by architectural historians show two other, quite distinct types of

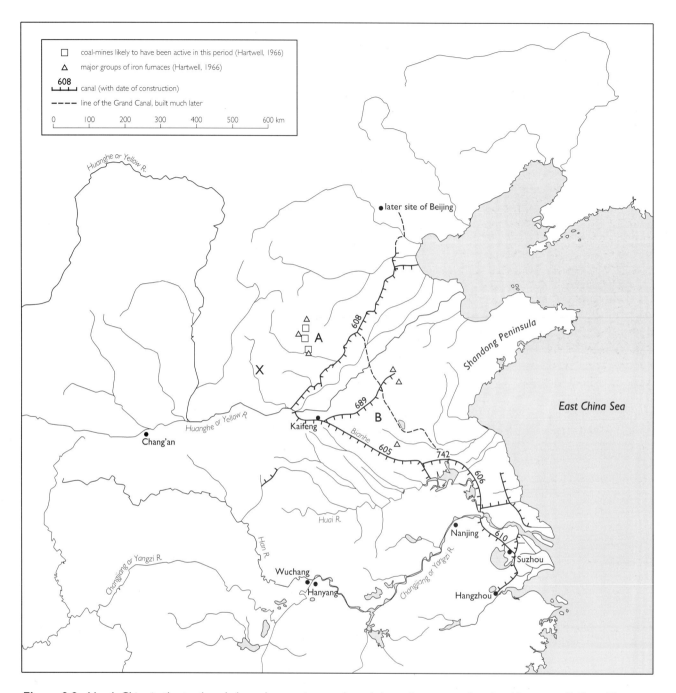

Figure 8.8 North China in the tenth and eleventh centuries: canals, and the main sources of coal and iron near Kaifeng. The area to the north of the city marked A was the most productive iron-producing area. Bituminous coal was also available here, and *may* have been made into coke for use in furnaces. The area marked B was somewhat less productive, while small items such as needles and knives were produced in area X. Anthracite coal was available in the latter area, but may not have been used for iron-smelting until later (adapted from Twitchett and Fairbank, 1979, p.136, map 2, with additional material from Hartwell, 1966)[1]

[1] I am indebted to Donald B. Wagner of Copenhagen and Peter J. Golas of the University of Denver for their advice, although neither is responsible for any detail of this map. The extent to which coal and coke were used remains uncertain, but the consensus is that at least some coke was made for use in furnaces.

framing, which may represent separate phases in a long historical development (Knapp, 1989, 1990). Continuity in architectural form was certainly valued, but that did not mean that carpentry techniques were frozen and unchanging.

The Tang dynasty emperors who reigned in Chang'an, like all rulers who aimed to keep China unified under a single government, were much preoccupied with transport links to the provinces in order to keep the capital supplied with food and fuel, to allow taxes to be remitted in kind (as will be explained later), and to move military supplies and sometimes whole armies. Under the preceding Sui dynasty, and mainly in the two decades prior to 618, canals extending to the north-east and south-east had been developed (see Figure 8.8). Construction proceeded rapidly, partly because a massive labour force was employed, but partly also because lengths of canal built by previous regimes were incorporated, and in some places, natural rivers were adapted to form part of the canal route. Moreover, most of the terrain was a flat alluvial plain, and river flows could be fairly easily diverted into the new canals, in many of which there was a considerable current. Few changes in level were necessary, and were mostly dealt with by 'double slipways'. Boats were hauled up a slipway from one length of canal by men working a windlass, then passed over a hump to slide down another slipway into a higher (or lower) length of canal. Sometimes, too, boats would be passed through a single lock gate in a weir. Pound locks more like the modern type, with a short stretch of water (a pound) enclosed between two sets of lock gates, are a well-documented invention whose first use can be dated to 983 CE (Needham *et al.*, 1971). Boats could then be floated into the pound at one level and floated out at a different level, after water was released from, or admitted to, the lock.

For two centuries after the Tang emperors came to power, in 618 CE, the canal system was well maintained, extensively used and occasionally extended, but then the Tang dynasty's military conquests in Central Asia began to be more fiercely opposed, and indeed rolled back, and there was internal dissension as well. Eventually, the sacking of Chang'an in 881 and again in 904 marked the effective end of the dynasty.

A time of fragmentation known as the period of Five Dynasties followed, and several decades elapsed before China was reunified, in 960, under the Song dynasty (pronunciation and old spelling, Sung). Even then, the Song failed to re-establish Chinese government in the north of the country, which was now ruled by peoples with a nomadic pastoralist culture who were sometimes referred to as Tartars or Eastern Turks. They had established what they called the Liao Empire in the partitioned parts of north China, and in areas further north extending into southern Mongolia (Franke and Twitchett, 1994).

8.4 *Kaifeng: an industrial centre as capital*

For the Song emperors, alien occupation of north China created several problems, one of which was to find a more secure location for the capital. The place they chose was an existing city, at a central point on the canal system, now known as Kaifeng. This small but busy city had a square plan, which was displaced from a strict north–south alignment in order to fit into a 'complicated web' of canals (Needham *et al.*, 1971, p.311). As the site developed on becoming the capital, a new city wall was built outside the original one, with four water-gates in it to allow access by canal and river-boats, but this time more precisely aligned on the north-point. To emphasize the improved alignment, a central street, the Imperial Way, linked the emperor's palace ('Imperial City' on Figure 8.9 overleaf) with the main south gate. Along this route were three or four temples and altars at which the emperor performed various ceremonies. For example, one ritual in the early spring sought to ensure that farmers would have good crops in the coming season.

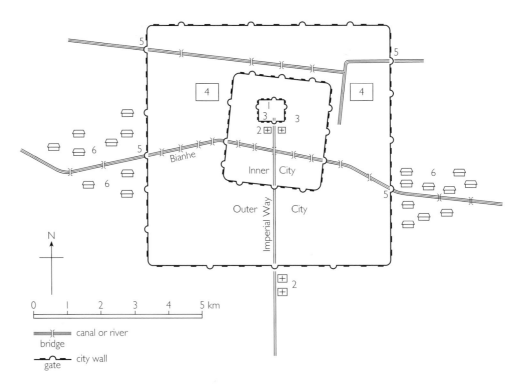

Figure 8.9 A speculative reconstruction of the plan of Kaifeng, *c.*1100. The Bianhe is the Bian canal, or canalized river

Key 1: Imperial City and palace; 2: altars or temples for imperial ceremonies; 3: possible locations of the astronomical clock; 4: warehouse and granary compounds; 5: water-gates in city wall; 6: suburbs, including the one with the bridge shown in Figure 8.11, p.278 (adapted from Wright, 1977, p.61; with the permission of the publishers, Stanford University Press. © 1977 by The Board of Trustees of the Leland Stanford Junior University)

Not only was agriculture critical for feeding the population, but it was also central to the political economy of the empire. Most taxes were paid in grain (by farmers) or in textiles (by their wives), and these goods were brought by canal from outlying regions. Twenty-three large warehouses were specially constructed in Kaifeng so that grain could be stored over long periods. Such buildings usually had substantial masonry foundations as protection against rising damp, and there were ventilators in their roofs (see Figure 8.10).

From here, grain was supplied to the imperial household and nearby army units. Some wages and salaries were paid in kind, and some grain was sold to provide funds for other government expenditures. The unit of currency was a copper coin with a hole in the centre. A thousand coins threaded on a cord constituted a 'string of cash', and Bray (1984, p.415) comments that the Song government's budget in the 1050s was in the region of 125 million strings of cash annually, which was probably equivalent to something like ten million tonnes of grain. This was more than the granaries could hold, and, undoubtedly, some taxes were remitted in copper cash or silver bullion.

Another feature of Kaifeng was its great astronomical clock. The Song emperors were as much concerned as their predecessors to issue an accurate calendar, and when it seemed that a discrepancy had arisen, the intensive astronomical research that ensued led to the construction of an elaborate clock within the Imperial City (see Figure 8.10). Built during the 1080s, this was a large calculating device designed to show the apparently changing positions of the stars and planets in real time during each night, and from month to month during the year. It was driven by a water-wheel, three and a half metres in diameter, which revolved very slowly – in fact, at 100 revolutions each day – under the control of a precise escapement mechanism (Needham *et al.*, 1960).

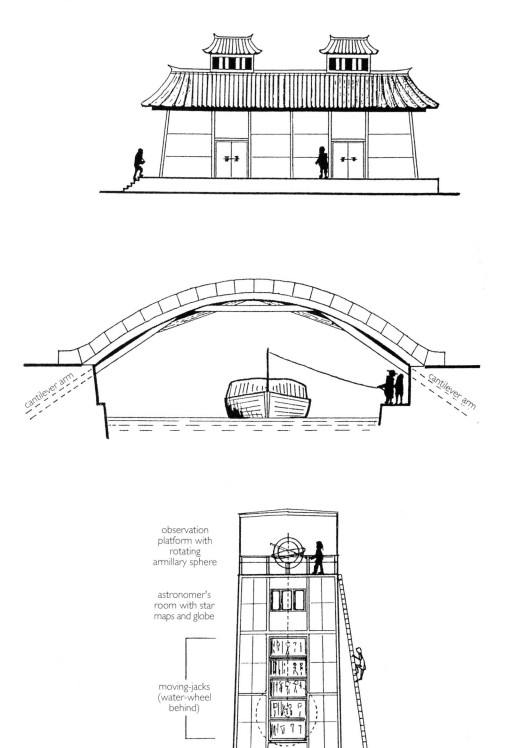

Figure 8.10 Three structures in Kaifeng, c.1100. *Top* granary (grain warehouse) with ventilators in roof; *middle* timber bridge using the 'rising cantilever' principle (see Figure 8.11); *bottom* astronomical clock driven by escapement-controlled water-wheel (dashed). The roof on the observation platform was removed when the astronomers were working. Displays of moving-jacks gave astronomical information as well as time and date (based on information in Bray, 1984, pp.415–22; Needham *et al.*, 1971, pp.162–3; Needham and Wang Ling, 1965, p.449; Needham and Yates, 1994, p.345)

Figure 8.11 Part of a large painting thought to date from about 1125, showing Kaifeng at the time of the Spring Festival. The waterway is presumed to be the Bianhe, a canal or canalized river, and the 'rising cantilever' bridge is built of timber. There is a towpath under the bridge, but the boat is being propelled with oars, its mast and sail having been lowered to allow it to go under the bridge. The timber-framed house in the foreground has a braced king-post supporting its roof. This form of construction went out of use later (photograph: Cheng Chen-To)

Hardly anything now remains of the Song capital at Kaifeng, but a remarkable painting dating from about 1125 shows canal-boats and a timber bridge of unusual construction (see Figure 8.11). The bridge is constructed using the principle of the 'soaring' or 'rising cantilever'. These terms are used by Needham *et al.* (1971, p.163) to describe a steeply inclined cantilever. The word 'cantilever' is here used in its ordinary sense, to mean a projecting beam counterweighted at one end, in this case by the mass of earth in which it is embedded. There were many wharfs where boats unloaded their cargoes, and it is said that much of the city developed in an 'unplanned and utilitarian' manner with numerous commercial and industrial premises (Wright, 1977). The latter included wheelwrights' workshops and premises where paper was made using bamboo, mulberry and straw as raw materials. There were printing shops, which produced books for private and commercial publication as well as for government issue, and there were retail bookshops (see Figure 8.12). One major publisher in the town was the National Academy, which published editions of the Confucian classics and official dynastic histories (Tsien, 1985). More surprisingly, there was also some 'heavy' industry in or near Kaifeng, including three privately owned workshops making iron products, and a government enterprise that made arms and armour. In mentioning these 'factories', Hartwell (1966) describes Kaifeng as 'the most important administrative, military, manufacturing and commercial centre' in China, noting how its population increased from 510,000 in 1021 to about 750,000 in 1078. Elsewhere, he describes the city as a 'multifunctional urban centre', and alleges that in this respect it was 'quite possibly unsurpassed by any metropolis in the world before the nineteenth century' (quoted in Wright, 1977, pp.60–63).

Figure 8.12 Detail of a bookshop in Kaifeng appearing in a long scroll painting, *Spring Festival on the River*, attributed to Chang Tse-Tuan, early twelfth century (Tsien, 1985, p.168, figure 1120)

Kaifeng's role as an industrial and commercial city as well as the imperial capital becomes less surprising when we learn how vigorously the Song government tackled various practical problems that faced it, notably with regard to food supplies. Agricultural land had been lost to the Liao Empire in the north, where the main crops were millet and wheat. The chief opportunity for increased food-production was further south, especially in the lower Yangzi region, where the main crop was rice. Changes in diet are implied, and there was emigration of farmers to rice-growing areas. In addition, a new, quick-growing rice variety was introduced from Champa, in what is now Vietnam, which made it possible for each plot of land to produce two crops a year. Deliberately promoting innovation, the emperor ordered several thousand samples of the Champa seed to be distributed in the lower Yangzi region and Fujian in 1012. Local officials were given precise (printed) instructions on sowing dates and cultivation, and were required to contact 'master farmers' and pass on the seed and the instructions. Widespread adoption of the new rice variety followed, and in subsequent decades, farmers bred higher-yielding strains. Output increased rapidly (Bray, 1984; Elvin, 1973).

There were contributions to these developments in agriculture from two technologies for which Kaifeng was an important centre: printing and iron-working. Apart from the printed instructions for the Champa rice (which officials had to relay orally to illiterate farmers), Bray (1984) has counted 105 books on agriculture that were printed during the three centuries of Song rule. Iron-working was important because of an earlier phase of innovation which had led to the introduction of ploughs with iron mould-boards as well as iron ploughshares, and also because new hand-tools and mills developed for use with rice culture had iron components.

The map of the canal system in Figure 8.8, p.274, shows the location of major ironworks within 200 kilometres of Kaifeng, and indicates that many of them were near to coal-mines (they were also near to quarries, which provided limestone flux for furnaces). Wood-fuel suitable for making charcoal in the quantities needed for iron-smelting was not available in all these places. Anthracite coal was used much later for iron-smelting to the north-west of Kaifeng in what is now Shanxi Province, an area known from the time of the Tang dynasty for small, high-quality products such as scissors and needles (Wagner, 1997, p.53). But in other iron-working areas shown on the map, the coal was not of this kind and could only be used if first converted into coke (which was referred to as 'purified coal'). There are references to coke being used in cooking-stoves as early as 870 CE, and Hartwell (1966) believes that it was used in some blast-furnaces north of Kaifeng (area A) from the beginning of the Song period.

The main reason for this large output of iron was a military one. In response to the occupation of north China by the Liao Empire, the Song emperors maintained an enormous army, and needed iron for blades, armour and arrowheads. Workshops producing these goods were located in the iron-producing areas (those marked A and B in Figure 8.8), and at Kaifeng itself. The most important weapon in current use was the crossbow and, at one stage, the imperial 'bow-and-arrow factory' was producing several million iron arrowheads per year (Hartwell, 1966). But apart from the use of iron in weapons and farm implements, efforts were made to develop new markets for cooking-pots (and woks), components for buildings, and ironwork for bridges (Needham, 1958). From 1079, further innovation made it possible to use iron for casting bells, such as were needed in city bell-towers (Rostoker *et al.*, 1984).

8.5 Hangzhou and the canal cities, c.1130–1280

For many years, relations between the Song government in Kaifeng and the Liao Empire to the north were relatively peaceful, or at least stable, but then the Liao rulers were overturned by a more aggressive Tartar group, whose leaders took the dynastic name of Jin. They soon began an attack on the Song Empire, and in 1126 besieged Kaifeng. When the city capitulated, it was looted, and some components of the great astronomical clock were carried off to the Jin capital (Needham *et al.*, 1960, p.137).

The emperor's son escaped this disaster, however, and with a remnant of the government eventually established what was regarded as a 'temporary' capital at Hangzhou (Hangchow), an old port and trading city at the southern end of the main canal system. Hangzhou occupied an elongated site beside an estuary. The whole area had originally been subject to flooding, but a long time before, a dike or sea wall had been built alongside the estuary, and then another dike was used to dam an inlet to create an artificial lake on the west side of the city. While this constrained growth of the built-up area, its scenic qualities were much appreciated, and in an area where many rivers and creeks were tidal, it was also important as a source of fresh water.

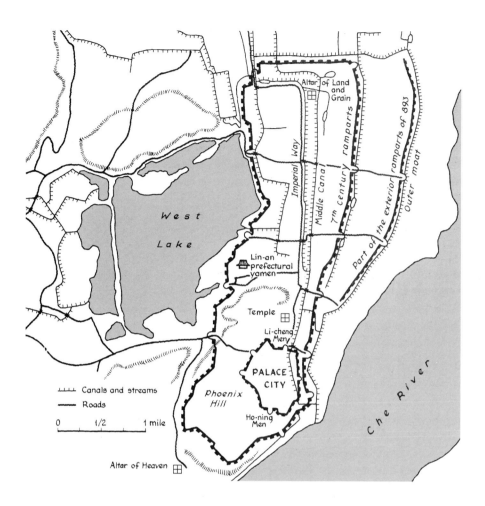

Figure 8.13 Plan of Hangzhou, *c.*1274, when it was still the Song dynasty's southern capital. Marco Polo supposedly visited the city in 1280 and used his Persian sources to describe it well, calling it 'Kinsai' (or 'Qinsai') (Skinner, 1977, p.65, map 3; © 1977 The Board of Trustees of the Leland Stanford Junior University)

Clearly, this capital city could never fit the cosmological planning principles that had been imposed on Chang'an and Kaifeng. The approach by canal entered the city from the north-west corner, and the main street, remodelled as a fairly grand Imperial Way, proceeded southwards (rather than northwards) from there to the only site available for the emperor's palace (marked 'Palace City' on Figure 8.13). This was the former site of a Buddhist temple. Contemporaries were appalled by the lack of symmetry in this plan. One compared living in Hangzhou to living in the 'side-room of a house', and another referred to it as a 'mean little place, lost in a corner of the empire' (Wright, 1977).

Nor was there much space between the lake and the river to accommodate without congestion the 1.2 million or more people who were living here around the year 1270 (Skinner, 1977), and there may have been more two- and three-storey buildings than was usual in Chinese cities.

Trade continued to be an important part of Hangzhou life after the city became the capital, and the population was markedly cosmopolitan, with visiting merchants from as far away as the Persian Gulf. One European merchant who supposedly came here in 1280 (just after the Song dynasty had been overthrown) was Marco Polo. However, it now seems that his description of the city, which he refers to as 'Kinsai' (or 'Qinsai'), may have been derived from Persian traders' accounts, with all the statistics and distances grossly inflated for dramatic effect.

Since he came from Venice, it is not surprising that Polo took note of what he was told about the canal system, though he exaggerated its size, claiming that there were thousands of bridges over the different waterways. In fact, a survey made by Song dynasty officials not long before had recorded 117 bridges within the city walls (and others outside), and had suggested that many needed rebuilding to allow more space for the passage of boats. There was a lock and tidal basin at the point where the canal system met the river estuary. This was operated in such a way that the basin filled at high tide and then water flushed out canals within the city, many of which carried sewage.

The city was traversed from north to south by the Middle Canal, and another main canal ran parallel with the Imperial Way (see Figure 8.13). Other canals served as a moat outside the city wall. The wall itself was nine metres high, and was faced with mortared limestone. Also built of stone to make them fireproof were some multi-storey warehouses whose rooms were let out individually to merchants. Branches of the canal ran right round the warehouses to enable boats to unload directly into them, and they are said to have looked like 'great towers of stone' (Boyd, 1962, pp.56–8; Needham *et al.*, 1971, p.69, note (b)).

Chinese capitals had usually been far inland, and it was something new for the emperor to be living in a port city. The experience altered governmental awareness of the sea, and led to the founding of a navy in 1132. There was much innovation in the design of sailing-ships for a defensive role, and vessels driven by paddle-wheels were also built. Water-wheels were commonplace in country mills. In these ships, whose movements were independent of wind direction, water-wheels worked in reverse, being driven by men on treadmills. For example, one boat of 100 tonnes displacement was driven by two wheels powered by twenty-eight men. There were also some 'stern-wheelers' (small vessels with one large paddle-wheel at the stern). All naval vessels were equipped to catapult gunpowder bombs into enemy ships, and the paddle-wheel boats owed some of their success to the fact that their decks could be roofed over to deflect such bombs, which was impossible for a ship with masts and sails. Needham *et al.* (1971, pp.688–9) quote successful use of these vessels in river battles against rebels (but they note that, seven centuries later, similar boats were not effective when deployed against armed paddle-steamers during the Opium War of 1840–42).

In this part of China, in the river delta lands south of the Yangzi, the canal system had many branches, and there were other canal cities apart from Hangzhou. Many smaller places – even villages – were located on the banks of the major artery (later known as the Grand Canal), and the relationship of surviving buildings to the canal in these smaller places has been analysed by the architectural writer Ronald Knapp (1989; see Figure 8.14). While the buildings he discusses are much later than the Song dynasty, a similar relationship of buildings to canals probably existed then. Houses were often set back from the side of the canal to leave space for a walkway but, in places, the house roofs were carried forward to provide cover for pedestrians in this high-rainfall area. Timber bridges, or high masonry arches, crossed the canals at intervals.

It is worth noticing, though, that an unusual design of arched bridge evolved in this region, because of the difficulty of finding a firm foundation in the deep alluvial soils for any structure heavier than an ordinary house. Bedrock was sometimes 100 metres below the surface. The solution was to build a bridge flexible enough to cope with movement in the foundations. The stones forming the arch-ring were not mortared but were interlocked, or held together by dovetail-shaped iron cramps. Another feature was a masonry 'shear wall' built through each abutment of the arch (see Figure 8.15). (An abutment is a structure built up from a foundation within the bank to support

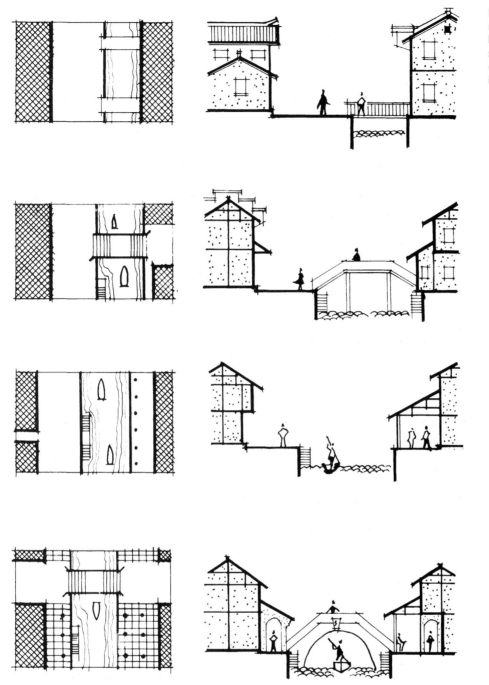

Figure 8.14 Some typical relationships of dwellings, bridges and canals in the smaller settlements around Hangzhou (Knapp, 1989, figure 1.6; illustration: Shen Dongqi)

Figure 8.15 Elevation of a single-span stone arch bridge, with the foundations shown in section. The foundations are on piles because of the soft subsoil in the Yangzi delta area near Suzhou. Shear walls, which allow for subsidence of the abutments, can be seen on either side of the arch (Fugl-Meyer, 1937)

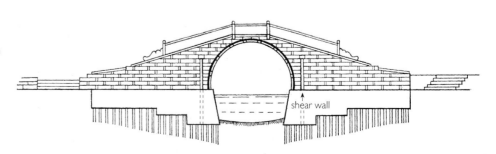

shear wall

and buttress the arch of a bridge.) The inclusion of this feature meant that if one abutment settled more than the other, the masonry of the arch could slide against the shear wall without anything collapsing (Needham *et al.*, 1971; Willetts, 1958).

While the bridge at Kaifeng illustrated in Figure 8.11 had space for a towpath to pass under it so that boats could be hauled by teams of men known as 'trackers' (rather than by draught-animals), the bridges shown in Figures 8.14 and 8.15 make no allowance for this means of moving boats. No pictures survive to show whether this was also true during the twelfth century, but it is worth noting that where canals lacked a towpath, different means of operation are implied. Thus it seems probable that boats within the canal cities were propelled from the stern using a large sculling oar, operated by a man or woman in a standing position. The oar slotted on to a pivot or bearing on the deck in such a way that the pull of a rope, also attached to the deck, caused the oar to twist (or 'feather') as it was pushed to and fro, thus lessening water-resistance. It could therefore be described as 'self-feathering' (Needham *et al.*, 1971). Heavy boats could be worked in this manner by three or four people, and on open stretches of canal, in the countryside, boats sometimes moved under sail.

The largest and best-known canal city apart from Hangzhou is Suzhou (Soochou), which is famous for its fine gardens. Marco Polo described the city with customary exaggeration as having 6,000 bridges crossing its many waterways. The distance from Hangzhou by canal is about 130 kilometres, and these two attractive places were bracketed together when people said: 'Above is Paradise; below are Suzhou and Hangzhou.' One way the cities differed, however, is that Suzhou was built to a classical plan – nearly square and enclosed by the usual city wall. It also had canals running parallel with nearly every street, often along the backs of buildings (see Figure 8.16).

Figure 8.16 Suzhou is served by a network of small canals, which are linked to the main canal system and what is now called the Grand Canal (Purvis, 1988, p.64)

Hangzhou and Suzhou were at the centre of an area of intensive agriculture producing food crops and also silkworms. Both cities had many silk-reeling and weaving shops. In addition, Hangzhou was an important centre for printing. Chinese printing did not involve a press, as in the West. Instead, a suitably carved wooden block was placed face upwards on a table. It was then inked, paper was placed face down on the block, and the back of the paper was gently brushed. By 1050, the carved block was replaced by movable type in experiments tried at Hangzhou. When the government moved to the city, eighty years later, the National Academy came too, and resumed its work as a publisher of classical literature. The government was also at this time printing paper money. With much paper also used by officials, writers and the many artists in the city, the demand was enormous, and it is said that the paper factory in Hangzhou employed 1,000 people in 1175 (Tsien, 1985, p.48).

8.6 *Perspectives on an age of conflict, 1250–1368*

During the 1270s, China finally succumbed to a long military campaign by Mongol armies, and their leader, Kublai Khan, became Emperor of China. Mongol military power was based on the special skills of mounted archers, but their conquest of China could only have succeeded with the help of Chinese defectors (Morgan, 1986). In their early campaigns, from 1211 to the 1230s, the Mongols conquered only the northern lands formerly ruled by the Liao and Jin emperors. But in 1251 they launched a serious campaign against the Song Empire. It was not until 1276 that Hangzhou fell and then, for over ninety years, China had to endure a deeply resented alien government, which brought other aliens with it – Arabs and Persians especially – to occupy key administrative positions in the civil service on which emperors depended for effective government. Although this is not intended to be the kind of history book that focuses on battles and the names of kings or emperors, it is essential to mention some of these events in order to make it clear that, contrary to what Westerners sometimes assume, China does not have a centuries-long history of smoothly continuous central government. There have been long periods of exceptional turbulence and destruction. There have also been episodes when the central government fell apart and the country was partitioned.

That needs to be said for two reasons. First, it helps to make clear why the capital of the Chinese empire was not always located in the same city. After serving as imperial capital, Chang'an and Kaifeng each suffered much damage and the areas around them were depopulated. More important, their circumstances were so altered by changes in frontiers, emigration and economic development that they were no longer convenient or appropriate places to be centres of government. Hangzhou escaped relatively lightly during the Mongol conquest, but the new rulers had chosen what is now Beijing as their capital some years earlier, when they still only controlled north China, and it was a more suitable centre for maintaining a connection with Mongolia.

A second reason why the periods of disruption in Chinese history need emphasis is that they may (arguably) have had a decisive influence on the development of technology. Wang Zhen (Wang Chen), a former civil servant, was an acute observer of the early years of Mongol government, and felt that he should document the agricultural and mechanical technology of the period. He thought that a book describing key technologies in detail would be useful when better times came and reconstruction was possible. The result was an 'encyclopaedia' of agriculture and a wide range of related subjects, which was printed in 1313 (Bray, 1984; Tsien, 1985).

Wang (surname, pronounced Wong) explicitly mentioned the devastated economy as a reason for writing his book, and he seems to have felt that some skills and techniques could die out through lack of use. As Bray (1984, p.60) comments, Wang was 'spurred by pity and a Confucian concern for the people's welfare'.

Comparing Wang's book with a similar survey of technology written three centuries later, in 1637, by Song Yingxing (Sung Ying-Hsing), one can certainly gain the impression that the range and scope of textile machinery had diminished. Song himself mentions a form of coloured porcelain, 'Hsuan red ware', for which the process of manufacture was 'lost' at the end of the Mongol period and later had to be rediscovered (Sung, 1966, p.154).

Pound locks in canals also went out of use at some stage during, or soon after, the Mongol occupation. According to Needham *et al.* (1971, p.360), this is because conditions had changed and smaller boats were being used. But it is also possible that repeated disruption of the canal administration had an effect. Another hydraulic device no longer used was a floating boom designed for diverting river currents and referred to as the 'wooden dragon'. This was re-invented during Qing (Ch'ing) times, and is still sometimes used in China today. But Elvin (1975, pp.101–3) quotes one of those concerned with revival of the technique as saying: 'It is a good method of river protection and should not be forgotten again.' With regard to more specialized technologies, it is quite clear that the knowledge and skill represented by the great astronomical clock in Kaifeng was entirely lost.

Losses of technology are not much discussed by historians, partly because habitual assumptions about technical progress as continuous and 'linear' make contrary instances seem freakish or uninteresting. But one striking example in the West was the loss of technologies concerned with mortar and concrete after the fall of the Roman Empire. In the 1750s, the English engineer John Smeaton developed some mortars with 'Roman' setting qualities, but in general it was not until the nineteenth century that 'Roman cement' and the uses of mass concrete were rediscovered.

To understand such losses, whether in China or the West, we need to be aware of the institutions on which technology always depends. They include systems of apprenticeship, training, education, production of technical books, methods of capital accumulation, business organization and government regulation or control. Some of these functions were located in cities, although it is hard to form a clear picture. We do not know much about how businesses were run in medieval China, either, and to what extent an urban base was necessary, although Eberhard (1967) and Hartwell (1966) have discussed entrepreneurs and, indeed, 'capitalists' during the late Song period. What we do know in much more detail is how the civil service operated, and hence explanations of technological development and decline tend to focus on that (Needham, 1954–, 1969).

For example, it can be seen as a weakness that civil servants with technical interests such as Wang Zhen seem to have been so few and far between. Linked to this was the influence of the examination system through which young men (never women) entered the civil service. When the examinations included some mathematics, as under the Song dynasty, mathematicians could get work as teachers, and for a time there were striking developments in algebra and other branches of the subject. Government officials trained this way had some competence in mensuration (the science of measurement), surveying and other subjects useful in communicating with practitioners of technology. But when the examinations became predominantly literary under the later Ming dynasty, interest and competence in these areas waned.

These are telling points, but they do not explain everything. Textile and pottery manufactures depended on the skills of large numbers of artisans. If major disruptions in economic and civil order threw them out of work and interrupted apprenticeships, and if the disturbance lasted through whole generations, skills could be lost for reasons that had little to do with government or civil service. Thus the extent of disruption during the Mongol period (and later upheavals) needs more careful assessment. On the other hand, desperate conditions during the Mongol occupation may have forced some technologies forward. The earliest metal gun barrel known to history, dated to around 1288, seems to be connected with resistance to the Mongol regime in the far north of China (Needham *et al.*, 1986, p.293).

One reason why the possible effects of this phase of disturbance have not been much discussed is related to a tradition of writing Chinese history in a way that seeks to stress continuity. The Mongol ruler, Kublai Khan, explicitly claimed to have founded a legitimate new dynasty, which he called the Yuan (Mote, 1994). Soon after, it became usual to slot the Yuan emperors neatly into the ordered list of imperial dynasties which provide the framework for every standard history of China (see Table 8.3) with no hint that they had disturbed that framework very considerably. But this is a matter of interpretation. In some respects the Mongols continued to behave like an alien occupying power.

Table 8.3 *Chinese dynasties during the period discussed in this chapter*

Conventional list of dynasties		Periods (CE)	Capital cities	Major periods of war, rebellion or disruption
	'second partition'*	479–589	no single capital	a long period of disunity and conflict ended by 'third unification'* in 589
隋	Sui	581–618	Luoyang	
唐	Tang (T'ang)	*618–906*	Chang'an†‡	
五代	Five Dynasties	906–960	no single capital	the empire was divided into several kingdoms
宋	Song (Sung)	*960–1126*	Kaifeng†	
				fall of Kaifeng, 1126, followed by a short period of disruption
		1126–1276	Hangzhou†	
元	Yuan	1260§–1368	Khan-baliq†¶	this period can also be interpreted as the 'Mongol occupation' of China
明	Ming	1368–1421	Nanjing	
		1421–1644	Beijing†	
清	Qing (Ch'ing)	1644–c.1660	Beijing	Manchu invasion and occupation, with much fighting
		c.1660–1840	Beijing	
		c.1840–1911	Beijing	episodes of European aggression; Taiping Rebellion, 1852–64
民國	Republic	1912 onwards	Nanjing preferred for a time	civil wars; Sino-Japanese war began 1937; revolution 1949

Periods known mainly for prosperity, economic growth and technological innovation are in italic type

* *Terms used by Needham (1954–)*

† *Cities discussed in this chapter*

‡ *The capital moved back to Luoyang during part of the Tang period*

§ *The Song dynasty had not been overthrown at this date, and controlled Hangzhou and parts of south China until 1276*

¶ *Khan-baliq was the Mongol name for what became Beijing*

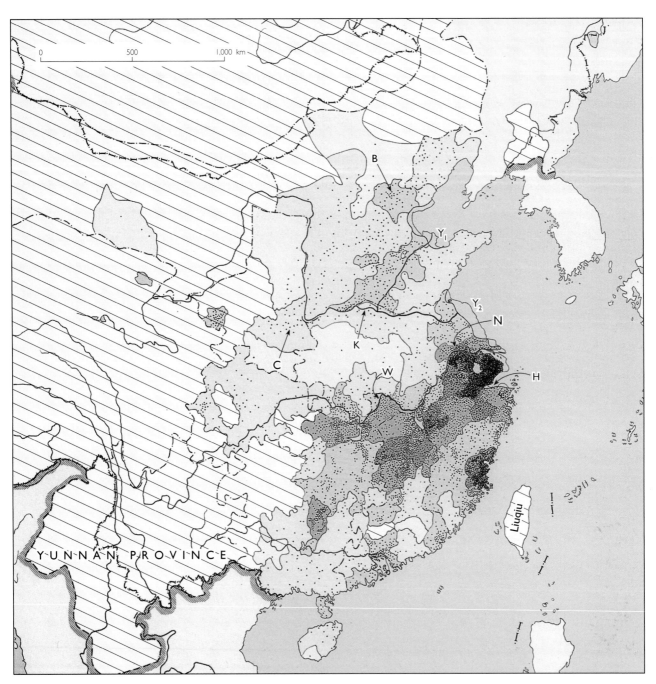

Figure 8.17 Population densities in China during the Yuan dynasty (late thirteenth and early to mid-fourteenth centuries). Each dot represents 10,000 people, and shading is darker in more densely populated areas. Hatching shows where the Yuan censuses failed to collect adequate data. Note how the area around the early capitals, Chang'an and Kaifeng, had been depopulated, and how the Yuan capital at what is now Beijing was even further from the densely populated areas and from most other urban centres such as Hangzhou and Wuchang. Hankou did not exist at this time: it was founded in 1465. Changes in course of the Yellow River are also indicated, and changes in coastline around its two principal mouths: Y_1 is the mouth of the Yellow River today, and prior to 1099, after which it developed a more northerly mouth; Y_2 is the mouth of the Yellow River between 1324 and 1855, and was the result of several changes of course during severe floods, most notably in 1194, 1288 and 1324

Key B: Beijing; C: Chang'an; H: Hangzhou; K: Kaifeng; N: Nanjing; W: Wuchang (adapted from Population Census Office of China, 1987; by kind permission of Professor Liu Yue of the Institute of Geography, Chinese Academy of Sciences, Beijing, People's Republic of China)

Another reason why this depressing period is often glossed over is controversy about whether it was really as black as it can be painted. The Mongols – or Yuan dynasty – built canals and provided a short period of orderly administration under Kublai Khan. They were unlucky in that China was badly affected by floods during this time, notably when the Yellow River devastatingly changed course and found a new outlet to the sea (Figure 8.17 shows both old and new courses). Outbreaks of bubonic plague also had disastrous results, and one that began in 1331 is thought by McNeill (1977) to have been the start of the plague that was carried across Asia by Mongol armies and entered Europe as the Black Death in 1347–8.

Military and economic disruption coupled with natural disaster are reflected in evidence of a decline in population (Mote, 1994, p.620). But again, the magnitude of the disaster is uncertain, in part because of deficiencies in Yuan efforts to take a census (see Figure 8.17). Some say that the population of the whole of China dropped from over 100 million to 60 million between 1200 and the end of the Yuan (or Mongol) dynasty in 1368 (McNeill, 1977; Morgan, 1986). Yet authoritative sources deny that the total population was ever as high as 100 million, indicating a less dramatic decline (Population Census Office of China, 1987).

After this difficult period, many cities must have had fewer residents. The quality of urban life probably changed in other ways as well. Some civil servants, disgusted by alien and what they often felt was corrupt government, withdrew from public life and lived in seclusion on their lands in the countryside. The mathematician Li Zhi, for example, though showered with honours by Kublai Khan and offered a government post, continued to live in his rural retreat (Ho, 1973). But his fame was such that people travelled into the country to visit him, and it is said that he responded by giving lectures on algebra in a barn. If such behaviour was at all typical, the cities must have lost some of their most intellectually lively residents, and were perhaps no longer so stimulating for science and invention.

On the positive side, though, the Yuan dynasty was able to build its new capital, sometimes called Khan-baliq, at what is now Beijing. This was not a virgin site, as it had been an administrative centre for the Liao Empire. But starting in 1266, the new rulers remodelled the city to make it their own. They planned it as a square, southward-facing walled city, following Chinese tradition, in order to give visible expression to their claim to be a legitimate Chinese dynasty. Concern with astronomy and the calendar was another tradition they maintained, and an astronomical observatory was established in the south-east corner, on a high part of the walls. In the centre of the city was a bell-tower, part of which survives, and south of that was the palace, in its own enclosure, some of which also survives in the foundations of buildings in the 'Forbidden City' (Morgan, 1986).

It was a big step for a Mongol leader, with his traditions of nomadic living, to commit himself to building so enduring a capital, but Kublai Khan made certain compromises. One was to migrate in nomadic style to a summer capital on the southern fringe of Mongolia in the appropriate season. This place was Shangdu, which Samuel Taylor Coleridge called 'Xanadu' when describing its 'stately pleasure-dome' in his poem 'Kubla Khan'. Another reminder of Kublai's nomadic origins was an area of steppe grassland, planted as a park in Beijing, and felt-wool Mongol tents erected in parts of the palace enclosure.

Once the capital of all China was established at Beijing, changes in the transport infrastructure were required so that the city could be adequately supplied, and so that tax-grain could be remitted to the government. One result of the Yellow River changing course was that many canals near Kaifeng had been destroyed. Rather than re-excavate them, the Yuan government decided to make a new channel connecting existing canals at Hangzhou and

Suzhou to Beijing via a route that was more direct, but still required 1,782 kilometres of waterway. Near Beijing, a short section of new canal with twenty lock gates was completed in 1293 under the direction of an astronomer skilled in surveying, Guo Shoujing. Next, some lengths of existing canal were used, until further south, higher ground between the Yellow River and the Yangzi was crossed by a summit-level canal planned in the 1280s by a military engineer. Large-scale summit-level canals were something new, and at first there were problems about keeping this one supplied with water (Needham *et al.*, 1971). Eventually, in 1411, construction of a long embankment created a reservoir fed by local streams which kept the summit-level full. The canal was then very successful.

8.7 Beijing brickwork and cosmology, 1368–1644

When a long period of rebellion against the Yuan government finally succeeded, and the Mongols were expelled, there followed a period of about seventy years, from 1368, when the direction China would take under its new dynasty, the Ming, seemed uncertain. This was reflected by a change of mind about where the capital would be. At first, the chosen location was Nanjing (Nanking), a city on the Yangzi River with extensive shipyards. Soon after 1400, however, the government decided to move from Nanjing (the name means 'southern capital') to Beijing ('northern capital'). At the same time, the Grand Canal was radically overhauled, which is when the summit-level water supply was improved.

In many ways, the decision to move the capital is hard to understand. Nanjing was at the centre of one of the most economically productive parts of the empire (see Figure 8.17). Beijing was remote from most other important centres, and some large part of its supply of food, fuel and building materials had to be transported over considerable distances by canal, or coastal shipping, or by caravans of camels and pack-ponies from the west and north. Historians have suggested that the cost of maintaining a capital city so far from the country's main centres of economic activity was a financial drain until the fall of the last dynasty in 1911 (Mote and Twitchett, 1988, p.244). It is easy to see why the Mongols wanted the capital to be close to their power base further north, but it is surprising that the Ming emperors, who hated most of what the Mongols had done, should wish to return to this site.

The decision reflects a brief period of civil war among the Ming, in which the rightful emperor lost his throne to a relative named Yong Le in 1403. This ruler had more political support in the north than elsewhere, and it was he who chose to relocate the capital. After Yong Le died, it seemed briefly that the capital would be moved back to Nanjing, but the decision finally taken to stay in Beijing signified a profound change in attitudes and orientation.

Yong Le's reign had been one of forceful activity, with military campaigns both to the north, in Mongolia, and far to the south, in what is now Vietnam. Prisoners from the latter region were used as forced labour in the building of Beijing. External trade was encouraged and a large naval force was sent on a series of expeditions into the Indian Ocean, to promote good diplomatic relations, and to do some exploring (notably along the coast of East Africa). Although the capital had been moved from Nanjing, for a time the shipyards and naval establishments in that city continued to play a prominent part in affairs.

But by 1433, the authorities had turned their backs on all this. Naval expeditions were stopped, shipyards were closed, and foreign trade was discouraged. China was beginning a long period of 'isolation from international affairs' (Mote and Twitchett, 1988, p.302). That this was a turning-point in world history is evident when we reflect that the first European explorers to sail into the Indian Ocean later in the fifteenth century were not seriously challenged, and indeed encountered a power vacuum where they might have been met by the formidable Chinese navy. The decline of naval shipbuilding, and of associated studies of navigation, may also have reinforced a neglect of mathematics and some aspects of astronomy within civil service circles. That in turn, we may tentatively surmise, may have added to the 'loss of technology' associated with the upheavals of the Yuan period. There were technological innovations in China under the Ming dynasty, including colour printing, improvements in pottery and brick manufacture, and developments in agriculture, but one hears little of developments in astronomy or mechanical invention.

Historians have documented new Mongol threats in the north and steep increases in the cost of the timber needed for shipbuilding to explain the end of naval exploration. They have noted political infighting associated with relocation of the capital, recording sharp differences of view between the civil service and the court (Mote and Twitchett, 1988). In discussing the Indian Ocean voyages, Needham *et al.* (1971) describe how a 'maritime party' in the government was defeated by more conservative 'neo-Confucians'. It is hard to escape the view, however, that something deeper than politics and the price of timber was at issue. For the governing class to turn its back so abruptly on the rest of the world, and also to lose interest in sciences such as mathematics, suggests a shift in values and a defensive, unadventurous outlook.

It is also possible to see the quality of urban life in the capital city as contributing to such change. Among earlier capitals, Chang'an and Hangzhou had been actively engaged in foreign trade and had been host to merchants from distant lands (India, Persia, Central Asia) and to several foreign religions (Christianity and later Islam, as well as Buddhism). Kaifeng had a Jewish community involved in textile trades as well as the entrepreneurs connected with its own industry. In leaving Nanjing, the Ming government was leaving a city with some of these same diverse trades and interests for a centre where, at first, nothing happened apart from administration and imperial ritual. The planning of the city, as we shall see, needed to make no compromises to accommodate transport canals or commerce, shipyards or iron-foundries. Politics and ritual, and the cosmological ideas on which these rested, could determine the plan of the city completely.

Beijing was formally inaugurated as the Ming dynasty's capital in 1421. It retained many features which the Yuan (Mongol) emperors had introduced, but the north and south walls were rebuilt on different alignments and the palace had a more northerly position relative to the walls and gates. The streets formed a grid of strictly aligned north–south and east–west routes, and Beijing became 'a city of rectangles within rectangles'. The outer city wall enclosed the rectangular Imperial City, which in turn enclosed another rectangle – the Palace City, also known as the Forbidden City. Within each of these enclosures were the smaller rectangles of the courtyards of private houses, temples or public institutions. It was a 'perfect example' of design by the assembly of elements of different sizes but standardized shape: 'its plan is thus perfectly harmonious and lucid' (Willetts, 1958, p.676) (see Figure 8.18 overleaf). Moreover, it brought together classical (Confucian) ideas, which were perhaps less perfectly expressed at Chang'an, and ideas from other traditional but less august sources, such as *feng shui*, or geomancy.

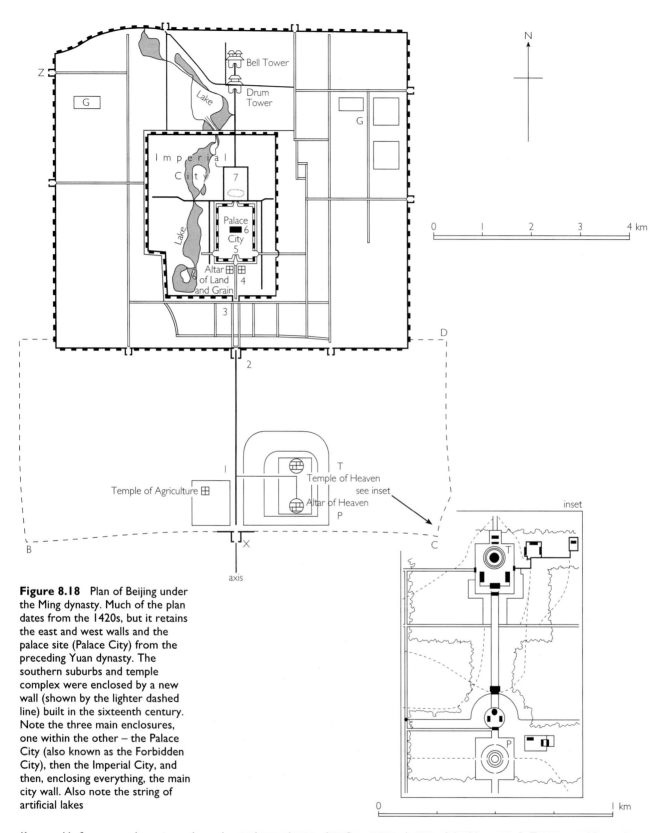

Figure 8.18 Plan of Beijing under the Ming dynasty. Much of the plan dates from the 1420s, but it retains the east and west walls and the palace site (Palace City) from the preceding Yuan dynasty. The southern suburbs and temple complex were enclosed by a new wall (shown by the lighter dashed line) built in the sixteenth century. Note the three main enclosures, one within the other – the Palace City (also known as the Forbidden City), then the Imperial City, and then, enclosing everything, the main city wall. Also note the string of artificial lakes

Key notable features on the main north–south axis 1: temple complex; 2: main south gate of the Ming city; 3: Tiananmen (Gate of Heavenly Peace); 4: temples, altars; 5: Meridian Gate (Gate of the Noonday Sun); 6: Hall of Supreme Harmony; 7: Prospect Hill (artificial hill)

Other features of interest G: granary compounds; T: circular 'Temple of Heaven' building (see Figure 8.19); P: Altar of Heaven – a circular, open-air platform approached from a series of circular terraces; B, C, D: new wall enclosing the southern suburbs; X: main south gate in the new wall; Z: north-west gate in the earlier Ming wall (see Figure 8.20, p.296) (compiled from maps in Boyd, 1962; Skinner, 1977, with the permission of the publishers, Stanford University Press. © 1977 by The Board of Trustees of the Leland Stanford Junior University; Struve, 1993)

There were a few departures from this concentration on rectangles, however, and they are all highly significant. Prominent but informal features of the plan were the artificial lakes in the western half of the city and an artificial hill formed from the spoil excavated in making the lakes. Figure 8.18 shows these features which had been begun in the Yuan period, but they were now redesigned and extended.

The replanning of Beijing also made provision for the various temples and altars of importance for imperial ritual. One of the most prominent and beautiful was the Temple of Heaven, then outside the city wall. The buildings here had (and still have) roofs of blue tiles to echo the blue of the heavens, and were complemented by a large altar, open to the skies, where the emperor would come to pray and sacrifice a bullock on the day of the winter solstice. Both the temple building and the altar were circular in plan because the circle itself symbolized heaven (see Figure 8.19).

All this geometry was centred on a north–south axial line which linked the various elements: the temple complex to the south, the rectangles of the Palace City, the artificial Prospect Hill (often also called Coal Hill), and to the north of that, the Drum Tower and Bell Tower. But this central axis was handled quite differently from any comparable axis in a European city, which might offer long views toward a prominent building – as in London The Mall offers a view toward Buckingham Palace.

Instead of creating an impression by means of a long view or distant prospect, what the central axis of Beijing offers is a series of forbidding gateways that give entrance to a sequence of enclosed spaces, some of them

Figure 8.19 The Temple of Heaven, Beijing. This round building which has blue-tiled roofs was designed during the Ming dynasty, but had to be rebuilt after a fire in about 1890 (Purvis, 1988, p.95)

very severe in the strictness of their geometry and absence of vegetation. Starting from the temple complex, the visitor passes through the great south gate of the city and then proceeds along a short avenue and through another gate to a space in front of the Gate of Heavenly Peace – the Tiananmen. This space was widened by planners under Soviet influence in the 1950s to create the great windswept area of Tiananmen Square, which is starkly non-traditional in its lack of a sense of enclosure. Beyond the Tiananmen are more gates penetrating more walls until the visitor is confronted by a flight of white marble steps leading up to the Hall of Supreme Harmony, which is where the emperor would receive visitors in audience.

Referring to the sequence of gateways and courtyards through which such a visitor must first have passed, Willetts (1958, p.679) comments that the impact on those who sought audience must have been overwhelming. This was an architectural device for ruling the world, and 'there can be little doubt of its capacity to reduce outsiders to a state of supplicatory awe'.

Looked at simply as a two-dimensional plan, however, Beijing can be read as a great cosmological diagram (see Figure 8.18), representing three contrasting shapes: the circle for heaven, the square (or rectangle) to depict the human world, and the irregular shapes of the lakes and hill to represent nature. The lakes, indeed, are a 'perfect foil' for the 'humourless rigidity' of the palace buildings and the processional axis (Willetts, 1958), and there are at least three reasons why this representation of nature has greater emphasis than it did at Chang'an. One is that the lakeside situation of Hangzhou had become something of a legend. Another is the philosophy of nature expressed by Daoism (Taoism) and a view of nature that stressed two elements in particular – mountains and water. Third, though, there were ideas coming from *feng shui* concerning the best location for buildings, having regard to weather and other forces of nature.[2] Practitioners of *feng shui* were not only referring to the desirability of sheltered, south-facing sites (which is what they usually chose), nor were they just considering a harmonious appearance of buildings in relation to landscape, but they were also considering less tangible forces, sometimes using a magnetic compass in their work (Knapp, 1990).

A view of nature that emphasized mountains and water meant that landscape paintings almost always included those features, and they were also represented in the design of small gardens by the inclusion of a token rock and pool. But on the grand scale of imperial Beijing, it was possible to introduce real lakes and the fifty-metre-high Prospect Hill, with its trees and several summits. According to Mennie (1922) and Wright (1977), the relationship of the hill to the palace buildings was decided on *feng shui* principles. Since the Grand Canal did not enter the city but terminated a few miles away, there were no practical functions to consider in planning water features. All could be part of the landscape design.

Western architectural writers express surprise that the major buildings in the Palace City are of timber-framed construction rather than built of more permanent materials, and note that they have been partly or wholly reconstructed several times since they were first erected during the Ming (or Yuan) period (Boyd, 1962). This comment seems to overlook the fact that much of the architectural impressiveness of the approach to the Palace City arises from the heavy brickwork of enclosing walls and the brick arches of gateways, as well as from the masonry platforms (raised bases for buildings) and marble terraces on which the buildings stand. Renewing timber structures

---

[2] There are many books on *feng shui* in British bookshops at the time of writing, to cater for the interests of 'new age' and 'alternative lifestyle' enthusiasts. One example is *The Feng Shui Handbook*, by Lam Kam Chuen (1995). These books can be helpful in giving an *impression* of the mythological world on which the subject is based, but they do not discuss *feng shui* adequately in its historical context, and cannot be relied on with regard to the planning of Beijing.

within this heavily monumental and enduring context could well be understood as a routine maintenance task in a permanent ensemble – rather like replacing wallpaper in a house that remains structurally unaltered for many years. It certainly does not mean that the builders lacked interest in the permanence of what they erected.

In no sense was technology a stimulus in the planning of a city such as this. Ideological concerns and cosmological concepts were of overwhelmingly greater significance. But the availability of technology set limits on what was possible. Marble balustrades, walls and stairways were used to front the terraces (or platforms) on which the main buildings of the Palace City stood. These materials could only be supplied from distant quarries given an efficient transport system. Some timber, especially in longer lengths or of special quality, was also brought via the canal and river system from distant forests. Fired brick, the preferred material for the many walls of the city, could not at first be used as extensively as desired. But at some date *after* the Ming brickworks near Beijing were set up in 1416, there were innovations that allowed production to be greatly increased, with costs reduced. Within the next century and a half, bricks became cheap enough for widespread use in walls of smaller cities, and even in ordinary houses (Knapp, 1989).

The book written in 1637 by Song Yingxing (Sung, 1966) vaguely implies that this improvement in brick-making may have been connected with increased use of coal to fire brick kilns.[3] It is possible that larger kilns became feasible, perhaps with better temperature control. Techniques for making coal-dust into briquettes allowed the fuel to be used economically. Perhaps also methods of heat economy common in pottery kilns were applied to brick-making. Song described a small 'clamp kiln' (meaning that bricks and coal were packed together). Much later, kilns were sometimes very large (Eberhard, 1967, plate 7), and probably worked differently – but the process and the details of innovation remain undocumented.

In this context, it is worth noting that the walls of Beijing were at first primarily of earthen construction, and only later were they faced with yellow-grey brick, which turned them into immensely impressive fortifications (see Figure 8.20). Similarly, when the southern suburbs of the city and the area around the Temple of Heaven were enclosed by a new wall in the sixteenth century, this was at first an earth wall, but later in the same century it was rebuilt in brick, with brick gatehouses (see Figure 8.18).

Many other cities throughout China rebuilt their walls in brick at this time. One of the best surviving examples is Xi'an, the small Ming city that occupied a fraction of the site of old Chang'an. Walls wide enough for a chariot to drive behind the parapets were constructed in grey brick with earth filling; Figure 8.22, p.298, shows the principle.

Another great Ming construction project was, of course, the Great Wall along the northern frontier of China. There had been earth walls on or near much of its alignment from early times, but these had not been maintained during Tang or Song eras. Indeed, Morgan (1986, p.66) comments that the main reason why this fortification did not stop the Mongol invasion in the thirteenth century is that 'the Great Wall as such was not there at the time'. The Ming governing class showed much evidence of feeling psychologically scarred by the experience of Mongol occupation, and wished to ensure that nothing like it should happen again. One reason for moving from Nanjing and turning their backs on the sea was a view that the northern frontier was the chief source of danger. Thus the Great Wall was rebuilt, beginning in about 1450. In many places, indeed, it was newly built. Both brick and stone were used in the most

---

[3] Song Yingxing's account of brick-making can be found in the Reader associated with this volume (Chant, 1999).

Figure 8.20 Xi-Chih Men, the northernmost gate in the west wall of Beijing (original Ming city) (Mennie, 1922, plate XX; photograph: courtesy of the Department of Special Collections, The Brotherton Library, University of Leeds)

important stretches, with more brick being employed in post-1550 parts of the work (see Figure 8.21).

It is useful here to consider a question raised by Frederick Mote (1977) about why so much effort went into rebuilding city walls in brick at a time when the country was more completely at peace than for several centuries. Military explanations miss the point, Mote suggests. Given the emphasis on enclosure in all Chinese architecture, city walls were a more positive means of architectural expression than Westerners usually realize. Even the Great Wall, he argues, which did have a military purpose, was more of a psychological barrier, at the geographic limit of the Chinese way of life. What city walls did was that they 'dignified cities'. They were symbolic of the 'presence of government'.

Others make this point by noting how the walls of the Palace City in Beijing were represented by artists. In 1754, for example, a Mongolian chieftain named Amursana who had been defeated in a local war came to Beijing to seek assistance. The court artist portrayed him being received in front of the walls of the Palace City at a point where a small square in front of the great Meridian Gate is enclosed by projections from the line of the wall. History records that the chieftain was actually received elsewhere, but it was regarded as important for the official picture to show his humility before the overpowering walls, and all they represented (Wu, 1963, pp.42, 119–20).

Similarly, when candidates who had taken the civil service examinations in Beijing were told that they had passed and would be awarded a degree, they approached the Meridian Gate in the wall of the Palace City to 'murmur their thanks' to the government which they were now qualified to serve (Wu, 1963, p.42).

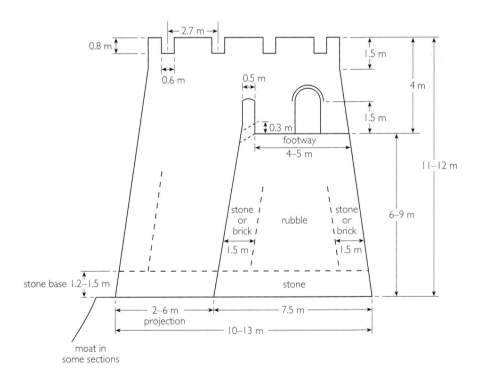

Figure 8.21 Typical cross-section of the Great Wall of China north of Beijing with a tower shown in elevation behind. The construction is characteristic of the Ming dynasty after 1450, or more probably, about 1550. Many Ming city walls are similar, but would often have tamped-earth filling rather than the stone rubble indicated here (Needham *et al.*, 1971, p.44, figure 726)

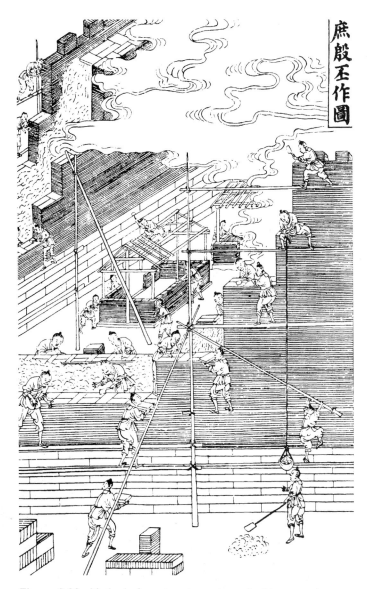

Figure 8.22 Method of constructing a city wall with a stone base, brick facing and a tamped-earth core. The artist expresses his sense of the walls being as high as mountains by showing them reaching up to the clouds (Needham *et al.*, 1971, p.51, figure 737)

The same connection between walls, gates and orderly government is indicated by the following anecdote. Confucius is said to have been in a horse-drawn carriage with one of his disciples when he encountered some children playing in the road. The children had used loose tiles and bricks to build a model city wall across the road and, as the carriage was about to scatter these playthings, one of them stood up and asked Confucius a question: 'Should the city wall be destroyed or ought the carriage to turn round if it cannot pass through the wall-gate?' Confucius apparently recognized an appeal to ideals of civil order and good government and asked his driver to turn the carriage around (quoted in Chiang Yee, 1938).

Walls and gates had other meanings, too, particularly about the privacy of the householder in his walled inner courtyard, or of the emperor in his 'forbidden' city.

Walls also provided the architecture of permanence and monumentality which those who ignore them say is missing from China (see Figure 8.20). The same word, *cheng*, means both 'city' and 'city wall', and while the timber houses within a city might be frequently replaced, it was the wall that endured and marked the permanence in the landscape of the urban place. A town that lacked an enclosing wall was not a real city. It would have no government offices, and would be administered from elsewhere (as was the case in Hankou, which for a long time had no wall).

But the symbolism of permanence could have a more spiritual meaning also. In Chinese culture, there was always a strong feeling for mountains because they had endured without change throughout history (Wickert, 1984, p.63). And not only did Beijing have its artificial hill, but also its mighty walls could excite feeling for the sublime as they seemed to reach into the clouds (see Figure 8.22). During the winter of 1643–4, a man named Liu Sangyou described his first view as he approached the city:

> the walls are so lofty, with parapets like sharp mountain peaks ... I murmured to myself: 'How beautiful – the fastness of mountains ... conferred by heaven!'
>
> (quoted in Struve, 1993, p.7)

But that is the view of an educated man. Working people and traders often found city walls oppressive. The great gates in them closed soon after sunset, and within the walls, watchmen could notice any untoward activity overnight. A walled administrative city might have the quiet of an English cathedral city. Its markets would often be in the suburbs, outside the walls, to which farmers could bring produce freely in the early mornings, or from which merchants could start on their journeys at whatever time they wished. Hankou, the great commercial city with which we began, flourished mightily from the late seventeenth century partly because it had no real walls and was not so restricted as many other cities. It actually had an all-night market, which would have been unthinkable in a walled city. In the twentieth century, for related reasons, many cities were eager to demolish their walls, often building ring roads along the same alignments. In Beijing, one of the few short lengths to be spared is the part on which the astronomical observatory has been located since Yuan times.

8.8   Conflict, commerce and natural resources

In 1644, Beijing was captured by armies from the region to the north-west known as Manchuria. This event followed two decades of rebellion and disorder, and it ended the Ming dynasty, replacing it by a new series of emperors of Manchu origin known as the Qing (pronounced 'Ching') dynasty. So devastating were the conflicts of these years that the period has been called 'the Ming–Qing cataclysm' (Struve, 1993). A picture showing the destruction of Yangzhou, a canal city in the lower Yangzi area, drawn some time after the event, shows a characteristic combination of solid walls and timber buildings after all had been taken apart by cannon-fire (see Figure 8.23). Hankou, further up the Yangzi, suffered two episodes of serious destruction: one when rebels captured the town in 1643, and another when the armies of the new dynasty recaptured it.

As in previous periods of severe conflict, there is evidence of some loss of technology. For example, about halfway between Hankou and the coast, a little south of the Yangzi River, were the potteries at Jingdezhen where the best Ming porcelain had been produced. These were seriously damaged in the fighting and, according to oral tradition recorded in the nineteenth century, when peace was restored, many attempts were made to rebuild the kilns, but there were certain details that nobody could get right. 'An experienced old man drew a design for a kiln which was to be 20 Chinese feet high. The kiln was built, but the firing of the pottery therein proved a failure' (Cornaby, 1925, p.271). In other words, once continuity of operation was interrupted for such a time that its resumption depended on the memories of elderly people, skills were lost and the subtleties of processes were forgotten. The oral tradition gives a fanciful account of how the problem was solved, and indeed, the Jingdezhen potteries did prosper again. But the finer qualities of Ming 'china' were never quite recaptured. Something of the earlier technology had been lost.

After peace and order had been re-established and, more specifically, after 1660, the Qing dynasty presided over a long period of prosperity. A main cause of the rising number of large cities was a rapid increase in population. Excluding remoter parts of its empire, such as Xinjiang, Tibet, Mongolia and Manchuria, China had a population of over 400 million by 1850, of which 25 million people lived in cities with 50,000 or more

Figure 8.23   During the fierce fighting that accompanied the fall of the Ming dynasty, the small canal city of Yangzhou was destroyed and its inhabitants massacred. This picture shows damage caused by cannon-fire, with collapsed timber-framed buildings and fragments of the brick walls that had enclosed them. A gate in the city wall is seen in the background (Struve, 1993, p.41, figure 4)

inhabitants (Elvin and Skinner, 1974). Whether this represents a *tripling* of overall numbers after a low point reached during the 1640s with the Ming–Qing cataclysm, or whether the increase was less dramatic, is uncertain (Elvin, 1973; Ho Ping-ti, 1959). Explanations are also controversial. A freeing of the rural peasantry from conditions of near serfdom (which delayed marriage) may have boosted population growth, and an increased take-up of smallpox immunization may also be relevant (Elvin, 1973, p.255). An old Asian technique known as *variolation* (from a word meaning 'pox' or 'pustule') was used, based on infecting the person to be immunized with matter taken from a recovered smallpox patient still carrying the disease in attenuated form. By contrast, the Western technique of *vaccination* (from *vacca*, the Latin for 'cow'), introduced by Edward Jenner in the 1790s, infected the person being immunized with cowpox, a milder but related disease. Variolation was just as effective as vaccination, but people who developed an adverse reaction were at greater risk.

Whether or not variolation had a demographic effect, what undoubtedly was important in China was agricultural innovation, including the introduction of new crops (discussed later), which made it possible to feed the increased population. Because so much of the land area of the empire was mountainous terrain, with desert and steppe and grassland to the north, the land available for agriculture was severely constrained (see Figure 8.24 top). Therefore, where good agricultural land occurred, as on the alluvial plains of the Yellow River basin and in the Yangzi delta, it was used very intensively, often leaving no room to graze animals or plant wood lots. For meat, the most important animal was the pig, because it could be fed on wastes from food-processing and cooking without need to graze. Draught-animals (the ox in the north and water buffalo in the south) were grazed on stubble after harvest, and at other seasons on uncultivated corners, roadside verges and cemeteries. Not enough draught-animals could be maintained in this way for them to be used as extensively as in the West. Hence, in many areas there were few wheeled carts or wagons, and many tillage and portering tasks were carried out by human labour. Pack-ponies or donkeys were used for overland transport, but sometimes again human porters carried loads for surprising distances. In north China, horses were obtained by trade with Mongolia in preference to taking up scarce farmland for their breeding.

By 1700, farming on the best land was probably approaching the maximum output possible without modern crop varieties and fertilizers. European visitors to the area around Guangzhou (Canton), the port in south China where Western merchant ships called, were impressed by the efficiency of agriculture there, and may have borrowed ideas for application in their own countries, particularly Holland. In some respects, Francesca Bray argues, 'The dazzling example of China shook the West into agricultural revolution' (Bray, 1984, p.555).

Be that as it may, Elvin (1973) confirms the high productivity of agriculture in the Guangzhou region, along the coast to the north, and in the Yangzi and Yellow River areas. But he notes that the Chinese themselves were aware that food-production was approaching the limit of what was possible, beyond which increasing population might outstrip resources. By 1750, the government was encouraging the import of rice as well as promoting new crops and the cultivation of marginal land.

These conditions also affected the availability of textile fibres, wood-fuel and timber, although the problem here could be lessened by importing raw cotton from India and by using more coal.

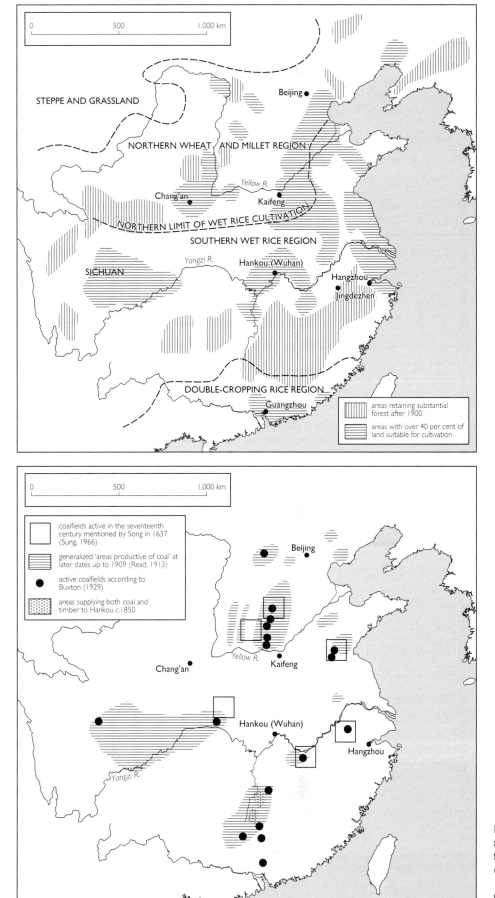

Figure 8.24 China: *top* agricultural land and forests; *bottom* coalfields (based on information in Bray, 1984; Buxton, 1929; Population Census Office of China, 1987; Read, 1913; Sung, 1966)

In detail, the situation varied greatly with climate, and hence with latitude. In the subtropical conditions of south China, two and sometimes three crops of rice could be grown on one plot of land each year if it were thoroughly manured. The area around Hankou, by contrast, could expect cold winters, sometimes with ice forming on the rivers, and here it would usually only be possible to produce a summer crop of rice. Cotton did well in this area, however, and was produced on a significant scale. Further north, in the Yellow River basin, were the cooler areas where wheat and millet rather than rice were grown (see Figure 8.24 top).

The areas of greatest fuel and timber scarcity were mostly to the north of the Yangzi, where population densities were high and there was a long history of intensive agriculture. Much of this area was alluvial land with no building stone. So rural houses were built with either mud-brick or rammed-earth walls, and might have bamboo rafters so that even the roof structure employed minimal timber. Earth walling, of course, need not be seen as an inferior building material – a point explored by Michael Bartholomew in Chapter 9 of this volume. In Britain, there are mud-walled houses which have lasted for two centuries, and there are researchers who advocate revival of the technique for precisely the reasons it was used in China.[4] It does not use scarce resources, nor is fuel needed for its manufacture.

In most of the Chinese cities discussed in this chapter, party walls between houses, or between shops, were often built in a combination of mud-brick and fired brick, the latter being employed on parts of the walls that showed on the streetfront or that were particularly exposed to the weather. But otherwise, urban houses continued to be built mainly of timber, which was transported by river or canal over considerable distances. For example, the bottom map in Figure 8.24 identifies one area from which building timber was supplied to Hankou in the nineteenth century. South of the Yangzi, in areas too hilly to support much agriculture, woodlands remained and forestry was practised, and this area provided much of the timber required by cities further north.

As to coal resources, Song Yingxing was writing in 1637 as if coal was available almost everywhere in China, especially where other fuels were scarce (Sung, 1966). However, when we plot the coalfields he names on a map (see Figure 8.24 bottom), this proves not to be entirely true. There had certainly been areas of wood-fuel scarcity for a long time, which is why some coke was used for iron-smelting and in cooking-stoves when the Song dynasty capital was at Kaifeng. The use of coal in brick kilns had probably expanded greatly during the century before Song wrote his book. But in the large areas of alluvial plain between the Yellow River and the Yangzi, there were often no local sources of coal and very little woodland.

Despite all the constraints, however, the Chinese economy expanded rapidly during the eighteenth century, and living standards were maintained or improved. There is evidence of considerable industrial activity. At Jingdezhen, the porcelain-making centre mentioned earlier, it was said that 'tens of thousands of pestles shake the ground', and at night, the heavens 'are alight' with glare from the kilns (Elvin, 1973, p.285). Ironworks were prospering, with blast-furnaces running twenty-four hours a day. In 1700, China probably still had 'the world's largest and most efficient iron industry' (Wagner, 1997, p.5), though it was soon to be overtaken by developments in the West. Elsewhere, there were water-powered paper mills, and sophisticated machinery was used at brine-wells, where salt was produced.

---

[4] Areas in Britain where earth walling was once widely used in rural building include Pembrokeshire, Devon, Dorset, parts of Oxfordshire and East Anglia, and the Solway plain in Cumbria. For modern earth-building techniques, see Norton (1997).

Under these conditions, cities such as Hankou, whose life centred on trade and transport, prospered greatly. Statistics are unreliable, not least because the number of people living in this city fluctuated with the seasons, and census enumerators failed to count the many merchants who lived in Hankou for a few months each summer, and the vast numbers of people who lived on boats (see Figure 8.31, p.319). However, as Table 8.4 shows, the population of Hankou probably passed the one million mark some time between 1800 and 1850 (Rowe, 1984, p.40; Skinner, 1977, p.236).

It is estimated that as many as 70,000 or 80,000 boats carrying freight called at the port each year around 1800. Even if some of these vessels were small, it is thought that the overall tonnage was greater than the Port of London dealt with annually in the early nineteenth century (Murphey, 1974, p.40). There were eight 'great trades' passing through the port as well as many minor ones. The most important included grain, salt, tea, oils and medicinal herbs. A little trade also came into the city overland on trains of pack-ponies, including consignments of shoes made in or near Beijing. Most of China had become an 'integrated market' by the middle of the eighteenth century, which helped ensure that limited resources were used efficiently. Individual regions specialized in whatever forms of production paid best, and exported extensively to other regions. Rowe (1984, pp.61–2) remarks that a 'uniquely efficient water-transport system ... allowed [China] to overcome the barriers of distance' in a way that was possible in Europe only with the advent of railways.

For example, the western province of Sichuan (Szechwan) was an important producer of medicinal herbs, rice and sugar, which came down the Yangzi by boat to Hankou. There, wholesale druggists repackaged and distributed the herbs to apothecaries throughout China, while many tonnes of rice went downriver to densely populated areas in the Yangzi delta. In the 1730s, the total amount of rice from all sources passing through the port of Hankou was about 600,000 tonnes annually.

Table 8.4 Estimates of the population of Hankou

Date	Resident population excluding summer residents and boat-dwellers (author's estimates; compare Skinner, 1977)	Estimates of population such as appear to include the summer influx and boat-dwellers (Rowe, 1984)
1800	less than 0.5 million (perhaps much less)	almost 1.0 million
1850	just over 0.5 million	almost 1.5 million

Merchants dealing in salt and tea became especially prosperous. An important export trade in tea to Russia went up the river and then overland, while ships carrying salt southwards, up tributaries of the Yangzi, often returned to the city with cargoes of coal. Some coal came from an area 400 kilometres to the south of Hankou which also supplied timber for building shops and houses (see Figure 8.24 bottom). Transport of the two commodities was often combined by building roughly made boats to carry the coal downstream with the intention that, after they had been unloaded, the boats would be dismantled and their planking sold to builders (Rattenbury, 1944, p.111).

Much of Hankou's trade was markedly seasonal. Tea, for example, was harvested from late spring onwards and came to market in the city from May until the end of the summer. Most tea merchants spent the winter in their home areas and came to Hankou for only a few months each year. River levels were much higher in summer than winter, rainfall being greater over large areas. So around Hankou, river margins and creeks offered moorings for more boats in summer, and some people who lived on their boats came to the city then to seek work, perhaps as porters in the tea market, or unloading the coal boats. Merchants and seasonal workers were mostly men, so not only did the population of the city increase substantially in summer, but the predominance of males in the urban population was more pronounced. Skinner (1977, p.533) notes a similarly skewed sex ratio in Beijing owing to men leaving their families at home in smaller rural settlements.

Most trade in Hankou was organized directly or indirectly by institutions that Western historians refer to as 'guilds'. The term is confusing because it refers to several kinds of organization, some operating like trade unions to defend the earnings of working people, and others representing the most prominent businessmen in the city. Some guilds were associations of people engaged in a single trade, such as the relatively humble Coal Boatmen's Guild in Hankou, or the more prosperous Tea Guild. On the other hand, several guilds are referred to more usefully as 'native-place associations' or 'place of origin' groups (Kapp, 1974), and were named after a locality, or after an academy they ran or a temple with which they were associated. Many guilds functioned as religious brotherhoods, and temple worship was a prominent part of activities within the guildhall (Rowe, 1984, p.290). What was especially important, though, was that the network of trade brought merchants to Hankou from many distant parts of China, and it was natural for those from the same place of origin to stick together. By forming a guild, they could erect buildings with accommodation where visiting merchants from the home area could stay, and where people from the area who had settled in Hankou could socialize. They sometimes also organized a school (or 'academy') for their children, and actively supported charities ('benevolent halls') which helped orphans or provided food for the poor. All this was in addition to doing all they could to develop trade between Hankou and their home area.

Two native-place guilds in Hankou are worth particular attention, not least for the buildings they erected. One was the Huizhou (Hui-chou) Guild formed by merchants who were trading mainly in salt. The other was the Shanxi–Shaanxi Guild, run by merchants from two northern provinces. This body built its first guildhall in 1683. It was typical of many other Chinese public buildings – and houses – in that its rooms were arranged around a series of courtyards. Although the main purpose was to provide rooms for 'commercial deliberation', the front courtyard contained a temple dedicated to a deity associated with the guild members' home area (rather as medieval guilds in Europe commemorated their patron saints in local churches). There were also stages for theatrical performances. Like other guildhalls in Hankou, the buildings were lavishly decorated with red-lacquered woodwork, carved stonework, ornate tile roofs 'and dazzling tile façades' (Rattenbury, 1944; Rowe, 1984, p.25). All this, however, was destroyed in 1854 during the Taiping Rebellion, and only after some hesitation did the Shanxi–Shaanxi Guild begin to rebuild (see Figure 8.25).

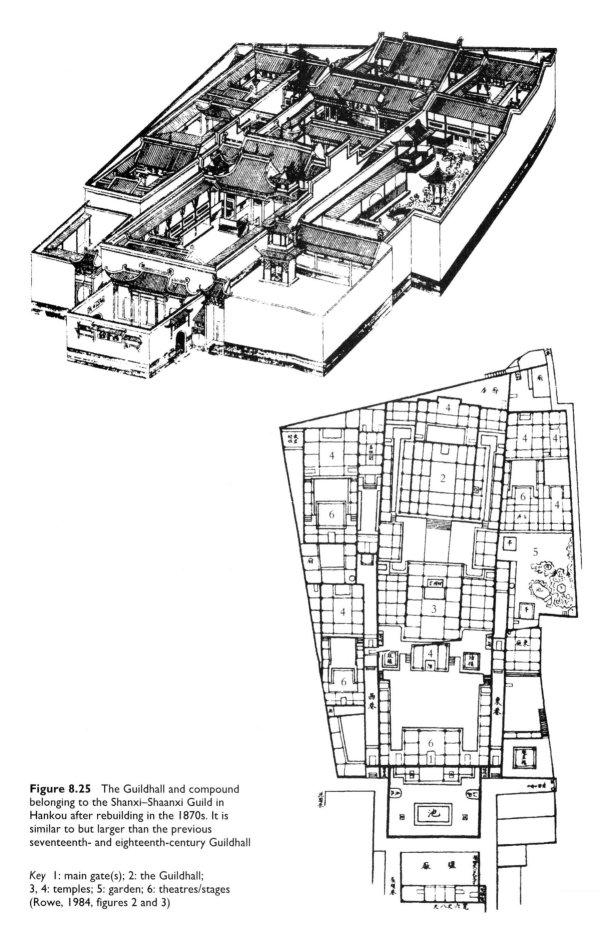

Figure 8.25 The Guildhall and compound belonging to the Shanxi–Shaanxi Guild in Hankou after rebuilding in the 1870s. It is similar to but larger than the previous seventeenth- and eighteenth-century Guildhall

Key 1: main gate(s); 2: the Guildhall; 3, 4: temples; 5: garden; 6: theatres/stages (Rowe, 1984, figures 2 and 3)

The Huizhou (or Hui-chou) Guild was rather different in that an important aspect of its activities was the academy it ran, and its guildhall buildings included a lecture hall for that purpose. The academy was mainly for children of its own members, but nobody from Hankou was denied its benefits. There was a high level of functional literacy within the city, and no doubt such academies were partly responsible. However, the main aim was to prepare students for the government civil service examinations, which were held every three years across the river at Wuchang.

As the Huizhou Guild prospered, it acquired much property within the city, and its enterprise in redeveloping shops and warehouses is recorded in surprising detail, telling us a good deal about eighteenth-century urban development. Indeed, a list of the properties owned by the guild in 1806 seems to mention almost every type of building which then existed in Hankou. It includes market buildings and piers, and also shops of the most fashionable kind, probably selling jade and silk. But there were also shops of a much humbler sort located in 'foundationless properties', and a few dwellings were also of this kind.[5]

Some of the poorer people in Hankou lived in houses built on stilts above the river's edge, occasionally with boats moored underneath (Rowe, 1984, p.26). However, 'foundationless' houses were another kind of dwelling in which the poor lived. The comparison to be made is with well-built houses erected on brick- or stone-paved platforms, which provided a solid and dry foundation for timber posts and walls. It was these platform foundations that were missing from the poorest buildings. Without them, timber would be in contact with damp ground and would quickly rot, so it is not surprising to find that 'foundationless' houses were not built of timber, but had earth walls with bamboo rafters supporting a roof of either tile or thatch. Some partitions would be of woven bamboo.

While many people used the river as their source of water, others obtained water from wells or by collecting rainwater running off roofs. A good courtyard house might have its own well, but well water here tasted bad and some people bought from water-carriers. The popularity of tea ensured that most water was boiled before it was drunk. Even so, there was little protection when cholera came to China in the 1820s, carried by people involved in ship-borne trade with Bengal (McNeill, 1977, p.261).

The type of building said to be most characteristic of Hankou was the 'shop-house' in which shopkeepers, their employees and their families lived above and behind their businesses. The whole of the front of a shop-house would comprise a folding partition – there was no window at ground-level (see Figure 8.26). When the shop was closed, it presented only blank woodwork to passers-by. When it was open, though, one could walk right into a narrow, arcade-like space, and the whole interior was visible from the street. Initially, shops were essentially single-storey with only attics above, where some of the owner's family or employees would sleep. As the population of the city increased, though, it was usual for shops to be two-storey, with the upper floor projecting over the street to gain a little extra space.

[5]   More detail is given in the first extract by William T. Rowe in the Reader associated with this volume (Chant, 1999).

Figure 8.26 Hankou shop-houses in perspective and cross-section. The buildings are timber-framed and have plank walls and tiled roofs. The overhang of the upper storey on the streetfront is clearly shown. The slightly raised foundation platforms have stone kerbs (1), and some walls between properties are of brick (2) for extra privacy, and to help contain any outbreak of fire. Shop-fronts consist of shutters (3) that can be folded right back when the shop is open. (They are shown partly closed in two of the shops in the perspective view; a third shop is displaying working clothes for sale.) Each house goes back a long way, but on a very narrow site. Inside, there are step-ladders (4) rather than staircases to the upper floor, and to the 'sun-drying platform' (*shaitai*) on the roof (5). There is usually also a tiny light-well (6) where large pots may be used to collect rainwater (7) (drawing: Arnold Pacey, from a photograph in Rowe, 1989, p.67, and incorporating information in Cornaby, 1925; Knapp, 1989, 1990; Rattenbury, 1944)

8.9　Hankou–Beijing comparisons, 1750–1840

It is hard to find clear illustrations of *urban* houses in China earlier than the latter half of the nineteenth century. Apart from the remarkable pictures of twelfth-century Kaifeng, with its houses, restaurants and bookshop (see Figure 8.12, p.279), we have to be content with relatively recent plans and photographs and check whether they fit what we know from other sources about pre-1840 buildings. However, with building design (like technology) evolving all the time, we should take nothing for granted.

For Beijing, however, a little more detail is provided by an accurate large-scale map of the city made about 1750 (Skinner, 1977). This demonstrates sharp contrasts between densely built-up areas in what Skinner calls the 'central business district', in the southern part of the walled city, and the more spacious layout of the areas where civil servants and members of the governing class lived. In other parts of the area within the walls, there was open land where crops were grown.

The map also shows that in the more spacious areas, very few buildings faced on to the streets, most being within walled courtyards. By contrast, in the business district, many buildings opened directly on to the street, as shop-houses needed to, and the courtyards behind were densely built-up. Two-storey shop-houses like those of Hankou (see Figure 8.26) were probably common in this area, although single-storey shops were also to be found, and on the wider boulevards, their temporary awnings and sales areas extended over the footways. If shops like the one illustrated in Figure 8.27 existed in 1750, they would be in a more exclusive district. This example is a single-storey building, with attics for storage, and decorative woodwork rising above the wide frontage to give it a taller, more impressive appearance. Such shops were not open-fronted, perhaps in response to the colder winter climate in Beijing, but also because shops selling books or porcelain and other quality goods were expected to provide more tranquil conditions, in which items could be examined and purchases made.

Using data from much later surveys, Skinner (1977, p.533) suggests that the population density of the most congested areas of Beijing could have reached 33,000 people per

Figure 8.27 The elaborate woodwork of a shop-front in Beijing. The party walls between one shop and the next are of brick. The photograph was taken in 1920 or just before, and the building may be nineteenth century. But note that there are gaslights high on the shop-front (Mennie, 1922, plate XXIX; photograph: courtesy of the Department of Special Collections, The Brotherton Library, University of Leeds)

square kilometre. These large numbers were accommodated without building above two storeys because many were unattached men living in what might be called hostel conditions. In shop-houses, Skinner believes, 'the salesrooms of stores, and workrooms of craftshops doubled as dining rooms and sleeping rooms' (ibid., p.533). Some of the men who lived like this were young apprentices. Others were migrant workers with families in some distant village. They accepted cramped living conditions because they were 'out to save as much as possible of their income' (ibid., p.533).

As for the owners of properties, they did not attempt to relieve congestion by constructing taller buildings because, as Skinner puns, hard-headed businessmen wished to keep overheads down. Had they wanted to construct multi-storey buildings, there were structures used as barracks and watch-towers above gates in the city wall that offered a ready model. Figure 8.20 (p.296), for example, shows a four-storey structure built on top of the solid mass of the city wall. This had outer walls of brick but an internal structure of timber, with timber posts or columns carrying the vertical loads from the upper floors. Other gatehouse structures had external timber-framing supporting verandas and pagoda-style roofs, but according to plans given by Boyd (1962), behind the timber exterior there were brick walls to the inner rooms. In all these buildings, while timber-framing appears to be doing all the work, the brick walls served to stiffen the structure.

One reason why multi-storey buildings of this kind were not erected by businessmen was undoubtedly their military associations and appearance. Possibly concerns about fire-risk would have made them unacceptable also. However, when we read of multi-storey buildings in thirteenth-century Hangzhou, we may surmise that some may have been like this, while the large warehouses there which were said to have been fireproof may have had vaulted basement storeys.

Figure 8.28 One of the main shopping streets in Hankou as it appeared in the 1880s. The streets were of similar width, and similarly congested, in the early nineteenth century. Some buildings had overhanging upper floors, and in places awnings could be extended across the whole width of the street. The garments suspended among the shop signs are not washing hanging out to dry, but an advertisement for a clothes shop. More garments are displayed on the extreme right. Important officials would travel in sedan chairs, as shown, and porters would carry loads suspended from shoulder yokes, like the one leaning precariously on the right (Cornaby, 1925, p.199)

Figure 8.20 also serves to make the point that carts hauled by mules or horses were extensively used for freight transport in Beijing, while Figure 8.27 indicates the role of donkeys. Two-humped Bactrian camels were also to be seen in the city, bringing goods from Mongolia (Mennie, 1922). In Hankou, by contrast, streets were so narrow that wagons were barely practicable (see Figure 8.28). Many loads were manually carried by porters from wharfs to warehouses.

As previously noted, dwellings of the better-off in Beijing reflected the same planning principles as the city as a whole, being based on rectangular walled courtyards aligned on a north–south axis, with the principal rooms facing south (see Figure 8.29, pp.310–11). Buildings within the courtyards were single-storey with trees and flowers in the courtyards, and often a pool. Even in the 1930s, photographs taken from the air show trees more dominant in the city than rooftops (Wright, 1977).

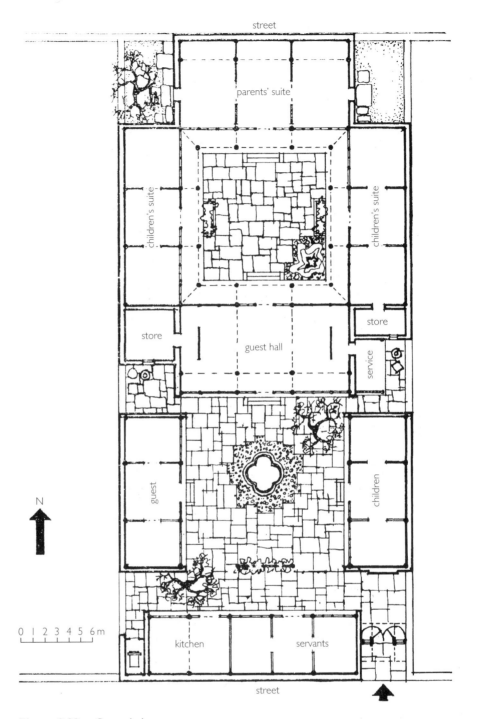

Figure 8.29a Ground plan

Figure 8.29 Plan of a typical Beijing house. All the buildings are single-storey, and there is a north–south alignment. The entrance is in the bottom right-hand corner, under its own roof, and is arranged so that there is no direct line into the house, to conform with *feng shui* principles as well as for privacy. The latrine, or privy, is next to the kitchen, unlabelled, in the opposite corner of the plan. From the street, all that a passer-by sees of the house are its tiled roofs, and a blank wall with one entrance. Inside the courtyards, the timber elevations of the buildings are often in shadow because of the wide eaves of the low roofs, which are shown in Figure 8.29b (Boyd, 1962, pp.80–81, figures 29–30)

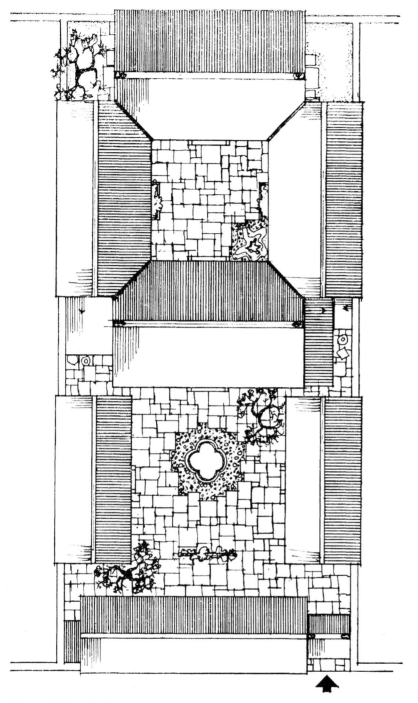

Figure 8.29b Roof plan

The side of each room that faced the courtyard of a house consisted mainly of windows and doors, which in summer were open to the outside air. Although glass was manufactured on a small scale to make lenses for spectacles and telescopes, window glass was not used. Windows were filled with wooden slats or decorative fretwork on which tough translucent paper was stuck in winter. This was also true of some shop windows (see Figure 8.27). Shutters might also be fitted to house windows, but rooms were not windproof.

Well-off families with extra garden or courtyard space could add new buildings as their sons grew up and married, so that large extended families with their servants might occupy one courtyard house. On the other hand, when people fell on hard times, the buildings of a courtyard house could be subdivided so that individual structures were occupied by different families, with the courtyard itself downgraded as a shared access area.

When occupied by a well-off family in the traditional way, such dwellings could be very pleasant, but they represent a form of building that had evolved in areas further south, and seem ill-suited to the dry but freezing winters of Beijing. During that season, people wore quilted clothes and thick felt-soled shoes. Inside a house, they used small portable charcoal braziers for warming their hands. Apart from that, the main source of warmth was the *kang*, a brick platform under which flues from a coal fire passed. People would use the *kang* as a divan for sitting or sleeping and parts of the palace had underfloor heating on the same principle (Boyd, 1962). In small houses, the *kang* would be warmed by the flue from the cooking-stove, which would be in the adjoining room. Such a stove had its coal fire largely enclosed within a brick box whose flat top contained two or three round apertures into which pans were fitted while their contents were cooking. The apertures were closed with covers or lids at other times. The grate, where the fuel was fed in, was low down at the front of the stove, with the flue passing through a brick wall at the back.

Less information is available about Hankou houses, but given the extremely dense population of the city (in striking contrast to parts of Beijing) it is not surprising to find that houses there had the 'tiniest of courtyards', barely adequate even as light-wells (Rattenbury, 1944). In two-storey shop-houses, as Figure 8.26 showed, the upper floor was usually reached by a ladder, and a 'sun-drying platform' on the roof provided an elevated equivalent of the more usual courtyard, with space for drying clothes.

Another feature of most houses was, of course, the privy, which was of the type where the user squatted rather than sat. In the more spacious cities, houses were often planned so that the privy was against an exterior wall, allowing the cavity underneath to be emptied via a small hatch from the street. However, the intricacies of privy design are not often recorded, and Figure 8.30 is an *impression* only, based on several travellers' experiences in a variety of cities, showing what is likely to have been a common arrangement in China and neighbouring countries. But plans of traditional houses in Beijing (Boyd, 1962) show that privies were placed so that they could be emptied without sewage-carriers passing through the main courtyards.

It was usual in most of China for night-soil removed from privies to be composted or otherwise used as fertilizer. In Hankou, it was carried by 'dung-boats' from the city to nearby agricultural districts (Rattenbury, 1944, p.14), and the city was

Figure 8.30 Sectional view of a household privy showing how the cavity underneath could be emptied from the street outside. Transport was first by buckets suspended from a yoke slung across the man's shoulder, then, for transport from cities out to farms, the buckets might be emptied into a small tank on a cart (or possibly into a boat) (Pacey, 1978, p.126, figure 8.2; reproduced with permission of John Wiley and Sons)

known, rather surprisingly, for its clean, well-paved streets. Paving and draining was one of the tasks undertaken by the guilds as part of their various property-development schemes. There was no town council, and government officials (such as the submagistrate and subprefect) were mainly concerned with law and order, and tax collection. Many of the functions we might expect a council to undertake were carried out by the guilds. This is nowhere more noticeable than in the organization of fire brigades. Fires in Chinese cities tended to be frequent, not only because houses were built of wood, but also because cooking-stoves were kept going most of the time so that tea could be brewed at short notice (Huc, 1859, p.139). But fire-fighting in Hankou remained rather casual until manually operated fire-engines, or 'water dragons', were introduced in the 1790s (Rowe, 1989, p.164). They consisted of small four-wheeled carts on which 'cylinder and piston' force-pumps were mounted. At first, soldiers worked these machines, but then government officers asked the guilds to sponsor fire-engines and their crews.

Fire-engines, weight-driven clocks, muskets, spectacles and telescopes were all manufactured in China during the seventeenth and eighteenth centuries using indigenous metalworking, millwrighting and gunsmithing technology often combined with ideas introduced from elsewhere. The earliest important infusion of Western technical ideas was associated with Jesuit missionaries, the first of whom, Mateo Ricci, gave a weight-driven clock to the emperor in 1601.

Another Jesuit, working with a Chinese author, produced a book in 1627 describing 'remarkable machines' from the 'Far West' (Elvin, 1975). Already in 1635, a workshop in Suzhou was building fire-engines that incorporated some of this technology (Needham and Wang Ling, 1965, p.222), and it was from somewhere near Suzhou that fire-engines were bought for Hankou.

Lifeboats on the storm-swept, kilometre-wide Yangzi River were another public service operated by guilds. Again, the government had failed to provide an adequate service, and around 1820, salt merchants began using a riverside pavilion as lifeboat headquarters and look-out point (see Figure 8.3, p.266). During the first sixteen years of this service, the lifeboats saved 4,132 lives and pulled 7,000 drowned corpses from the river – one casualty for every 100 vessels passing through the port (Rowe, 1989, p.108).

It is often said that the official bureaucracy of imperial China weighed heavily on commerce and choked enterprise. In 1850, the government employed 30,000 civil servants. That sounds a lot, but it represents only one bureaucrat for every 15,000 inhabitants, whereas a German author (Wickert, 1984, p.158) comments that in his country, there is one civil servant for every twenty-seven inhabitants. In other words, he observes, Chinese government officials were 'surprisingly thin on the ground'. That may explain why they lacked the capacity to run lifeboat or fire services in Hankou. Indeed, one might surmise that the business community suffered from too little government rather than too much. They paid very little tax (Elvin, 1973, p.92; Rowe, 1989, p.211), but were distracted from commercial activity by the need to manage schools, roadworks and lifeboats.

However, a more important point is also made about Chinese bureaucracy. Although many aspects of urban development were left to the initiative of merchants, officials of the central government were in ultimate control, and tended to run the towns as merely one part of the larger 'counties' to which they belonged.[6] The absence of distinctive forms of urban administration is regarded as having delayed 'the emergence of a "modern" urban society',

---

[6] The English word 'county' is used in official Chinese publications to translate *xian*, the term for an administrative area considerably smaller than an English county. *Xian* are often named after, and can be confused with, their principal towns. Today, however, the 236 largest cities are separate administrative units, and are not included within counties (Population Census Office of China, 1987, p.xiii).

inhibiting the economic dynamic that such a society might have generated. Hankou had an unusual degree of autonomy, however, and by all accounts was an unusually dynamic mercantile community in the eighteenth and nineteenth centuries (Rowe, 1984, p.38).

8.10   The role of industry

The physical appearance of the cities we have studied here varied greatly according to their relationships to waterways and the monumentality (or otherwise) of walls, gates and pagodas. But the underlying contrast throughout has been between spacious, formally planned administrative cities (Chang'an, Beijing), and the crowded, informal planning of commercial centres, such as Hangzhou in its original form, and Hankou. A rather separate issue has been the extent to which industrial or manufacturing activity was to be found within cities of either kind, notably Kaifeng, the Song dynasty capital, with its nearby iron industry and paper and printing workshops.

Hankou was primarily a port and commercial centre rather than a manufacturing town, and its characteristic industries were a reflection of this. Thus sacks, boxes and baskets were manufactured to service the repackaging of goods in its warehouses, and ships were built and repaired to support its function as a port. However, with coal and iron-ore mines not far away, it is not surprising to find some ironworking there before 1800, such that 'massive quantities' of knives, scissors, padlocks and cooking-pots were being made (Rowe, 1989, p.46).

There was also a textile industry, for not only were there large markets for cotton and silk, serviced by hundreds of boats, but Hankou was the market-town for a local cotton-growing area where there were also cottage industries producing cloth. It was typical of the textile industries that silk-reeling, cotton-spinning and weaving were rural occupations (though Hankou also had some weaving shops), while cloth was sent to town for dyeing and finishing. In a small town, cloth straight from the dye-vats might be spread out to dry on the city walls (L. Pacey, 1935, unpublished journals, and letters from China). In Hankou, dyeing, calendering (polishing cloth with stone rollers), trimming and packaging of cotton cloth were large, well-organized industries (Rowe, 1984), although with minimal mechanization.

There is, of course, the greatest possible contrast between the organization of these industries and the very large scale of major works such as canals and city walls. In contrast to the many small enterprises and independent producers in the textile industry, linked together by merchants' transactions and by self-governing guilds, canals and city walls were state projects.

Their construction was managed by civil servants who had powers to assemble the necessary labour force by conscription of both women and men. Soldiers might be employed on some projects, too, and convicts and prisoners of war also. But the basic understanding was that everybody owed the state a specified number of days' work each year (which the better-off could usually commute by paying extra tax). The majority of labourers on many projects were farmers and farmworkers, and it was usually arranged that their obligations could be discharged at seasons when agricultural work was slack. Thus they were not taken away from their land at seed-time or harvest.

A detailed account of works on a new canal near Kaifeng in the 1830s shows what was involved. There are hundreds of people with hoes, shovels and wheelbarrows (the wheelbarrow being a much earlier Chinese invention). Some people are operating pumps to keep water out of the excavation. A red flag is flying to indicate that a competition is in progress between different gangs to see which can complete their allocation of work first. Prizes on offer include wine, meat and items of clothing. The civil servant responsible for this project had decided that prizes were better than threats as a way of keeping people at work, since the latter only led to strikes (Needham *et al.*, 1971, p.262).

This method of constructing major works had a very long history in China, the largest projects, involving hundreds of thousands of workers at a time, being the canal-building schemes of the Sui and Tang dynasties, and the many city walls and the Great Wall of the Ming. The point is made, though, that conscription was not slavery; people gave only a limited number of days' work at a time and could not be permanently uprooted from their homes. Thus conscripts were not much used in remote quarries, according to Needham *et al.* (1971), who claim that this restricted the use of stone in building.

Comment was made earlier about the limited extent of good farmland in China and the consequent shortage of draught-animals. The equipment used on the canal project of the 1830s just quoted – the wheelbarrows and pumps – was in principle similar to what would be used on an English canal project of the period, but what was strikingly different was the lack of draught-animals, either to assist in earth-moving operations or to power pumps. Human muscle power did everything.

Land shortages affected the textile industries also. Cotton was the main textile worked in Hankou, reflecting its increased importance during the seventeenth and eighteenth centuries. And the rise of cotton in place of traditional coarse fibres such as hemp was related to the fact that *ten* times as much fibre could be produced from the same area of land (Elvin, 1973), and some cotton could be obtained by import from abroad also. Moreover, there were useful by-products, as oil could be produced by pressing cottonseed, just as it could be obtained from soya and rapeseed. Indeed, an oil-pressing industry would probably have developed at Hankou except that *feng shui* experts advised against it on the grounds that Hankou was built on shoals of river sediment, which were allegedly too insubstantial to withstand the forces generated by oil-seed presses. Only at the end of the nineteenth century were such objections overcome (Cornaby, 1925).

Not only with regard to draught-animals and cotton production, but in many other technologies, the central theme of eighteenth-century innovation was how to gain more output from land. Thus technological innovation in China took a different path from innovation in Europe. Although travellers in the countryside around Hankou reported many water-mills at work, mechanization aimed at raising labour productivity was less important than innovation aimed at raising the productivity of land and natural resources (Elvin, 1975).

The most significant innovations of the period, indeed, were new types of fertilizer, and new crops such as groundnuts, sweet potatoes, white potatoes and maize. Chinese officials had noted these growing in the Philippines during the sixteenth century, where the Spanish had introduced them from the Americas (Herklots, 1972, p.436). They had tried them out in China around 1600, and now that populations were rising, these crops provided a means of greatly increasing food-production on land where rice could not be grown, especially in north China. Fertilizers were produced by processing waste products from many industries, such as bean-curd and soy-sauce manufacture, vegetable-oil production, and the cleaning of raw cotton by ginning. Bean-cake and cottonseed fertilizers were said to be especially effective. Elvin (1975) also writes of refinements in crop rotations and the evolution of 'symbiotic farming', in which every plant grown on a farm complemented some other aspect of plant or animal production.

One difficulty for a Westerner in understanding the significance of these developments is that, in the West, 'technology' is identified with machines rather than with agricultural ecology. But in China, there was evident enthusiasm for machines as well. Crowds turned out around 1800 to see a new

Archimedean-type screw pump demonstrated, and poems contained references to machines. Commenting on this, Elvin (1975, p.107) suggests that the Chinese were too practical in their attitude to machines, and not sufficiently imaginative. There was no tradition of fantasy-invention such as one finds in the notebooks of Leonardo da Vinci or in the Western tradition of perpetual-motion machines. This outlook could have a negative aspect, as when labour-saving devices were assessed, very practically, from the point of view of the labourer rather than the entrepreneur. Did the new device make his or her work easier, or would it lead to a reduction of earnings? During periods of civil unrest from the late eighteenth century onwards, there were episodes of machine-breaking comparable to those of the Luddites in England (Elvin, 1975; Hsieh, 1974). One observer in the mid-nineteenth century, who commented on resistance to the more complex types of textile machinery, illustrated a machine for processing silk thread on eleven spindles simultaneously, which he said was the most complex mechanism he could find. He was aware that bigger machines of the same kind had existed previously (Thomson, 1977).

8.11 Conclusion

Taking a longer perspective, though, and looking beyond Hankou, we have noticed here a good deal of technological innovation which was a response to the requirements of city-builders, including new ways of casting bells for city bell-towers; the canals, sewers and tidal lock in Hangzhou; the improved brick-making techniques of the Ming period, and the introduction of fire-engines in Hankou. We have also noticed innovatory changes in building design, such as the two-storey Hankou houses with their 'sun-drying platforms'. However, Elvin (1975) also notes instances where innovators failed to respond. The price of coal in Beijing was rising steeply in the eighteenth century because of problems due to flooding of mines, and yet little was done to improve the simple suction-pump already used in some mines or to develop other techniques for removing water.

A second example of missing innovation might be the lack of multi-storey buildings in congested cities, although some relevant technology was available, and a third instance is the nearly total absence of aqueducts for the supply of water for drinking, laundry and washing. Needham *et al.* (1971, p.376n) comment that 'the abundance of waterways surrounding and penetrating most of the principal cities' made a separate system of aqueducts seem unnecessary. With regard to the contamination of canal water, these authors imply that people were aware of the desirability of boiling water prior to drinking, even apart from their liking for tea. Elsewhere, however, Needham and Wang Ling (1965, p.129) point out that Hangzhou had a piped water supply of sorts installed in 1089 and renewed in 1270 using bamboo pipes. Then, in Beijing, there was copper or bronze piping in the palace water system from the fifteenth century.

When Western commentators on the Chinese scene feel that desirable and easily achievable innovations are missing, whether they be pumps, aqueducts or high-rise buildings, they rarely ask how local circumstances affected the matter, or how the people involved thought about them. Instead, they are tempted into sweeping generalizations about how China 'failed' to produce an 'industrial revolution' because of 'technological stagnation' during the empire's later history. But having researched one sector with exceptional rigour, Bray emphatically insists that there was no 'technological stagnation'. In agriculture, there was continuous innovation, but it did not have the results that Westerners expect because of technical peculiarities of local crops and conditions (Bray, 1984, p.613).

More generally, though, Western historiography of China has suffered from interpreters who devote much energy to explaining why there were no industrial or scientific revolutions in China, or else to claiming that actually there were. Eberhard (1967, pp.45, 68) discusses the iron industry during the Song period and asks: 'What prevented the Chinese from going further?' Describing the career of one entrepreneur in detail, he points to problems of law and order, and extortion by tax officials. An industrialist, he suggests, had to spend too much time and money simply defending his assets. In addition, without much explanation, he asserts that another obstacle was 'the very structure of the Chinese city'. It is with that comment in mind that Rowe (1984) discusses the lack of effective local government in Hankou, and the role of the guilds in performing some of its functions.

Yet Hartwell (1966) sees the Kaifeng iron industry as comparable at least to the early stages of the European Industrial Revolution, and Elvin (1973) speaks of an 'economic revolution' affecting most of China under the Tang and Song dynasties. It is even possible to think that there were scientific revolutions of a sort in the late Song period, as well as in the seventeenth century. Sivin (1984) supports the latter view, though with tongue in cheek, since he also argues that it is really a mistake to expect China to have gone through the same sort of intellectual evolution as the West. In Europe, after all, the scientific revolution gained much of its initial impetus from the Renaissance and Reformation, movements that are clearly specific to European cultural and religious history. If we were to find a scientific revolution in another civilization, it would be related to a quite different pattern of cultural change, and would therefore present only the most approximate parallels with the European experience. One very broad parallel might be the stimulating effect of urban life on intellectual and scientific pursuits in Hangzhou during the early thirteenth century as compared with, say, Florence in the fifteenth and sixteenth centuries.

However, historians of China are inclined to pay less attention to supposed 'revolutions' than to an alleged cyclic pattern in the history of each dynasty. They point out that periods of peace and prosperity in the early part of a dynasty's rule often gave way to peasant rebellions and other upheavals later on. Such problems tended to mount until the dynasty was swept away, often to the accompaniment of cataclysmic disorder. Boyd (1962, p.13) describes this cyclic pattern as part of the background of Chinese architecture and town-planning. Thus it is possible to see old ideals of city-planning revived in the ascendant phases of several dynastic cycles, as at Chang'an, Kaifeng and Beijing (compare Figures 8.4, 8.9 and 8.18, noting the square plan, south-facing orientation and palace on the central axis in each case).

If the historian of technology takes note of this view, however tentatively, what he or she will notice is that, in each successive cycle, a somewhat different kind of technology and style of innovation emerged. Thus the Song dynasty was notable for its iron industry and agriculture, and its use of printing, while the Ming dynasty had its brick-making technology and its wall-building projects. The Qing dynasty took a different direction, impelled in part by population growth, supported by river-borne and canal-borne commerce, and focused on technologies related to productivity of land (with a strong iron industry also).

This latter pattern of development during the Qing period, and the story of Hankou to which it is closely linked, was interrupted when China's defeat at the hands of the Western powers allowed foreigners to take over some parts of the trading system, with the British concession at Hankou a part of this process. Even worse was the catastrophic Taiping Rebellion in the 1850s, which led to death and destruction within China on a scale comparable to the First World War in Europe during 1914–18. Hankou was engulfed by conflagrations during this episode on no fewer than three occasions, and had to be rebuilt from almost nothing in the 1860s and 1870s. What historians interested in revolutionary patterns of change should consider more fully is why these periodic outbreaks of destruction occurred.

The foregoing suggestions add up to the recommendation that instead of trying to apply Western concepts of industrial or scientific 'revolution', historians of technology should look at China in terms of that country's own distinctive history. Then new and different questions arise, among which three have been emphasized here. One is whether we need to recognize some 'loss of technology' as occurring during cataclysmic episodes. The issue has been broached rather tentatively because so few historians have recognized the problem and there is little well-researched evidence or analysis.

A second question concerns the need for a better understanding of the period of rapid commercial expansion which began in the late seventeenth century and continued to the end of the eighteenth. During this expansion, many previously small cities grew rapidly, Hankou being a prime example. There was a growth of organizations supportive of trade, including guilds and banks. Government policies were of a kind that today would be described as 'economic liberalization', designed to encourage trade and the free operation of markets. There was also much technological innovation, especially in agriculture. Western historians have tended to overlook or misunderstand the latter, however.

A third question concerns the complex relationships between cities and technology. One point is that cities were a stimulus to innovation, partly through the specialist demands they generated – for better bricks, or for fire-engines – and partly because they brought together people with different skills and backgrounds. Capital cities such as Chang'an, Kaifeng and Hangzhou included merchants, craftworkers and foreign traders within their populations as well as courtiers and civil servants. Once the transfer of the capital to Beijing had been completed in 1421, however, the government operated in comparative isolation from the main areas of economic activity, and hence from that kind of stimulus. It is from this time onwards that historians who focus on life in the capital speak of declining interest in science and missed opportunities for innovation, while historians who look at China's commercial cities and at the life of the countryside share Bray's (1984) insistence that there was no technological stagnation (especially not in the eighteenth century; see Elvin, 1973, 1975).

Another point, made strongly by Skinner (1977), is relevant to the converse influence that technology may have had on the development of cities, and on urbanization generally. Thus Skinner considers *six* influences that may have affected the number of people living in cities. They include population pressure in the countryside, division of labour in handicraft industries, and various developments in commerce and administration. Finally he asks: could the application of technology be regarded as a 'determinant' of the extent of urbanization?

Skinner assesses these six possible influences for several of the very varied regions of China. One is the middle Yangzi region, which includes Hankou, where he estimates that long-distance trade was by far the most powerful stimulus for urbanization (Skinner, 1977, p.235). This is in contrast to the lower Yangzi region, where Hangzhou and Suzhou are situated, as well as other canal cities (and the growing but still small city of Shanghai). During the seventeenth to nineteenth centuries, commerce and trade are admitted to have had a very powerful influence on urbanization, but the application of technology is scored as almost equally influential. One aspect of this is that the building of canals, land-drainage works and sea walls created the spaces in which cities could grow. Another aspect was the development of urban industries, such as printing in Suzhou and the workshops where fire-engines were made. These would draw workers into the cities in search of higher wages.

It is also possible to make comparisons with a global perspective. In the eighteenth century, when very few cities anywhere had populations in excess of one million, China's biggest cities must be counted as large. If we ask what gave them such distinction in world terms – what contributed to their size – then the efficiency of water transport and the high volume of traded goods must be seen as important. Few cities elsewhere could cope with such traffic flows. However, there is a difference between conditions that *make large cities possible* (among which canal engineering must be included), and influences that *stimulate* urban growth, such as a boom in trade. Neither can be regarded as rigidly *determining* the size or planning of a city.

Hankou during the early nineteenth century illustrates some of this in the oddest possible way. At the peak of the summer business season it may have ranked as one of the three most populous cities in the world, along with London and Paris (see Table 8.4, p.303, right-hand column). But it seems likely that as many as one-third of the summer population was living on boats (see Figure 8.31), and as winter approached, they sailed away in search of other work or, perhaps, warmer weather. We might say that boat-building technology *made it possible* for the city to support a much larger population than could be accommodated on its restricted site, given that high-rise buildings were not considered. But it would be rash to say that boat-building technology *determined* the character of Hankou. A sounder judgement would take more account of human purposes. People living on boats were usually owner-occupiers and were free to move on at different seasons to search for work. Boats were something they could control and use for their own purposes. The technology of building houses and shops, however, was largely controlled by the guilds and private landlords. And the question of who controls a technology is often more important for determining what happens than the nature of the technology itself.

Figure 8.31 Hankou boat-dwellings. A photograph dating from the nineteenth (or early twentieth) century at the junction of the Han and Yangzi Rivers, which helps to explain difficulties in assessing the population of Hankou: the inhabitants of such boats were often not counted (Hellier, 1906)

References

BOYD, A. (1962) *Chinese Architecture and Town Planning 1500 BC–AD 1911*, London, Tiranti.

BRAY, F.[7] (1984) *Science and Civilization in China*, vol.6, part 2 'Agriculture', Cambridge, Cambridge University Press.

BUXTON, L.H.D. (1929) *China, the Land and the People: a human geography*, Oxford, Clarendon Press.

CHANG SEN-DOU[8] (1977) 'The morphology of walled capitals' in G.W. Skinner (ed.) *The City in Late Imperial China*, Stanford, Stanford University Press, pp.75–100.

CHANT, C. (ed.) (1999) *The Pre-industrial Cities and Technology Reader*, London, Routledge, in association with The Open University.

CHIANG YEE (1938) *The Silent Traveller in London*, London, Country Life Books.

CORNABY, W.A. (1925) *A Necklace of Peach-stones*, Shanghai, North China Daily News and Herald.

CULLEN, C. (1990) 'The science/technology interface in seventeenth-century China', *Bulletin of the School of Oriental and African Studies*, vol.53, pp.295–318.

EBERHARD, W. (1967) *Settlement and Social Change in Asia*, Hong Kong, Hong Kong University Press (Collected Papers vol.1).

ELVIN, M. (1973) *The Pattern of the Chinese Past*, London, Eyre Methuen.

ELVIN, M. (1975) 'Skills and resources in late traditional China' in Dwight Perkins (ed.) *China's Modern Economy in Historical Perspective*, Stanford, Stanford University Press, pp.85–113.

ELVIN, M. and SKINNER, W.G. (eds) (1974) *The Chinese City between Two Worlds*, Stanford, Stanford University Press.

FUGL-MEYER, H. (1937) *Chinese Bridges*, Shanghai, Kelly and Walsh.

FRANKE, H. and TWITCHETT, D. (eds) (1994) *The Cambridge History of China*, vol.6 'Alien regimes and border states, 907–1368', Cambridge, Cambridge University Press.

HARTWELL, R. (1966) 'Markets, technology and the structure of enterprise in the eleventh-century Chinese iron and steel industry', *Journal of Economic History*, vol.26, pp.29–58.

HELLIER, J.E. (1906) *Life of David Hill*, London, Morgan and Scott.

HERKLOTS, G.A.C. (1972) *Vegetables in South-East Asia*, London, Allen and Unwin.

HO PENG-YOKE (1973) 'Li Chih' in C.C. Gillispie (ed.) *Dictionary of Scientific Biography*, New York, Scribner, vol.8.

HO PING-TI (1959) *Studies on the Population of China 1368–1953*, Cambridge, Mass., Harvard University Press.

HSIEH, W. (1974) 'Peasant insurrection and the marketing hierarchy in the Canton delta' in M. Elvin and G.W. Skinner (eds) *The Chinese City between Two Worlds*, Stanford, Stanford University Press, pp.119–42.

HUC, M. (E.-R. HUC) (1859) *The Chinese Empire*, London, Longman.

KAPP, R.A. (1974) 'Chungking' in M. Elvin and G.W. Skinner (eds) *The Chinese City between Two Worlds*, Stanford, Stanford University Press, pp.143–70.

KNAPP, R.G. (1989) *China's Vernacular Architecture: house, form and culture*, Honolulu, University of Hawaii Press.

KNAPP, R.G. (1990) *The Chinese House: craft, symbol and the folk tradition*, Hong Kong, Oxford University Press (China).

LAM KAM CHUEN (1995) *The Feng Shui Handbook*, London, Gaia.

LIU GUOJUN and ZHENG RUSI (1985) *The Story of Chinese Books*, Beijing, Foreign Language Press.

McNEILL, W.H (1977) *Plagues and People*, Oxford, Blackwell.

MENNIE, D. (1922) *The Pageant of Peking*, Shanghai, A.S. Watson.

MORGAN, D. (1986) *The Mongols*, Oxford, Blackwell.

MOTE, F.W. (1977) 'The transformation of Nanking' in G.W. Skinner (ed.) *The City in Late Imperial China*, Stanford, Stanford University Press, pp.138–47.

MOTE, F.W. (1994) 'Chinese society under Mongol rule' in H. Franke and D. Twitchett (eds) *The Cambridge History of China*, vol.6 'Alien regimes and border states', Cambridge, Cambridge University Press, pp.616–64.

---

[7] Individuals contributing to Needham's series *Science and Civilization in China* are treated as co-authors or, in volumes to which Needham himself did not contribute, as sole authors.

[8] In Chinese names, the surname comes first; for this reason the word order does not have to be reversed in an alphabetical list.

MOTE, F.W. and TWITCHETT, D. (eds) (1988) *The Cambridge History of China*, vol.7 'The Ming Dynasty, 1368–1644, part 1', Cambridge, Cambridge University Press.

MOULE, A.C. (1957) *Quinsai and other Notes on Marco Polo*, Cambridge, Cambridge University Press.

MURPHEY, R. (1974) 'The Treaty Ports and China's modernization' in M. Elvin and G.W. Skinner (eds) *The Chinese City between Two Worlds*, Stanford, Stanford University Press.

NEEDHAM, J. (1954–) *Science and Civilization in China*, 7 vols, Cambridge, Cambridge University Press.

NEEDHAM, J. (1958) *The Development of Iron and Steel Technology in China*, London, Newcomen Society.

NEEDHAM, J. (1969) *The Grand Titration: science and society in East and West*, London, Allen and Unwin.

NEEDHAM, J., HO PING-YU, LU GWEI-DJEN and WANG LING (1986) *Science and Civilization in China*, vol.5, part 7 'The gunpowder epic', Cambridge, Cambridge University Press.

NEEDHAM, J. and WANG LING (1959) *Science and Civilization in China*, vol.3 'Mathematics and the sciences of the heavens and the earth', Cambridge, Cambridge University Press.

NEEDHAM, J. and WANG LING (1965) *Science and Civilization in China*, vol.4, part 2 'Mechanical engineering', Cambridge, Cambridge University Press.

NEEDHAM, J., WANG LING and LU GWEI-DJEN (1971) *Science and Civilization in China*, vol.4, part 3 'Civil engineering and nautics', Cambridge, Cambridge University Press.

NEEDHAM, J., WANG LING and PRICE, D. DE SOLLA (1960) *Heavenly Clockwork: the great astronomical clocks of medieval China*, Cambridge, Cambridge University Press.

NEEDHAM, J. and YATES, R. (1994) *Science and Civilization in China*, vol.5, part 6 'Military technology', Cambridge, Cambridge University Press.

NORTON, J. (1997, 2nd edn) *Building with Earth: a handbook*, London, Intermediate Technology Publications.

PACEY, A. (ed.) (1978) *Sanitation in Developing Countries*, Chichester and New York, John Wiley.

PACEY, A. (1990) *Technology in World Civilization*, Cambridge, Mass., MIT Press.

POPULATION CENSUS OFFICE OF CHINA (1987) *The Population Atlas of China*, Hong Kong, Oxford University Press (China).

PURVIS, M. (1988) *China Journey*, London, Merehurst Press.

RATTENBURY, H.B. (1944) *China, My China*, London, Frederick Muller.

READ, T.T. (1913) 'The mineral production and resources of China', *Transactions of the American Institute of Mining Engineers*, vol.43, pp.3–53.

ROSTOKER, W., BRONSON, B. and DVORAK, J. (1984) 'The cast iron bells of China', *Technology and Culture*, vol.25, pp.750–67.

ROWE, W.T. (1984) *Hankow*, Stanford, Stanford University Press, vol.1.

ROWE, W.T. (1989) *Hankow*, Stanford, Stanford University Press, vol.2.

SIVIN, N. (1984) 'Why the scientific revolution did not take place in China – or didn't it?' in E. Mendelssohn (ed.) *Transformation and Tradition in the Sciences*, Cambridge, Cambridge University Press.

SKINNER, G.W. (ed.) (1977) *The City in Late Imperial China*, Stanford, Stanford University Press.

SUNG YING-HSING (1966) *T'ien-kung K'ai-wu: Chinese technology in the seventeenth century* (trans. and ed. E.T.Z. Sun and H.C. Sun), Pennsylvania, Pennsylvania State University Press.

STRUVE, L.A. (1993) *Voices from the Ming–Qing Cataclysm: China in tiger's jaws*, New Haven and London, Yale University Press.

THOMSON, J. (1977) *China, the Land and its People*, Hong Kong, John Warner (first published 1873).

TSIEN TSUEN-HSUIN (1985) *Science and Civilization in China*, vol.5, part 1 'Paper and printing', Cambridge, Cambridge University Press.

TWITCHETT, D. and FAIRBANK, J.K. (eds) (1979) *The Cambridge History of China*, vol.3 'Sui and T'ang China, 589–906, part 1', Cambridge, Cambridge University Press.

WAGNER, D.B. (1997) *The Traditional Chinese Iron Industry and its Modern Fate*, Curzon Press, Richmond for Nordic Institute of Asian Studies.

WICKERT, E. (1984) *The Middle Kingdom* (trans. J. Maxwell Brownjohn), London, Pan.

WILLETTS, W. (1958) *Chinese Art*, Harmondsworth, Penguin Books.

WRIGHT, A.F. (1977) 'The cosmology of the Chinese city' in G.W. Skinner (ed.) *The City in Late Imperial China*, Stanford, Stanford University Press.

WU, N.I. (1963) *Chinese and Indian Architecture*, London, Prentice-Hall.

Chapter 9: THE CITY IN PRE-COLONIAL AFRICA

by Michael Bartholomew

9.1 Introduction

'The Middle Ages', 'the Renaissance', 'the Early Modern period', 'the period of Industrialization' … these are familiar terms, invented by European historians for the purpose of differentiating phases of their continent's past. It was never intended that the terms should apply universally: nobody envisaged talk of medieval New Zealand, or of Renaissance Alaska. None the less, the terms have had a baleful effect on the presentation of the history of countries for whose past the categories of European periodization are inappropriate. Historians of China, for example, have had to struggle to keep at bay European terms such as 'the Scientific Revolution' and 'the Industrial Revolution' in order to free themselves for the project of telling Chinese history in its own terms.

The problem is acute when we turn to African history. As the writers of a survey of African history bluntly put it, 'The old sequence of ancient, medieval and modern is obvious nonsense for Africa'. Furthermore, in Africa, the seemingly more basic epochs, such as the beginning of agriculture, or the Iron Age, came at different times in different regions of the continent (Curtin *et al.*, 1995, p.xiv). It is therefore not obvious whereabouts – in this broadly chronological series on the city and technology – this chapter on African cities should go.

There are topographical and cultural problems also. First, what actually corresponds to the term 'Africa'? Obviously we can find the continent in an atlas. But the range of languages, religions, traditions, climates and ethnic groups within the continent is, and always has been, colossal; and it is not at all clear that, collectively, these elements constitute a cultural and historical entity that we can unproblematically signal with the word 'Africa'.

Second, and coming closer to the issue of cities and technology, African history is plagued by the connotations of the word 'primitive'. Because the continent was thinly populated and has left few written records, and because most of its human settlements were never intended by their builders to be permanent, it is tempting simply to make do with two stereotypes of African towns and cities. The first is characterized by 'primitive', perishable, insubstantial huts of the type 'surpassed' by cultures elsewhere in the world centuries ago. The second is characterized by European-style buildings implanted by colonizers, around which have gathered slums of rickety shanties built by the descendants of the aboriginal inhabitants. Neither stereotype encourages us to look closely at the distinctive technologies that were – and still are – used by Africans when they constructed large settlements.

We have to be careful, therefore, neither to casually extend inappropriate notions of period and of progress, nor to apply the term 'city' only to large human settlements that happened to have reared large heaps of durable masonry. It might be helpful here to give a couple of examples of the sort of approach we need to take when considering African cities. The first concerns fortifications. Historians of the European city have long been familiar with the Renaissance theory and practice of fortification, especially as it was applied in Italy. In Chapter 5 of this volume, David Goodman surveyed the topic. In her

Figure 9.1 The locations discussed in this chapter

study *African Traditional Architecture,* Susan Denyer gives an account of an indigenous African solution to the problem of attack on a settlement by enemy cavalry:

> The splendidly conceived mazes of the plateau area of Nigeria are a fascinating example of an ingenious defence. Complex entrance tunnels to villages were formed out of live cactus hedges and planned on the principle of a maze to frustrate any attempts by horsemen to enter either by storm or stealth. The tunnel … began sometimes over 1 km from the village. The sides were too close to allow a horse to be turned and there were many blind alleys and passages which took one back to the place just passed. So the front sections of a hostile column of mounted men would be brought back face to face with the rear columns of the force – a perfect ambush. Once trapped the enemy would have considerable difficulty in breaking out even with the aid of axes, as the type of cactus used exuded a white juice which was so caustic that it took the skin off an arm or a leg or blinded an eye splashed with it (and it has recently been found to be carcinogenic).
>
> (Denyer, 1978, pp.66–7)

To emphasize the point that Denyer makes: this defensive maze depends on a very highly developed technological application of botanical knowledge, yet it is impossible to date the practice; its devisers have left no documentation, and it has left scarcely a trace on the landscape (see Figure 9.2 overleaf).

The second example, also drawn from Denyer, is chosen in order to dispel impressions that African human settlement is casual or random. Figure 9.3

Figure 9.2 Defensive maze
of cactus hedging; Birom village,
Bukuru, Nigeria, c.1920
(Denyer, 1978, p.9)

Figure 9.3 Dogon village, Ireli, Mali, c.1935 (Collection Musée de l'Homme, Paris; photograph: Marcel Griaule)

shows the extraordinary density of a settlement in Mali. This may appear to be a random accumulation of individual small buildings, but in fact the layout of this settlement is governed by a set of extremely complicated cultural rules:

> The Dogon, who live on the rocky Bandiagara escarpment in what is now the Mali Republic, base their philosophy, it is claimed, on the idea of germinating cells vibrating along a spiral path to break out of a 'world egg'. The spiral and the egg shape therefore had special significance. Each village was laid out either in a square to represent the first field cultivated by man or in an oval with a hole at one end to represent the world egg broken open by the spirally vibrating cells. In either case it also represented a person lying in a north–south direction with the smithy placed at the head and certain shrines at the feet, while huts used by women during their menstrual period were the hands and the family homesteads the chest. The anthropomorphic nature of the village was further expressed by a conical foundation shrine and by a hollowed stone for grinding which signified the male and female sexual organs. The surrounding fields fitted into the system.
>
> (Denyer, 1978, p.20)

Figure 9.4 Dogon granaries near Mopti, Mali, c.1935. These multi-storey granaries, made of mud and small pieces of stone, were built under overhanging cliffs at the highest point in the villages (Collection Musée de l'Homme, Paris; photograph: Marcel Griaule)

Technology always operates within particular cultural constraints, and here the constraints are evidently very tight. But lest it be thought that cultural rules have inhibited the efficient functioning of the Dogon buildings, an examination of the granaries in Figure 9.4 indicates that a weatherproof and pestproof solution to the vital problem of storage of grain has been achieved.

Clearly, vivid as they are, hard-to-date examples such as these cannot be threaded on to a single chronological string that joins the cities of Mesopotamia in the fourth millennium BCE with Singapore in the twenty-first century. But examples of this sort can add substantially to our understanding of the multifarious, ingenious ways in which humans have packed themselves into the small areas we call cities. Within this book, this chapter on African cities locates itself in the part labelled 'Pre-industrial Cities in China and Africa'. For the reasons given above, this location is pretty arbitrary. Many of the examples that I shall present arose entirely independent of influences from beyond Africa, and they do not fit neatly on to any chronology. But sometimes, particularly in the case of sub-Saharan cities, the settlements were part of trading networks that reached well beyond the borders of Africa, and that connected Africa to large movements in world history. Parts of Africa were in routine contact with other cultures, centuries before European colonization.

9.2   Environmental constraints

To Europeans, the climatic range between the south and the north of their continent seems large, and they can rightly point to features of northern and southern cities that have been governed by climate. In the north, buildings are designed to catch the sun; in the south, they are designed to ward it off. In the north, a prime consideration for builders is the shedding of rainwater; in the south it's not so important. But the differences can easily be exaggerated. Residents of medieval Florence would have seen few unfamiliar sorts of building had they been transported to medieval York. Indeed, if the visit had been made a century or so later, the visitors would have seen that the new Renaissance styles that were transforming their own city were being applied, with no significant structural modification, to chilly Yorkshire. Moreover, the materials from which the buildings in both cities were made were not fundamentally different. There were more thatched roofs in York, but certain features were found throughout Europe – hewn and unhewn stone, brick, timber and tile. Styles that originate in one part of Europe tend to travel well to other parts. The style, for instance, that was devised for the Parthenon on the Athens Acropolis serves admirably for the entrance to Huddersfield railway station. A few types of building do not travel so well. Flat-roofed houses in northern Europe, derived from village houses in Greece, tend to let in the rain. But by and large, in Europe, the range of climate and the range of readily available building materials are not so extreme as either to prohibit or to constrain the building of cities, or to produce fundamentally different sorts of city as we move from south to north. The character of European cities derives less from the brute facts of terrain and climate than from the cultural history of the continent. The style of Huddersfield railway station tells us more about the span of cultural history connecting classical Greece and Victorian Britain than it does about the climate of Yorkshire. Only when the tundra and subarctic regions of the far north of Europe are reached do we meet climatic and environmental conditions that set limits on the very possibility that a city, of any sort, could be built.

In contrast, environmental barriers to city-building, equivalent to the tundra on the very edge of Europe, are common throughout Africa. Large, permanent settlements in the middle of the vast areas of desert within the continent are simply not possible – unless by the importation of high-tech infrastructures for the purposes of exploiting valuable resources such as oil. Large, permanent settlements are scarcely more possible in the middle of tropical rain forests. Savannah regions whose soil is not rich enough for settled agriculture, but which will support nomadic hunting or pastoral tribes, are unlikely to yield city-building traditions. And overall, where populations are spread thinly across vast and inhospitable terrains, there will be little impetus for city-building. The environment in Africa sets much fiercer constraints on human culture than does the environment in Europe. The range of climates, and the associated terrains and vegetation that are crossed as a traveller moves from north to south in Africa, is extreme – ranging from vast desert to vast rain forest. Maybe this range sets up barriers to cultural transmission more formidable than those that lie between, say, Greece and the UK. Culture perhaps travels more readily along lines of latitude than along lines of longitude.

The African terrain, the temperature and the rainfall impose gross constraints, but there are other, less obvious but equally powerful environmental forces in play. For example, in Chapter 1 of this volume Colin Chant drew attention to the part played by the north wind in the emergence and flourishing of North African civilization in the Nile Valley (pp.34–5). You may recall that it just happens that the prevailing wind in Egypt blows from the north, and that the River Nile flows *towards* the north. This happy environmental conjunction means that craft can use the current when travelling downstream, but can *sail* upstream, pushed by the northerly breeze. Navigation on rivers where the prevailing wind blows in the same direction as the current is much more difficult. (Getting craft up to ports such as London, for example, where an east-flowing river is accompanied by a predominantly south-westerly wind, was not easy. London-bound sailing ships routinely had to anchor way down the estuary, and wait for a favourable windshift.)

A further environmental constraint in inland Africa was the range of the tsetse fly. This fly attacks cattle, horses and camels (as well as humans). No culture, therefore, which depends either on cattle husbandry or on transport based on draught- or pack-animals can flourish in an area infested with these flies. Some local control of the fly was possible: patches of the vegetation that formed its habitat were cleared, and smoke from fires was used for fumigation. But in tsetse-ridden parts of the continent, goods were transported between settlements by human porters, not animals. It is easy to see that such a transport system alone, all other considerations apart, sets severe limits on the growth of trading cities. As we shall see presently, the sub-Saharan zone of Africa, where there did develop a chain of permanent trading cities which depended entirely on camels and donkeys, is at the far edge of the tsetse's northern limits, and thus just out of range of its depredations. Summing up the significance of this particular determinant of African settlement, Philip Curtin, a historian of Africa, concludes:

> insects could sometimes influence the course of history more than vegetation did. Tsetse flies made a fundamental difference in what human societies could do.
>
> (Curtin *et al.*, 1995, p.66)

The environments of Africa, therefore – especially those far from the coast – are not propitious for the emergence of cities like Florence and York, with monumental civic or religious buildings like the Duomo or York Minster. African settlements and buildings tend to be smaller than European, and to have a much closer, more intimate relationship with the environments that governed their establishment.

9.3   Building materials

Stone

We can usefully distinguish four sorts of material from which African buildings
and settlements were constructed. Let us start with stone – remembering,
though, that its durability and permanence, and therefore its visibility to
archaeologists and historians, give it a prominence in the history of building
that its actual use in the past may not justify. (History is not exactly about the
past: rather, it is about those bits of the past that have left records.) Building
with stone was not common in Africa, but there are some important
exceptions. In Ethiopia, a region to which Christianity spread very early and
directly from the Middle East (at the same time as its spread northwards into
Europe), there were established, by the tenth century CE, traditions of building
churches in stone – or of hewing and carving them out, in one piece, from
hillsides (Denyer, 1978, p.53). Somewhat later, under Islamic influence, a string
of trading ports sprang up along the East African coast (for example, Mombasa
and Zanzibar). At one of these ports, Kilwa, there are the remains of an
elaborate palace complex, built during the mid-thirteenth century from blocks
of coral rock (Hull, 1976, pp.113–14). But the most remarkable stone-building
traditions, entirely indigenous to Africa and its native cultures, flourished away
from the coast – in eastern and southern Africa. Their most celebrated
archaeological remains are those of the city of Great Zimbabwe, which lies
260 kilometres south of Harare. 'Zimbabwe' means 'houses of stone' or
'venerated houses' (Garlake, 1973, p.11). ('Zimbabwe' was also used to signify
the whole country when the name was changed from 'Rhodesia'.)

The Europeans who discovered the ruins of Great Zimbabwe were not
predisposed to believe that such an extensive city, built in stone by evidently
highly skilled masons, could be indigenously African. The first, prejudiced and
amateurish colonial archaeologists indiscriminately shovelled out layer after
layer of what they dismissively called 'Kaffir' remains, in search of evidence
that would show that the city was really European, or at least non-African, in
origin. Their depredations made it difficult for later, fully competent
archaeologists to establish the exact sequence of the development of the site.
Much is still unclear, but it is almost certain that Great Zimbabwe was built
between 1000 and 1500 CE by local people, drawing on native traditions.

The city is located in a temperate, fertile, well-wooded, tsetse-free region
studded with granite outcrops. As these outcrops erode, they degrade into
regular-sided pieces of rock which lend themselves to being laid in courses
(like bricks). The suggestion of P.S. Garlake, who has written a detailed survey
of the site, is that the earliest walls were simply small rocks heaped up to fill
the gaps between large, immovable boulders. By linking up rings of boulders
with these rough walls, enclosures were established. As the city developed,
wall-building techniques became more sophisticated. Eventually, unmortared
walls ten metres high were built of carefully selected granite rocks, laid in
regular, bonded courses, some incorporating vigorous decorative patterns (see
Figure 9.5).

The site is difficult to interpret because it does not have a single, continuous
wall defining the city's perimeter, and within which the city's buildings were
laid out. Rather, a collection of enclosures – most with walls much lower than
those of the chief's palace – is spread across the landscape, and the purpose of
each enclosure is not clear (see Figures 9.6–9.8). Despite the massive walls, the
enclosures may not even have been primarily defensive, for they do not take
systematic advantage of naturally occurring defensive features of the terrain.

Within the enclosures, buildings made of timber, thatch and *daga* were
constructed. (*Daga* is one variant of what is unhelpfully known in common

Figure 9.5 Great Zimbabwe: wall of Zimbabwe or chief's palace. The palace was surrounded by this massive wall about ten metres high and, on average, three metres thick, enclosing an area ninety metres by sixty-five. The wall was constructed of unmortared, dressed granite blocks enclosing a rubble core. Near the top, thinner slabs of stone were worked into a double-chevron pattern. Large soapstone bird sculptures once looked outwards from the flat top (photograph: The Ancient Art and Architecture Collection)

Figure 9.6 Great Zimbabwe: the Great Enclosure, as in Figure 9.5, with smaller enclosures in the foreground. The Great Enclosure has a number of narrow entrances – 'smoothed off' breaks in the wall; these are reconstructions, now thought to be inaccurate (photograph: The Ancient Art and Architecture Collection)

Figure 9.7 Great Zimbabwe, the Great Enclosure: these curved steps show the masonry at its most accomplished (photograph: Peter Chèze-Brown)

Figure 9.8 Great Zimbabwe: the Conical Tower – the focus of the Great Enclosure. The tower testifies to a settled and elaborate civic or religious culture, but its exact purpose is unknown (photograph: Peter Chèze-Brown)

parlance as 'mud' – a building material whose range of properties will be discussed later in this chapter.) *Daga* was used to plaster the stone walls wherever the wooden buildings abutted them, and to supply smooth, hard floors.

What went on in the city? Archaeological evidence, painstakingly reassembled after the first clumsy diggings, shows that Great Zimbabwe was a centre for metalworking. Crucibles containing copper, bronze and gold have been found. Iron-smelting produced arrow- and spear-heads. Garlake estimates that the population was somewhere between 1,000 and 2,500. The city must have built up an extensive export and import trade before it went into mysterious decline in the late fifteenth century, for fragments of Chinese pottery have been discovered. It seems likely that the city was exporting gold, via port cities, probably to Arab countries or even to China, and importing luxury goods. This in turn indicates that Great Zimbabwe was a fully-fledged, though small, city. It had emancipated itself sufficiently from subsistence activities to develop specialist metalworking trades and to set up extensive trading networks. (The nearest coast is 480 kilometres away; and Kilwa, a trading port which flourished at the same time as Great Zimbabwe, is 1,440 kilometres away.)

The gold-mining connection with Great Zimbabwe indicates the possibility of the transfer of a technology from mining to building.[1] Ore-bearing rocks in the mining areas were split by being heated in fires and then doused with cold water. The granite rock from which the walls of Great Zimbabwe are made tends to erode naturally into flat slabs but, in some walls, it appears to have been broken down into more precisely shaped blocks by the same method of heating and rapid cooling.

Great Zimbabwe is by far the biggest stone ruin in the region, but there are fifty or so smaller sites. At a far humbler level, wherever it occurred locally, uncut stone rubble was used fairly widely in Africa for the foundations and lower walls of buildings that were built chiefly of timber and thatch.

Timber and vegetation

This brings us to the second group of materials – timber and vegetation. Until fairly recently, such perishable materials, and the buildings constructed from them, had a rather lowly place in the history of architecture. But now, driven by modes of cultural study which privilege no particular way of living and building, and by an emerging concern in the industrialized countries for the construction of 'green' buildings,[2] interest in traditional African architecture has grown. In a book upon which I have drawn extensively in this chapter, Denyer has meticulously catalogued and analysed traditional African buildings, establishing a taxonomy of thirty-two distinct types, most of which were built from timber – usually thin laths and poles rather than hefty beams – with thatched roofs and woven fibre or mud-plastered walls (Denyer, 1978, pp.133–42).

Each style exhibits a slightly different adaptational response to the environment in which it developed. In some, for instance, the thatched roof does not sit directly on top of the walls but is supported from outside the walls on free-standing posts. This gives the building deep eaves, valuable both for shade and for the shedding of rain; it also allows for a gap between the top of the walls and the inside of the roof, through which a cooling breeze can pass. Contrary to myth, not all traditional African dwellings were small round huts.

---

[1] I owe this point to Arnold Pacey.

[2] In cereal-growing areas of the industrialized world, for example, successful experiments have been conducted in the building of houses from straw bales. Such houses, even those built for the fierce climates of the North American prairies, are thermally efficient, strong and weatherproof, once waterproof renderings and roofs have been applied to them.

Figure 9.9 Mangbettu assembly hall, Zaire, c.1910. In contrast to the round houses, the Mangbettu assembly halls (and chiefs' houses) were immense rectangular buildings (Denyer, 1978, p.45)

In Zaire, for example, there was a tradition of large, rectangular buildings, and Denyer gives an illustration of one (see Figure 9.9; the photograph is dated around 1910, so we cannot be absolutely certain that the building type has an unbroken lineage back to earlier centuries). Buildings of this sort were assembly halls – often fifteen metres high, with ground plans ninety metres by fifty – built entirely from palm-wood poles, thatched with palm fronds, and walled with woven palm-frond matting. Their immense size testifies to a civic life that produced appropriate official buildings, albeit buildings made from highly perishable materials. But to focus straight away on large civic buildings of this sort is perhaps to miss one of the chief points about African timber and thatch buildings, for overwhelmingly they were small. And it is a moot point whether they commonly clustered together in densities sufficient to justify the designation 'urban' – and whether their general character, when they did, is best conveyed in the vocabularies often used in studies of urbanization. Examples drawn from southern Africa in the early nineteenth century, and described by Richard Hull in his book on pre-colonial African cities and towns, make the point.

Hull describes the way in which, in the area known now as the Transvaal, raids from the north had prompted the local populations to urbanize themselves. They constructed stone-built towns with defensive walls. There were hundreds of these little towns, sited at the bases of hills. Eventually, in the early nineteenth century, they were overwhelmed by Zulu invaders. The inhabitants retreated, re-founding their stone towns on inhospitable, inaccessible mountaintops in what is now Lesotho. Hull describes one such town, Thaba Bosiu, sited on a hilltop near the Orange River:

> Its half-square-mile summit provided for the accommodation in times of siege of large numbers of people with livestock and other provisions. Its steep cliffs and narrow passes were excellent natural defences; and the surrounding countryside could support a substantial population in times of peace. A French missionary spent several decades in Basutoland [now Lesotho] and described Thaba Bosiu in 1833 as a jumble of low huts, separated only by narrow lanes, crowded with children and dogs. In the town center was a vast space where cattle were penned at night in well-constructed, perfectly round stone enclosures. Adjoining the square was a courtyard devoted to business transactions and public speeches.
>
> (Hull, 1976, p.23)

Even in its degraded state, then, here was evidence of what was perhaps an urban culture. The French missionary is clearly describing a town of some sort,

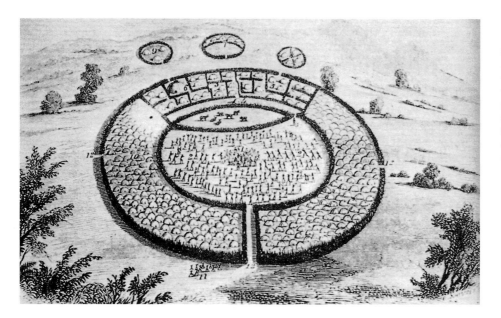

Figure 9.10 Royal elliptical kraal at Umgungundhlovu, *c.*1830; all Zulu royal cities followed the same design. These were essentially predatory cities, which thrived not on market or craft activity but on pillage of the surrounding countryside. At its zenith, Umgungundhlovu contained over 1,700 dwellings (Killie Cambell African Library)

but maybe these settlements were simply fortified agricultural villages – farmers driven into close communities by the threat of Zulu raids. They may not have achieved the measure of emancipation from the basic processes of food production that is one of the indicators of truly urban life.

What sorts of settlement did the invading Zulus build, down on the plains? Their culture was one of raiding and moving on, not one of settled city-building, yet maybe we need to expand our notions of what a town or city can be, for the Zulus built kraals – enclosures – within which were erected large numbers of dwellings. In the largest, Umgungundhlovu, over 1,700 dwellings – each accommodating about twenty soldiers – were built within a timber palisade (see Figure 9.10). Additionally, there were cattle pens and a large parade ground. Plainly, Umgungundhlovu was a big place, but it existed solely for the purpose of pillage and defence, and thus might fail a strict test of the truly urban. Its buildings were intended to be temporary, and were constructed of light woven and thatched materials (see Figure 9.11 overleaf for an example of a Zulu house). Garrison towns built of stone survive to this day in most parts of the world, often forming the heart of later urban developments, but in the Zulu Transvaal this enormous concentration of highly combustible dwellings was effaced utterly when Boer commandos set fire to it in 1838.

Modern colonial imports

The third group of materials can be covered only fleetingly. The impact on Africa of European colonization, and the attendant importation into the continent of alien buildings and cities, is a vast subject: I have deliberately excluded it from this chapter in order to make way for the work of scholars who have been uncovering the indigenous traditions of African building. The skyscrapers, oil refineries, motorways and urban shanty towns of modern Africa would require a separate chapter – or book. But relationships between traditional and imported technologies exist at humbler levels than the more obvious levels of the railway, or the high-rise concrete office block. Studies, for example, of the consequences of the introduction of corrugated iron for roofs, or glass for windows, would be instructive. Even this low-level approach cannot be pursued in this chapter; but as an emblem of it, I record the observation of two fieldworkers who were studying the traditional houses of Botswana in 1974 (Farrar and Pacey, 1974). These houses are circular, with conical thatched roofs. The very apex of the cone is a vulnerable point in the weatherproofing and windproofing of the roof. Farrar and Pacey found that

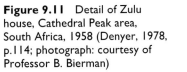

Figure 9.11 Detail of Zulu house, Cathedral Peak area, South Africa, 1958 (Denyer, 1978, p.114; photograph: courtesy of Professor B. Bierman)

local metalworkers had tackled this problem by fabricating small cone-shaped caps from galvanized sheet metal. This ingenious and limited application of an imported technology is contrasted by the authors with what they see as the less successful transfer of another material – corrugated iron sheets, for roofs. These roofs confer high status on their owners, even though the thermal properties of thin iron sheet are inferior to those of traditional thatch. (Thatch keeps the house cool in the day and warm at night; sheet metal has very little insulating effect – although it is better for the collection, via guttering, of rainwater.)

Earth

Returning to the main subject of this chapter – the pre-colonial period in African urban history – the last building material to be considered is mud, one of a number of building materials derived from earth, but not modified by having been fired in a kiln. Despite its inevitable connotations of the primitive, earth is a material which has given rise to some highly developed technologies and to some sophisticated buildings and cities. Earth building is by no means uniquely African. In Chapter 1 of this volume (Sections 1.4 and 1.5), Colin Chant discussed the use of mud-bricks in ancient Near Eastern cities.

Earth can be used in a variety of ways. It can be left in place, and then rammed ('tamped') or trodden until it is so compacted that it will serve as foundations for buildings, or even as the floors of the buildings. Sometimes earth is carried to the site and heaped in a mound which is then tamped to provide a solid base for a building slightly higher than the surrounding ground. In Chapter 8 Arnold Pacey described the rammed, or tamped, earth bases of Chinese buildings (p.269). Some earths, especially clays, can be dug up, mixed with water and binders (straw, animal hair, gravel, animal dung or blood), and used as a plaster to coat walls or floors. This is the earth known popularly as 'mud' or 'daub' (as in the term 'wattle and daub', where light panels of woven vegetable material – 'wattles' – are plastered with a carefully prepared 'daub' of clay and dung). Mud plaster can be brought to a high finish: in Great Zimbabwe, for example, the *daga*, while wet, was patted gently with the palms of the hand to bring the finest constituents floating to the surface to give a smooth, hard finish (Garlake, 1973, p.18). Clay mixtures are not always used as plasters: the soft mixture can be put in moulds, dried and hardened in the sun, and then used as bricks that are cemented together with liquid clay. In this form, or when rammed between parallel shuttering-boards half a metre apart, mud can form thick, load-bearing walls. (Surprisingly large numbers of still perfectly sound, mud-walled houses remain in use, in Europe. In the English West Country, the mud walls are known as 'cob': Clifton-Taylor, 1972, pp.287–93.)

The focus of the rest of this chapter will be the mud-walled cities of sub-Saharan Africa, cities in which the techniques of building with earth were brought to an extraordinarily high level of sophistication, as indicated in Figures 9.12 and 9.13.

Figure 9.12 Hausa mud-built mosque (Moughtin, 1985, p.124; photograph: courtesy of Emeritus Professor J.C. Moughtin)

Figure 9.13 Djenne house, Mali, c.1950 (photograph: Archives La Documentation Française)

Of course, not all African mud buildings aspired to the grandeur of the mosque and the house in Figures 9.12 and 9.13. Less spectacularly and more routinely, mud was used for floors and for the plastering of the walls of small huts. Here is Denyer's description of the construction of floors – which, she explains,

> could be almost as hard as cement and quite smooth. A good, hard floor was obtained by beating the mud with a wooden beater while it was setting. The mud was mixed with charcoal, as in Zulu houses, or some other small aggregate, or with cowdung and then smeared with ashes. In parts of the Sudan the Raik Dinka found that the mud from low ant-hills was particularly good for making a hard practically waterproof, bluish cement.
>
> (Denyer, 1978, p.94)

Mud is the material from which a number of substantial cities were built, and it is to these cities that we now turn.

9.4   *The cities of sub-Saharan Africa*

The Sahara Desert set a barrier, as formidable as a great ocean, between the cultures of the Mediterranean rim and the cultures of equatorial Africa. Influences from the Levant and the Roman Empire, and later from Islam, could spread fairly easily westwards along the North African coast and its fertile hinterland, but not very far southwards – save along the Indian Ocean coast. In Chapter 4, David Goodman described the way in which the Islamic 'urban ideal' spread westwards along the Mediterranean, establishing towns (Fez, for example) with highly developed civic amenities (p.137). At the end of the westward expansion, Islamic influence turned north, crossed the Straits of Gibraltar and entered Spain. By contrast, travel southwards, across the Sahara Desert, was formidably difficult. But it was not impossible, and there had long been camel caravan routes which had kept open lines of communication between the cultures on either side of the desert, especially when trade in highly desirable commodities such as gold and slaves was in prospect. Gold had been imported from sub-Saharan Africa into the Roman Empire from around 300 CE. There grew up a chain of cities stretching along the southern rim of the desert, across the continent from west to east, sited at places where there was sufficient water and agricultural land to supply a permanent settlement, and from where caravan routes headed north across the sand to Mediterranean cities such as Marrakech or Tripoli, or to Nile Valley cities such as Cairo. This cross-desert trade was in gold, salt, textiles and slaves.

Figure 9.14 Plan of Timbuktu in the early nineteenth century (Caillié, 1830)

Timbuktu is the classic example of such a city (see Figure 9.14). It is sited where the River Niger, quite early in its course to the Nigerian coast, curves northwards to the edge of the desert (see Figure 9.1). Communication from Timbuktu downstream to the equatorial zones of what is now Nigeria can be made by boat. The city stands, therefore, at the junction of river traffic and camel caravan traffic. (Cities typically spring up at places where modes of transport change.) A Moorish traveller, born in Fez, described Timbuktu, which he visited in around 1510:

> Here are many shops of artificers and merchants, and especially of such as weave linen and cotton cloth. And hither do the Barbary [North African] merchants bring the cloth of Europe. All the women of this region except maid-servants go with their faces covered, and sell all necessary victuals. The inhabitants and especially strangers are exceedingly rich. Here are many wells containing most sweet water; and so often as the River Niger overfloweth they convey the water thereof by certain sluices into the town. Corn, cattle, milk and butter this region yieldeth in great abundance … The inhabitants are people of a gentle and cheerful disposition, and spend a great deal of the night in singing and dancing through all the streets of the city. Here are great store of doctors, judges, priests and other learned men, that are bountifully maintained at the king's cost and charges. And hither are brought diverse manuscripts or written books out of Barbary, which are sold for more than any other merchandise.
>
> (quoted in Hull, 1976, pp.13–14)

Here then was a city built firmly on trade, regulating its water supply technologically by means of sluices, and sustaining a literate and educated culture. The reference to veiled women indicates that Timbuktu was predominantly a Muslim city. It might be tempting to see cities of this sort simply as colonial intrusions into Africa – in the way, perhaps, that Roman towns were intruded into Britain. Undoubtedly, the influence of dynamic Islam in sub-Saharan Africa was great. For example, Muslims, with their reverence for scripture, brought the practices of reading and writing into Africa. In the early fourteenth century, the African ruler of Mali went on pilgrimage to Mecca, bringing back with him Muslim scholars, including architects. And during the nineteenth century a jihad, or holy war, consolidated the Muslim control of the entire region. But inevitably, local customs played their part in the fabric and culture of the cities that were emerging. Some cities, for example, were oriented and laid out according to indigenous African cosmological principles, and the variety of African family structure traditions influenced the way in which housing compounds were laid out. Timbuktu was already a substantial city long before the fourteenth century, when it fell under Islamic influence. It is best, therefore, to see these cities as cosmopolitan trading and craft-manufacturing centres, certainly heavily Islamic, but embodying also the traditions of all the peoples who traded back and forth. The structure of the cities was determined also by the needs of defence – some of them have defensive walls – and by the constraints imposed by the rugged environments in which they were built.

Hausa cities

One particular group of sub-Saharan cities has been studied in great detail. J.C. Moughtin's *Hausa Architecture* (1985) surveys the mud-built cities of Hausaland, an area in what is now the north of Nigeria, bordered in the west by the River Niger, and stretching in the east towards Lake Chad (see Figure 9.1). The principal cities are Kano, Zaria and Sokoto. Although Moughtin supplies a good deal of information about the history of these cities, and especially about the interplay between indigenous African and incoming Muslim cultures, dates of buildings are not particularly important. This is because he is dealing with a building tradition that is unbroken: he can illustrate techniques of wall-building, for instance, by reference to the way it was done in the 1980s, and yet justifiably suggest that this was the way it was done centuries ago. Moughtin's study is tailor-made for the student of cities and technology, for he describes – in great and sympathetic detail – the technologies that were, and to a diminishing extent still are, employed in the construction of these cities. Students of architecture familiar with the rhapsodies of architectural historians on such materials as Cotswold limestone or Dutch brick will be intrigued by Moughtin's reverence for the structural properties of mud and his respect for the skill of the city-builders who used it. Furthermore, he explores the relationship between the technology of the cities and Hausa culture, framing both with a description of the landscape and climate that have shaped them.

It is a commonplace in the history of building that, throughout the world, towns and cities built according to ancient traditions, and from local materials, achieve unique harmonies of form and function. The buildings bear intimate relationships with the landscapes from which they rise, and their forms have been steadily adapted over the centuries to serve the exact needs of their occupants. This way of thinking can, however, lead to problems: we can be tempted to *start* by assuming that there is always a good match between traditional buildings and their environments, rather than by investigating whether there indeed is one. Hausa cities – literally built from the mud, vegetation and animal products of the land they lie on, adapted to the particular social needs of the Hausa people, responsive to the extremes of the climate, and poised always, were they to be abandoned, to erode swiftly back into the landscape – seem to be a dream come true for the devotee of sustainable, appropriate technology. Plainly, Moughtin is full of admiration for these cities, but quite properly he does not assume, in advance, what he would like to be able to prove. Indeed he tempers his enthusiasm for the cities by indicating that – especially in relation to climate – they are a less than ideal response to the environment.

The climate in Hausaland is extreme. A hot, dry, dusty season is followed by a season of torrential rain. No building will be perfectly adapted to both seasons, and Hausa cities are a necessary compromise. But the compromise seems to have been struck in favour of buildings that perform best in hot, dry conditions. This indicates perhaps that the building types themselves originated in the desert and were brought in on the wave of Muslim influence. Dotted about Hausaland are the remains of ancient, indigenous stone-walled buildings. Seemingly, such a building tradition offers greater potential than mud if permanent cities, subject to periods of intense rainfall, are to be built. But the Hausa people have not drawn on it. Instead they have built with mud derived from laterite – the residue of eroded, mineral-rich rocks. Obviously, mud walls and roofs will wash away in heavy rain. Consequently the Hausa have developed elaborate techniques for waterproofing mud surfaces and channelling water away down slopes and into spouts. But in the absence of the standard solution to a wet climate – a pitched roof with overhanging eaves – a good deal of time in Hausa cities has to be spent in maintaining rain-damaged buildings.

At the other climatic extreme, the fierce heat of the dry season would seem to prompt the placing of buildings in orientations that maximize the possibility of the circulation through them of the prevailing breezes, and minimize the area of wall facing directly towards the sun. Yet, Moughtin argues, Hausa buildings are not typically arranged in this way. The cosmological system which governs the layout of cities, and which directs that buildings should be oriented north–south, takes precedence over orientations that might be more climate-friendly. Here, culture regulates technology.

In other respects, however, the structural properties of mud as a building material work extremely harmoniously and conveniently with the cultural conventions of the Hausa people. The cities themselves are ancient and permanent, and within them highways and public spaces, and certain monumental buildings (mosques, for example), are well established. But at the level of the housing compound, each of which is defined by a continuous wall, there is considerable flexibility. Compounds are large enough to enclose not only houses, but plots of agricultural land and accommodation for animals. Within the Hausa compound wall, accommodation will be subdivided or knocked down, and new units will be built, all in response to routine changes imposed by the elaborate Hausa family and kinship structure, and by Muslim regulations concerning the segregation of the sexes. In this respect, culture and technology work hand in hand. Here is Moughtin's account:

> The Hausa house plan for both Muslim and non-Muslim families follows the traditional African pattern, having rooms arranged within or surrounding a courtyard. In common with other types of African housing, the compound wall is an important feature of the Hausa house. It may have been developed initially as a wind-break or for reasons of defence and to a certain extent it still fulfils these functions, but whatever its original purpose, there are now other very good reasons to justify the cost of its construction. The wall demarcates an area within which members of the family may withdraw from society and remain in privacy. It also serves the same defensive function for the family as does the city wall for the community as a whole, providing security for the family, preventing the escape of small farm animals and discouraging thieves from entering the compound.

> The compound contains the main economic unit, the extended family, which works the same fields, shares the same grain store and eats from the same pot. The marital units which make up the extended family occupy separate areas within the compound ... The structure of the family is in a process of constant change leading to growth, subdivision or decline ... within the compounds of the extended families there is evidence of constant adaptation to changing family circumstances ... huts are added where and when they become necessary; later, screen walls are built to subdivide the house into *sassa* [marital unit areas] for married sons. When the family shrinks through death or the loss of a breakaway family group, the land soon returns to agricultural use. Land within the compound has a high agricultural potential and since the maintenance cost for repairs to buildings is high, the tendency is for unused huts to be demolished, the site put to other uses and the building materials recycled ... the process of demolition is completed during the heavy rains. It has been the custom for the dead occupant to be buried beneath the floor of his hut and the building permitted to decay, eventually forming a mound over the grave. Like the settlement itself, the house is not an accumulation of accommodation which remains static for long periods ... This process of growth and decay is facilitated by the short life of the building materials, but is given form and continuity through the cellular nature of the architecture where the unit of growth and decay is the hut.
>
> (Moughtin, 1985, pp.57–8)

The Hausa city is permanent, but most of its constituent buildings are impermanent. But this does not lead to randomness in the overall pattern of the city, for the compound walls define enduring spaces. Furthermore, as in most cities, there are zones, or administrative wards, associated with particular craft guilds or families. Moughtin again:

> Both the ward and the city are based on a pattern of street blocks. Each street block, irregular in shape, is enclosed by a high mud wall and contains the main social and economic units, either the extended family or the simple unit, based on marriage. Family compounds are linked by a complex system of pedestrian routes. The pathways connect the important elements in the city, the *dendal* [a large, public open area in front of the emir's palace], palace and Friday Mosque, the market place and the gateways. These routes are sequences of spaces and passages, some of which are barely wide enough to permit the passage of a fully-laden donkey. The passages widen into outdoor praying places, shaded sitting areas in front of the houses, Koranic teaching areas and the spaces left after the recent filling in of the one-time borrow pits [pits from which material has been taken for building walls].
>
> (Moughtin, 1985, p.47)

It is important to stress the logic that underlies cities of this sort, for to a foreign pedestrian, away from the open public spaces and ceremonial buildings, the urban experience would be of walking down twisting narrow alleys formed by blank walls, with just an occasional gateway. This would give the visitor little sense of the vitality and orderliness of the private housing compounds behind the walls.

For the fascinating details of the actual construction of Hausa cities, including the way in which builders met the challenge of constructing what would in Europe be termed Gothic arches, the reader is encouraged to consult Moughtin's book, or study the excerpts in Chant (1999), the Reader associated with this book.

9.5 Conclusion

The cities of sub-Saharan Africa are not typical of the continent as a whole. Africa does not have cities and extensive city-building traditions equivalent to those in Europe or China. But I hope that this chapter has indicated that Africa can shed light from particular angles on the relationship between cities and technology. First, the climates and terrains in Africa frequently set limits on the very possibility of a city; and second, the cultures that flourished in Africa did not characteristically set a high premium on either permanence or monumentality. But in certain places Africans did congregate in the tightly packed areas we designate 'urban'; and where and when they did, their technical solutions tended to be distinctively African.

References

CAILLIÉ, R. (1830) *Travels through Central Africa to Timbuctoo; and across the Great Desert, to Morocco, performed in the years 1824–1828*, London, Colburn and Bentley.

CHANT, C. (ed.) (1999) *The Pre-industrial Cities and Technology Reader*, London, Routledge, in association with The Open University.

CLIFTON-TAYLOR, A. (1972) *The Pattern of English Building,* London, Faber.

CURTIN, P., FEIERMAN, S., THOMPSON, L. and VANSINA, J. (1995, 2nd edn) *African History*, London, Longman (first published 1978).

DENYER, S. (1978) *African Traditional Architecture*, London, Heinemann.

FARRAR, D.M. and PACEY, A.J. (1974) 'Africa fieldwork and technology: Report 4' (unpublished research report).

GARLAKE, P.S. (1973) *Great Zimbabwe*, London, Thames and Hudson.

HULL, R.W. (1976) *African Cities and Towns before the European Conquest*, New York, Norton.

MOUGHTIN, J.C. (1985) *Hausa Architecture*, London, Ethnographica.

Conclusion:
THE SJOBERG MODEL

by David Goodman

This volume has discussed a large number of cities over a vast time-span and scattered over much of the world. Taken together, they exhibit enormous diversity. But is it possible to discern some underlying features that all of these cities have in common – some essence of urbanism? A bold assertion that this was indeed the case attracted attention when it was developed in a book written by a US academic and published in 1960. According to Gideon Sjoberg, a sociologist, all cities before the onset of industrialization were (and in some cases still are) of the same type. He therefore sees nothing of importance distinguishing ancient Babylon, Nineveh, Rome, medieval Florence, Early Modern Paris, Guangzhou of the early twentieth century, present-day Kathmandu and Timbuktu. Their 'structure, or form' – spatial, political, social and economic – is everywhere the same. In all pre-industrial cities – whether in Europe, Africa, India or China – the city centre is the hub of government and religion, and the residence of the elite. Manual labour is despised. Craft production is achieved through a simple division of labour. The underlying reason for these and all other common characteristics is, he argues, the undeveloped state of technology. Unlike the industrial city whose technology is based on the inanimate energy of steam or electricity, on advanced tools and 'know-how', the pre-industrial city is throughout fashioned by dependence on animal or human power and simple tools, and by lack of technical knowledge. Why did the craft specialists in pre-industrial cities of the past (and in those of the present) concentrate in particular streets or districts? Because of rudimentary technology: primitive transport and the dependence of illiterate craft specialists on oral communication forced them to group together (Sjoberg, 1960).

British historians have been reluctant to welcome the introduction of sociological models such as Sjoberg's to interpret the past. The British tradition in history is empirical and not theoretical. And the more conservative of British historians – including such outstanding scholars as the late Geoffrey Elton, a constitutional historian – have declared that sociologists have much more to learn from history than historians from sociology. These are wise words, well worth heeding. But we must give Sjoberg a fair hearing. After all, in the United States and, above all, in France, historians have held sociology in higher esteem. And in Britain, Peter Burke, well known for his work on the history of popular culture in Early Modern Europe, has campaigned for a greater appreciation of the importance of sociology by British historians. He therefore read Sjoberg's book without any prejudicial hostility. What would he think of it?

Burke found serious flaws in Sjoberg's model. First, it suffered from the weakness of the sources he used for his historical data – old-fashioned, antiquarian or anecdotal works, long since outmoded by modern, well-researched, analytical studies of urban history. Consequently Sjoberg was led to inaccurate generalizations for the Early Modern period. The evidence of paving, drainage and street lighting in Early Modern Paris is enough to undermine Sjoberg's caricature of the pre-industrial city. Burke also criticizes Sjoberg for his far too general assumptions of pre-industrial illiteracy and the absence of a sense of time; both literacy and consciousness of time were well developed in some Early Modern cities (even, in the case of time, in medieval cities, as we have indicated in Chapter 4). For Burke, a second basic weakness in Sjoberg's model was the sharp distinction between pre-industrial and industrial cities. Is not the segregation of ethnic minorities, a supposedly distinctive characteristic of the pre-industrial city, to be found in New York's Harlem and in many other cities today? And was not a ruling elite dominant in industrial Birmingham just as it was in pre-industrial cities?

Burke concedes that models simplify and so tend to distort the past. One can therefore well ask why they should be used at all, since the purpose of history is to understand the past in all its complexity. Nevertheless, Burke will not reject urban models, but seeks a better one than Sjoberg offers. While he hints that technology may be at times a useful criterion – for example, in distinguishing the sanitary structures of Renaissance Rome and Victorian Birmingham – he warns that mechanized transport technology will not easily separate the pre-industrial from the industrial city. He tends to look instead to a model based on the sheer size of cities, suggesting that a city of, say, 100,000 may bring together cities of various periods into a common type (Burke, 1975).

Clearly, any attempt to invent a model to embrace cities across the world risks unacceptable neglect of important, distinctive characteristics of individual cities. Sjoberg's 'pre-industrial city' is far too broad a model to withstand even cursory historical scrutiny; not even the much smaller scope of the 'Islamic city' model can do that.

References

BURKE, P. (1975) 'Some reflections on the pre-industrial city', *Urban History Yearbook 1975*, Leicester, Leicester University Press, pp.13–21.

SJOBERG, G. (1960) *The Pre-industrial City Past and Present*, New York and London, Free Press and Collier-Macmillan.

Index

Acknowledgements

Grateful acknowledgement is made to the following for permission to reproduce material in this book:

p.43 Oates, J. (1986, rev. edn) *Babylon*, Thames and Hudson Ltd.

p.44 Flannery, K.V. (1994) 'Childe the evolutionist' in D.R. Harris (ed.) *The Archaeology of V. Gordon Childe: contemporary perspectives*, UCL Press.

p.45 Butzer, K.W. (1996) 'Irrigation, raised fields and state management: Wittfogel redux?', *Antiquity*, vol.70, no.267; by permission of the author.

p.78 Aristotle (1962) *Politics* (trans. T.A. Sinclair), Penguin Books Ltd.

p.158 Bulliet, R.W. (1989) *The Camel and the Wheel*, Harvard University Press.

pp.260–61 Hernando Cortés, *Five Letters of Cortés to the Emporer* (trans. J. Bayard Morris); translation copyright © 1969 J. Bayard Morris; reprinted by permission of W.W. Norton & Company, Inc. and Routledge.

Figures 7.1, 7.13, 7.14 Padron, F.M. (1988) *Atlas Histórico Cultura de América*, by permission of Gobierno de Canarias.

Figure 7.3 Grosser (1997) *Westermanns Atlas zur Weltgeschichte*, by permission of Georg Westermann Verlag GmbH.

Figure 7.6 © HarperCollins Publishers 1991; MM–0398–123.

Every effort has been made to trace all copyright owners, but if any has been inadvertently overlooked, the publishers will be pleased to make the necessary arrangements at the first opportunity.